Thermodynamics

K. Rathnam
Professor of Mechanical Engineering (Retired)
Annamalai University
Annamalai Nagar

PEARSON

Chennai • Delhi

ISBN 978-81-317-9550-7

First Impression

Published by Dorling Kindersley (India) Pvt. Ltd, licensees of Pearson Education in South Asia.

Head Office: 7th Floor, Knowledge Boulevard, A-8(A), Sector 62, Noida 201 309, UP, India.
Registered Office: 11 Community Centre, Panchsheel Park, New Delhi 110 017, India.

Compositor: Mukesh Technologies Pvt. Ltd., Puducherry.
Printed in India by Saurabh Printers Pvt. Ltd.

Contents

Foreword

Whenever a new book on subjects such as Thermodynamics or Fluid Mechanics appears, the question 'Why another book?' arises. However, after reading the manuscript of 'Thermodynamics' by Professor K. Rathnam, I am convinced that it is written with a specific purpose that is not fulfilled by a standard treatise on Thermodynamics. The choice of the title is an expression of modesty on the part of the author because the book contains adequate, if not ample, theory needed to solve the problems.

One would notice the absence of historical and philosophical details, but I feel that is not an act of omission. Professor Rathnam has a specific objective in mind and this objective is fulfilled without including such details. He has adhered to the B.E./B.Tech. syllabus for Thermodynamics while compiling the questions in this book.

The number, quality and gradation of problems must have taken an enormous amount of hard work for which the author has to be highly commended.

To conclude, this is a book with a specific purpose that will be highly appreciated by undergraduate college students as well as those preparing for competitive examinations.

M. M. Oberai
Doctor d'Universite' de Paris (Sorbonne)
Former Head, Department of Mechanical Engineering
Indian Institute of Technology Kanpur

Preface

A sound knowledge in the fundamental principles of Engineering Thermodynamics is needed to understand the working of thermal power plants, internal combustion engines, refrigeration and air conditioning. In the study of theory behind the transformation of thermal energy to mechanical energy and vice versa, one encounters situations relating to the type of gadgets employed and the basic relations between heat and work and the properties of the system. A characteristic feature of thermodynamics is that these relations can often be stated without the details of the process undergone by the system. Another feature of thermodynamics is that it furnishes simple tests to validate the feasibility of an inventor's proposal for a new power plant. This book, containing around 300 objective type questions, 160 solved numerical problems and an equal number of practice problems, aims to kindle the interest of students who are yet to start learning the subject. The concepts in thermodynamics are difficult to understand in one reading. The questions are arranged topic wise, with emphasis on elucidating the concepts. The questions framed are of varying degrees of difficulty. Finding answers to these questions will help the reader to assess his/her understanding of the subject. The reader will have to consult standard textbooks before using the keys provided in the book. To this end, the bibliography will be of some help to the reader. I believe that the readers who read the booklet conscientiously will gain the knowledge necessary for solving exercise problems later. I hope that the questions provided in this book help to kindle the imagination of its readers and inspire them to pursue this subject further.

The contents of this book is an offshoot of questions I had prepared for the course on thermodynamics for undergraduate students at Annamalai University. I had an opportunity of teaching this subject consistently for quite some time. I express my sincere thanks to Professor V. R. Muthuveerappan, B.E. (Hons), M.S. (Cornell), former Head of the Department of Mechanical Engineering, who was instrumental in allotting this subject during the formative years of my teaching career and also for his guidance. I cherish, even now, the confidence the Professor showed in letting me handle one of his favourite subjects. I am indebted to my Professor M. M. Oberai, D.Sc. (Sorbonne), an authority in high-speed viscous flows, and the first author of the book 'A First Course in Thermodynamics', for his foreword. Last but not the least, I thank the editorial team at Pearson Education for their timely help in bringing out the manuscript in book form accessible to the students and others.

K. Rathnam

1

Basic Concept

Thermodynamics is a branch of science that deals with energy transformations and consequent changes in the properties of a system. A thermodynamic study that is undertaken without recourse to the nature of individual particles falls into the domain of classical thermodynamics. Classical thermodynamics in conjunction with the fields of fluid dynamics, heat transfer and the mechanics and dynamics of solids provides fundamental knowledge for the analysis of a wide variety of problems. Hence, a firm foundation in thermodynamics is a prerequisite to the understandings of modern scientific and engineering studies.

Many of the contrivances employed in thermodynamics study are conceptual in nature. A conceptual motion will be considered real or fictitious depending upon whether or not there are actual systems that conform to its specifications. All parts of theoretical physics have their own fictions. These include point masses, frictionless inclined planes and pistons, the ideal physical pendulum in mechanics and point charges in electromagnetic field. All of these are idealizations of non-thermodynamic systems. We shall soon be faced with important thermodynamic fictions. These fictions will be merely postulations of idealized behaviour that never exist in reality, but that can be approximated to a high degree. The most important fictions are those that constitute the limits of extrapolation from the actual behaviour.

1. Explain the concept of system in thermodynamics and also define the boundary and surroundings.

Ans: The idea of a system plays an important part in thermodynamics and it may be defined as a region in space containing a quantity of matter whose behaviour is being investigated. The quantity of matter is separated from its surroundings by a boundary, which may be a physical boundary, such as the walls of a vessel or some imaginary surface enveloping the region. The boundary of the system has to be defined as through this boundary the energy transformations take place with the system. The term *surroundings* is restricted to those portions of matter external to the system that are affected by the changes occurring within the system. The surroundings are frequently referred to as the environment.

Two examples of systems and the representative boundaries are shown in Figure 1.1. Figure 1.1(a) illustrates the flow of a fluid through a pipe or duct. The dashed line represents the boundary, fixed in space, which outlines the region of space for which a thermodynamic study is to be made. It is to be noted that a portion of the boundary is imaginary at the open ends designated by points 1 and 2. These boundaries are selected arbitrarily for accounting purposes as inlet and outlet and have no significance in the actual physical processes. Thus, we see that it is not necessary for any or the entire boundary to be physically distinguishable when making

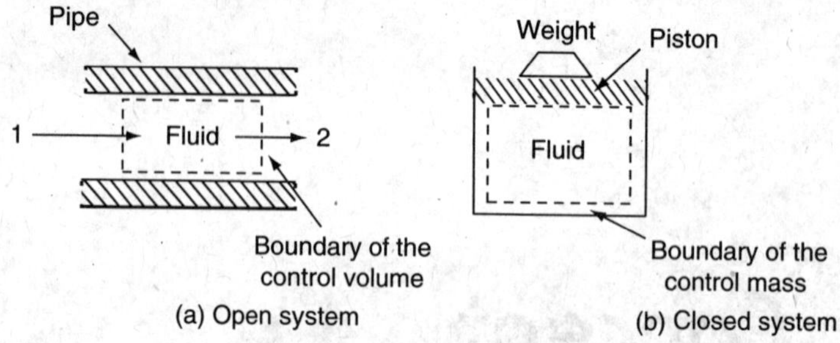

Figure 1.1 Thermodynamic Systems

thermodynamic analyses. Nevertheless, it is extremely important to establish the boundaries of a system clearly before beginning any form of analysis, in order to incorporate heat and work interactions.

In Figure 1.1(b), a piston-cylinder assembly is shown. Again, the selected boundary is represented by the dashed line. In this case, the boundary is physically well established, for the envelope lies just inside the walls of the cylinder and the piston. It is to be noted here that the shape of the boundary will change when the position of the piston is altered, if the boundary is selected to surround the fluid within the assembly. The change in shape or position of the boundary is always permissible as long as such changes are recognized in subsequent calculations.

2. What are the types of system used in thermodynamics? Explain.

Ans: In the study of thermodynamic systems, one of the major differences as viewed by the observer is between those systems that contain a fixed quantity of matter within the boundaries and those that experience a transfer of matter across the boundaries. When the transfer of mass across the bounding surface of a system is prohibited, the system under consideration is a closed system. Although the quantity of matter is fixed, energy is allowed to cross its boundaries in various forms. The quantity of matter under observation is called the control mass. The control mass within the piston-cylinder assembly in Figure 1.1(b) is in closed system.

It is appropriate to define an open system as a system for which mass is permitted to cross the selected boundaries and energy. Such a passage of matter is not restricted to one region of the boundary; it may occur simultaneously at several sections of the imaginary or real surface. When attention is focused on a region of space that acts as an open system, the region is called a control volume, and the boundary of the control volume is the control surface. The region of space described in Figure 1.1(a) is a control volume, and this region would be analysed as an open system.

Systems for which all interactions with the surroundings are excluded are called isolated systems. Such a system has a fixed mass, and no heat or work interactions occur across its boundaries.

3. What is meant by thermodynamic equilibrium? Discuss the three conditions to be satisfied to achieve thermodynamic equilibrium.

Ans: The concept of equilibrium is an important one, since it is only when a system is in a state of equilibrium that the thermodynamic properties have any real meaning. There are three major subclasses of equilibrium; all of which are necessary to fulfil the condition of thermodynamic equilibrium. These are usually called thermal, mechanical and chemical equilibriums. If a force acts on (the boundaries of) a system, the pressure within the system will start changing so as to balance this force. If the system is now left to itself, the change will ultimately come to an end and certain value of pressure will be attained which will stay constant with the passage of time. This condition or state in which there is no unbalanced force within the system or at its boundaries is defined to be one of the mechanical equilibrium. In general, the mechanical equilibrium does not ensure uniformity of pressure within the system, for example, still water contained in a tank is in mechanical equilibrium but the pressure increases as we move downwards from the surface, and fluid moving through an open system will have different pressure at inlet and outlet regions. Again, if the condition of a system is such that all chemical reactions in it have ceased and there is no diffusion (transfer of mass from one part to another), the system is said to be in chemical equilibrium. Let us consider a system consisting of two bodies A and B such that B is hotter than A. The bodies are placed in thermal contact (contact through a diathermic wall) and the system is isolated from the surroundings. After some time the two bodies attain thermal equilibrium with each other.

4. State and explain the significance of the Zeroth law of thermodynamics.

Ans: 'If each of the two given bodies A and B is in thermal equilibrium with a third body C then the bodies A and B are in thermal equilibrium with each other.'

The importance of the Zeroth law of thermodynamics lies in the conclusion drawn from it that all bodies such that every pair among them is in thermal equilibrium must have a common characteristic; this characteristic is called temperature (symbol T). Thus, temperature is defined as the characteristic that has the same value for all bodies that are in thermal equilibrium (taken two at a time).

5. What are the two important temperature scales that are used for measuring the temperatures?

Ans: Centigrade and Fahrenheit scales are used for the measurement of temperatures. For the quantitative measurement of temperature, we must fix the zero point and the size of the unit. For this purpose, two fixed points are chosen: the freezing and the boiling points of pure water under one standard atmospheric pressure (760 mm of Hg or 1.013 bar). In the centigrade scale (due to Celsius), the freezing point is taken as zero and the range between the two (freezing and boiling) points is divided into 100 degrees. In the Fahrenheit scale, the freezing point is 32°F and the range is 180°. Thus, the relation is

$$\frac{C}{100} = \frac{F-32}{180}.$$

6. Describe how perfect gas scales are introduced in thermodynamics.

Ans: If a glass bulb containing a perfect gas is placed in thermal contact with a constant temperature bath, then after a while, thermal equilibrium will be attained. As the volume of the gas is fixed, the pressure will reach a certain definite value depending on the temperature of the bath.

We start with a certain quantity of the gas in the bulb and measure the pressure p_i and p_s corresponding to the freezing and boiling point of pure water under one atmospheric pressure and plot the ratio (p_s/p_i) versus p_i. We repeat this procedure, starting every time with a different quantity of the gas. It is observed that

(i) for each gas, the points $\left\{p_i, \frac{p_s}{p_i}\right\}$ lie on a straight line

(ii) the straight line corresponding to various gases intersect in a point on the $\frac{p_s}{p_i}$ axis, the ordinate of the point being 1.366.

Kelvin's and Rankine's scales are based on the postulate

$$\frac{T_s}{T_i} = \lim_{p_i \to 0}\left(\frac{p_s}{p_i}\right) = 1.366.$$

In the Kelvin scale, we set $T_s - T_i = 100$ and obtain $T_i = 273$ K and $T_s = 373$ K.
In the Rankine scale, we put $T_s - T_i = 180$ and obtain $T_i = 492$ R and $T_s = 672$ R.
The Kelvin, Celsius, Rankine and Fahrenheit scales, which are linear, are related by the equations

$$T_C = T_K - 273; \quad T_R = 1.8\, T_K; \quad T_F = T_R - 460;$$

7. Define property of a thermodynamic system.

Ans: The property is any observable characteristic that describes the condition of the system, for example, pressure, temperature, volume and energy. These properties are uniform (single-valued ones) only if the system is in thermodynamic equilibrium. For our study, we need to consider only a few of the properties.

8. Explain what do you understand by extrinsic property and intrinsic property.

Ans: A property of the system that involves a relationship between the system and the surroundings is called extrinsic, such as weight, potential energy and kinetic energy.

A property of the system that depends on the system without recourse to the surroundings is called intrinsic, such as volume, mass, pressure and temperature.

9. Distinguish between intensive property and extensive property giving two examples for each.

Ans: A property whose value remains the same whether we refer to the whole system or a portion thereof is called intensive property, for example, temperature and pressure. Properties such as volume and energy depend on the mass of the system being considered and hence they are called extensive properties. The extensive properties can be intensified by expressing their values per unit mass; these values are called specific values.

Thus, specific energy, $e = E/m$.

10. Explain the state of system.

Ans: The state of a system is the condition of the system characterized by the values of all the properties that help to describe the system. Specifying the values of the properties is termed specifying the state of the system. The term *state* will refer to an equilibrium state.

11. Define process and path as applied to a thermodynamic system.

Ans: A process is any transformation of a system from one equilibrium state to another over an interval of time, with the possibility of heat or work interactions at the boundaries of the system. Of special significance in thermodynamics is a cyclic process, which is defined as a process with identical end states.

The *path* of any process is the specification of the series of states through which the system passes during the process.

12. State the equation of state for a perfect gas.

Ans: All the properties are not necessary for a particular study. Further, all the (relevant) properties cannot take independent values; they are connected by certain relations. One such relation is called the equation of state and is generally written in the form

$$f(p, v, T) = 0$$

where v is the specific volume.

A substance obeying the special equation of state

$$p\,v = R\,T$$

or

$$p\,V = m\,R\,T$$

(where m is the mass and R a constant) is called a perfect gas.

Solved Problems

1. What is the reading on the Centigrade scale when that on the Fahrenheit scale is 110°?

   We know $\dfrac{C}{100} = \dfrac{F-32}{180}$; $\dfrac{t_C}{100} = \dfrac{110-32}{180}$;

   Hence, $t_C = 43.3$°C.

2. At what temperature are the Centigrade and the Fahrenheit readings identical?

   We know $\dfrac{C}{100} = \dfrac{F-32}{180}$; $\dfrac{x}{100} = \dfrac{x-32}{180}$;

   Hence, $x = -40°$.

3. A Fahrenheit and a Celsius thermometer are in equilibrium with a fluid. If the reading on the Fahrenheit thermometer is twice that on the Celsius thermometer, what is the absolute Rankine temperature of the fluid?

   We know $\dfrac{C}{100} = \dfrac{F-32}{180}$; $t_f = 2\,t_c$ (given); on substitution, we get $\dfrac{t_c}{100} = \dfrac{2t_c - 32}{180}$.

   Hence, $t_c = 160°$, $t_f = 320°$ and $T_R = 320 + 460 = 780$.

4. Define a new temperature scale, say °N, in which the boiling and freezing points of water (steam and ice points) are 400°N and 100°N, respectively, and correlate this with the centigrade scale. The °N reading on this scale is a certain number degrees on a corresponding absolute temperature scale. What is this absolute temperature corresponding to 0°N?

   $\dfrac{t_C}{100} = \dfrac{t_N - 100}{300}$, solving this, we get $3\,t_C = t_N - 100$.

   We know $\dfrac{T_S}{T_i} = 1.366$; $T_S - T_i = 300$, solving these two equations, we get

   $$T_i = 819.6 \text{ N(abs) and } T_S = 1119.6 \text{ N(abs)}$$

   Absolute temperature corresponding to 0°N = 719.6 N(abs).

5. One kilogram of helium is placed in a rigid tank of 0.5 m^3 volume. If a thermometer in the tank reads 77°C and the barometric pressure is 76 cm Hg, what would a pressure gauge read that is inserted into the tank, in bar?

   $$\text{Gas constant, } R = \frac{8314.4}{4} = 2078.6\frac{\text{J}}{\text{kg K}};\ T = 273 + 77 = 350$$

$$pV = mRT;\ p = \frac{1 \times 2078.6 \times 350}{0.5} = 14.55 \text{ bar; } 1 \text{ atm} = 76 \text{ cm Hg} = 1.0132 \text{ bar.}$$

Pressure gauge reading = 14.55 – 1.0132

= 13.5368 bar.

6. A balloon is filled with methane (CH_4) gas at 30°C and 1.0132 bar (abs) pressure until the volume is 50 m³. Calculate the mass of methane in kg.

$$\text{Gas constant, } R = \frac{8314.4}{16} = 519.65 \frac{\text{J}}{\text{kg K}},$$

$$m = \frac{pV}{RT} = \frac{1.0132 \times 10^5 \times 50}{519.65 \times 303} = 32.17 \text{ kg.}$$

7. A tank contains helium at 6 atm and 40°C. A kilogram of the gas is removed, which causes the pressure and temperature to change to 4 atm and 30°C, respectively. Determine the volume of the tank in litres.

Gas constant for helium, $R = 8314.4/4$

$= 2078.6$ J/kg K

$$p_1 V_1 = m_1 R T_1 \qquad (1)$$

$$\text{and } p_2 V_2 = m_2 R T_2 \qquad (2)$$

Given $m_2 = m_1 - 1$ and $V_1 = V_2$

Dividing the equations (1) and (2), we get, $\frac{p_1}{p_2} = \frac{m_1}{m_2}\frac{T_1}{T_2} \Rightarrow \frac{6}{4} = \frac{m_1}{(m_1 - 1)}\frac{313}{303}$

Solving this, we get $m_1 = 3.212$ kg,

Hence, volume of the tank, $V = \frac{3.212 \times 2078.6 \times 313}{6 \times 1.0132 \times 10^5} = 3.438 \text{ m}^3$ or 3438 litres.

8. A tank contains carbon dioxide gas at 5.5 bar and 38°C. A leak occurs in the tank which is not detected until the pressure has dropped to 4.5 bar. If the temperature of the gas at the time of detection of the leak is 21°C, compute the mass of carbon dioxide that has leaked out if the original mass was 27.5 kg.

Gas constant for carbon dioxide, $R = 8314.4/44 = 188.96$ J/kg K

$$p_1 V = m_1 R T_1 \text{ and } p_2 V = m_2 R T_2$$

$$V = \frac{27.5 \times 188.96 \times 311}{5.5 \times 10^5} = 2.938 \text{ m}^3$$

$$m_2 = \frac{4.5 \times 10^5 \times 2.938}{188.96 \times 294} = 23.798 \text{ kg}$$

Mass of CO_2 leaked out = 27.5 – 23.798

= 3.702 kg.

9. A turbine is supplied with steam at a gauge pressure of 14 bar. After expansion in the turbine, the steam passes to a condenser which is maintained at a vacuum of 710 mm Hg by means of pumps (wet and dry air pumps). The barometric pressure is 772 mm Hg. Express the inlet and exhaust steam (absolute) pressures in N/m^2. Assume density of mercury as 13.6×10^3 kg/m^3.

   The barometric pressure $= 13.6 \times 10^3 \times 9.81 \times 0.772 = 1.02997 \times 10^5$ N/m^2.

   Steam pressure at inlet to the turbine $= 14 \times 10^5 + 1.02997 \times 10^5 = 15.02997 \times 10^5$ N/m^2.

   Steam pressure at exhaust $= 13.6 \times 10^3 \times 9.81 \times (0.772 - 0.710) = 0.0827179 \times 10^5$ N/m^2.

10. The temperature t on a thermometric scale is defined in terms of a property P by the relation

$$t = a\,\ell n(P) + b, \text{ where } a \text{ and } b \text{ are constants.}$$

   The temperatures of the ice-point and the steam-point are assigned the numbers 0 and 100, respectively. Experiment gives values of P of 1.86 and 6.81 at the ice-point and steam-point, respectively.

   Evaluate the temperature corresponding to a reading of $P = 2.50$ on the thermometer.

$$0 = a\,\ell n(1.86) + b \tag{1}$$

$$100 = a\,\ell n(6.81) + b \tag{2}$$

   Solving the above equations, we obtain the values $a = 77.05$ and $b = -47.817$.

   Hence, the temperature corresponding to a reading $P = 2.50$.

$$t = 77.05\,\ell n(2.50) - 47.817 = 22.78°\text{C}.$$

Practice Problems

1. Specify whether the following properties are extrinsic or intrinsic and whether they are extensive or intensive.

(i) mass	(v) density	(ix) molar density	(xiii) temperature
(ii) weight	(vi) energy	(x) mass concentration	(xiv) surface area
(iii) volume	(vii) specific weight	(xi) mole fraction	(xv) elevation
(iv) velocity	(viii) strain	(xii) pressure	(xvi) potential energy.

2. Specify whether the following systems are open or closed when in operation.
   (i) a water wheel
   (ii) an automotive engine
   (iii) a manually operated ice-cream freezer
   (iv) a pressure cooker
   (v) a rigid vessel containing air and an oily rag, which ignites spontaneously
   (vi) the liquid cooling system of an automobile
   (vii) a boiler in a steam power plant
   (viii) the steam flowing through a system composed of a turbine coupled with an electric generator.

3. Carry out the following temperature conversions:

(i) 100°F to °C
(ii) 45°C to K
(iii) 745 R to °F
(iv) 290 K to R
(v) 330 K to °F

(37.7°C, 318 K, 285°F, 522.6 R, 134.6°F)

4. It is desired to store methane gas (CH_4) underground at a pressure of 14 bar and a temperature of 13°C. Estimate the volume required in m³ to store 50 kg of the gas under these conditions.

($5.3078\ m^3$)

5. Two tanks A and B are connected by suitable pipes by a valve which initially is closed. Tank A initially contains $0.28\ m^3$ of nitrogen at 7 bar and 60°C, and tank B is evacuated. The valve is then opened and nitrogen flows into tank B until the pressure in this tank reaches 1.4 bar and the temperature is 16°C. As a result, the pressure in the tank A drops to 4.8 bar and the temperature changes to 43°C. Determine the volume of the tank B in m³.

($0.337\ m^3$)

Objective Type Questions

1. The pressure of the system could be categorized as belonging to

(i) the geometric state
(ii) the hydrodynamic state
(iii) the kinematics state
(iv) the quantum mechanical state

2. The energy form that cannot be transformed directly into work is

(i) chemical
(ii) internal
(iii) kinetic
(iv) potential

3. A Fahrenheit and a Celsius thermometer are in equilibrium with a fluid. If the reading on the Fahrenheit scale is twice that on the Celsius scale, then the absolute temperature of the fluid on the Kelvin scale is

(i) 313
(ii) 373
(iii) 433
(iv) 493

4. If the pressure of oxygen ($R = 260$ J/kg K) contained in a tank drops from 0.4 bar to 0.3 bar on removal of 1 kg of the gas with negligible change in temperature at 27°C, then the volume of the tank, in m³, is

(i) 7.2
(ii) 7.4
(iii) 7.6
(iv) 7.8

5. Which one of the following let-outs by an automobile diesel engine has one use to the road users?

(i) Carbon dioxide
(ii) Carbon monoxide
(iii) Noise/vibration
(iv) Unburned hydrocarbon

6. The integral $\int (x\,dy + y\,dx)$, on evaluation along the path $y = x^2$ connecting the points A(2, 4) and B(3, 9), gives the value
   (i) 6 (ii) 7
   (iii) 19 (iv) 30
7. Which one of the following pairs is not correctly matched regarding the thermodynamic forces encountered in the systems?

System	*Thermodynamic force*
(i) Fluids	Pressure
(ii) Elastic solids	Stress
(iii) Surface films	Viscosity
(iv) Dielectrics	Electric current

8. The contents of a thermodynamic system is described geometrically, dynamically and chemically by
   (i) its mass, moles and volume
   (ii) its volume, moles and mass
   (iii) its pressure, mass and volume
   (iv) its volume, pressure and moles
9. Which one of the following items of the IC engine exhaust gases if considered as a thermodynamic property is false?
   (i) Pressure (ii) Temperature
   (iii) Colour (iv) Viscosity
10. Which one of the following changes is responsible for the measurement of temperature in a bimetallic dial thermometer?
   (i) Volumetric expansion of metals
   (ii) Relative strain of dissimilar metals
   (iii) Emf production at the junction
   (iv) Resistance of the metals
11. Intensive properties alone are listed in group
   (i) temperature, enthalpy, pressure
   (ii) volume, entropy, density
   (iii) temperature, density, pressure
   (iv) volume, enthalpy, entropy
12. Identify the thermodynamic system that has neither work interaction nor heat interaction.
   (i) Steam condenser (ii) Automobile engine
   (iii) Throttle valve (iv) Air compressor
13. Extensive properties alone are listed in group
   (i) pressure, volume, temperature (ii) volume, enthalpy, entropy
   (iii) temperature, mass, density (iv) enthalpy, pressure, density

14. The volume of air leaking out of a building when air temperature increases uniformly from 27°C to 39°C, expressed in per cent, is
    (i) 2
    (ii) 4
    (iii) 6
    (iv) 8
15. A tank is divided by rigid wall into two sections as shown in the given figure. Pressure gauge B reads 1.75 bar (g) and gauge C reads 1.5 bar (g). If the barometric pressure is 1 bar, then the reading on gauge A, in bar (g), shall be

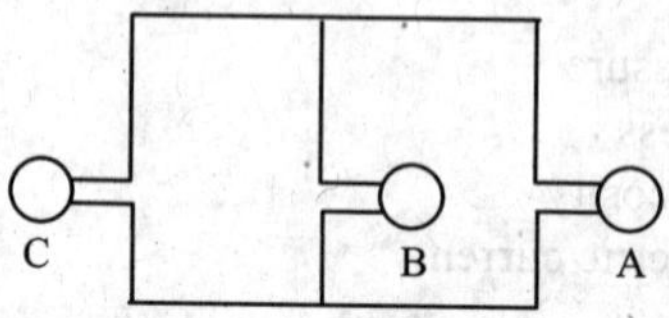

Figure 1.2 Composite Tank

    (i) –0.50
    (ii) –0.25
    (iii) +0.25
    (iv) +0.50
16. Which event is an example of an irreversible process?
    (i) Fast adiabatic expansion of gas
    (ii) Slow isothermal compression of steam
    (iii) Extension of spring
    (iv) Plastic deformation of material
17. Isentropic flow is
    (i) an irreversible flow
    (ii) an ideal fluid flow
    (iii) a reversible flow
    (iv) a frictionless reversible flow
18. How many independent properties are required to fix the internal state of water flowing steadily through a horizontal pipe?
    (i) One
    (ii) Two
    (iii) Three
    (iv) Four
19. The heat absorption or rejection by a thermal reservoir or just heat reservoir with a thermodynamic system is an
    (i) isometric process.
    (ii) isobaric process.
    (iii) isothermal process.
    (iv) isenthalpic process.
20. Identify the process in which there is absorption of heat.
    (i) Diathermic
    (ii) Endothermic
    (iii) Exothermic
    (iv) Hygroscopic

Answers

1. (ii) 2. (i) 3. (iii) 4. (iv) 5. (iii) 6. (iii) 7. (iii) 8. (iv) 9. (iii) 10. (ii)

11. (iii) 12. (iii) 13. (ii) 14. (ii) 15. (ii) 16. (iv) 17. (iv) 18. (iii) 19. (iii) 20. (ii)

2

Work and Heat Interactions and First Law of Thermodynamics

1. Explain the concept of quasi-static process and derive the expression for work interaction.

Ans: The concept of work is usually introduced in the study of mechanics. Mechanical work is defined as the product of a force F and the displacement S of the force when both are measured in the same direction. In thermodynamics, it is convenient to consider work, which is an interaction between a system and its surroundings, as energy in transition and is done by the system or is on the system. Before proceeding with the specific equations for work interactions, it is appropriate to define a quasi-static process. During such a process, the substance within the system must be infinitesimally close to a state of equilibrium at all times. The path of quasi-static process is a series of equilibrium steps. There are two important reasons why the quasi-static process is important in the study of thermodynamics. The equations for the work interactions given in the following sections require that the interrelationships between properties be known at all times during the process. The second major consequence of the quasi-static concept deals with the value of work interactions under such a condition. The work output is found to be a maximum and the work input a minimum, for a given quasi-static process. In the presence of non-equilibrium states, the work output is less than the maximum and the work input is greater than the minimum required. Thus, the quasi-static process is a standard to which all other processes may be compared.

Consider a piston-cylinder assembly, as shown in Figure 2.1, containing a substance enclosed in it. The system boundaries are indicated by the dashed line in the diagram. The mechanical force exerted uniformly by the substance on the piston is given by pA.

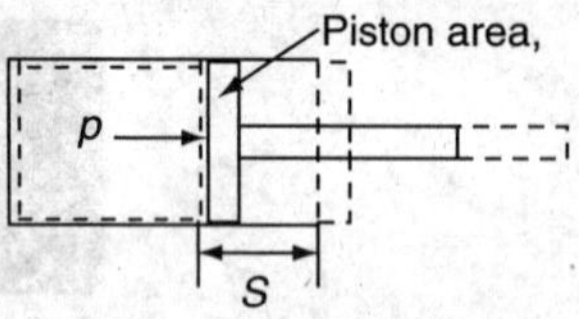

Figure 2.1 Cylinder and Piston Enclosing a Fluid System

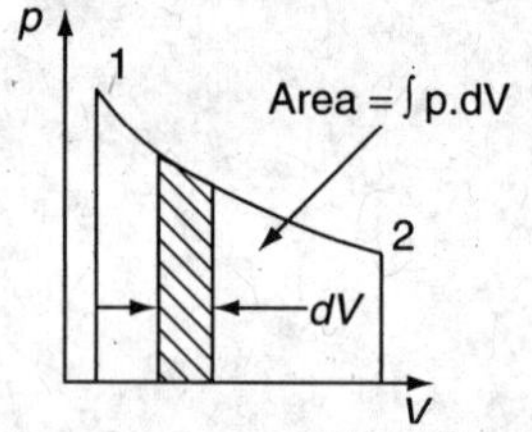

Figure 2.2 State Diagram for the Fluid System

The mechanical work done on the piston by the substance is then expressed as

$$\delta W = F \cdot dS = pA \cdot dS.$$

Since $A \cdot dS$ is the change in the volume dV due to a deformation of the boundary, the differential quantity of work done is simply,

$$\delta W = p \cdot dV.$$

If the system is passing through a series of equilibrium states during a process from an initial state 1 to a final state 2, then we may write for the work done during the process as

$$(W)_{\text{by the system}} = \int p \cdot dV \quad \text{(or)} \quad (w)_{\text{on the system}} = -\int p \cdot dV.$$

A plot of the quasi-static process on the pressure volume co-ordinates is quite useful for describing graphically the boundary work of a process involving a closed system. A typical diagram for the evaluation of the work is shown in Figure 2.2. The cross-hatched area on the diagram represents the work done by the gas within the cylinder when the volume is increased by an amount dV. The entire area beneath the curve from point 1 to point 2 represents the total work done when the gas expands from point 1 to point 2. A large number of paths could be drawn between the same two end states. In general, the area beneath each of these paths will be different. This emphasizes the fact that work is a path function.

2. What is the sign convention for positive work and negative work? How are positive work and negative work represented on pressure-volume diagram?

Ans: It is necessary to adopt a sign convention to indicate the direction in which the work flows. If the work is done by the system on the surroundings, the work is said to be positive. If work is done on the system by the surroundings, the work is said to be negative.

If the area lies to the right of the process as indicated in Figure 2.3, then the work done is positive. If it lies to the left of the process, then the work done is negative.

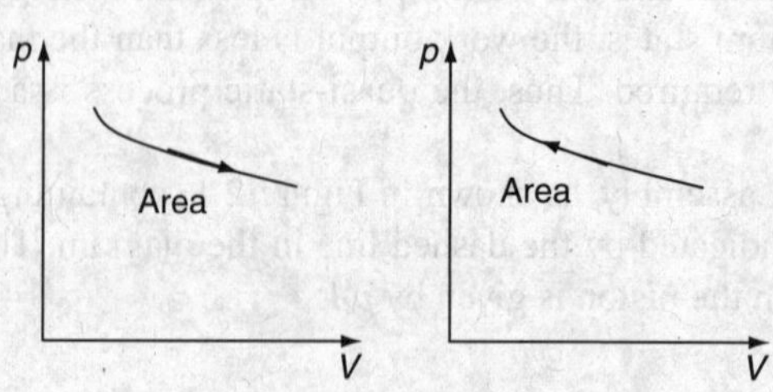

Figure 2.3 Area Representation of Boundary Work

3. Derive an expression for work interaction during an isothermal process.

Ans: The integration of the equation for the work requires knowledge of the functional relationship between p and V. A gas is contained within a piston-cylinder assembly. The ideal gas equation is $pV = mRT$. It is allowed to expand quasi-statically and isothermally from state 1 to state 2. The work done by the gas on the piston is given by the integral of $p.dV$

$$W = \int pdV = \int_{V_1}^{V_2} mRT\frac{dV}{V} = mRT\ln\frac{V_2}{V_1}$$

By measurement of the temperature of the process and the initial and final volumes, the maximum work done by the system can easily be calculated. If the work interaction is to be expressed per unit mass, the lower case letter will be employed.

4. Derive an expression for work interaction during a polytropic process.

Ans: Consider now the evaluation of work for a system which undergoes a quasi-static polytropic process, pV^n = constant

$$W = \int p\,dV = \int_{V_1}^{V_2} \frac{C}{V^n}dV = \frac{p_1V_1 - p_2V_2}{(n-1)}.$$

Knowing the pressure and volume of the initial and final states, the work interaction during the process can be calculated from the aforesaid equation.

5. Explain the heat interaction with the system and its sign convention.

Ans: Heat, denoted by the symbol Q, may be defined in an analogous way to work as follows. Heat is something that appears at the boundary when a system changes its state due to a difference in temperature between the system and its surroundings. Heat, like work, is a transient quantity which only appears at the boundary while a change is taking place within the system. Heat is the energy transferred by virtue of a temperature difference. If heat flows into a system from the surroundings, the quantity is said to be positive. If heat flows from the system to the surroundings, it is said to be negative.

Summarizing neither heat nor work are properties of a system, but both are transient quantities, only appearing at the boundary while a change of state occurs within the system. Consider a vessel of constant volume containing a gas. Let the gas undergo a change of state in such a way that the pressure and temperature are increased. This change could be obtained, as in Figure 2.4a, by placing the vessel in contact with a body at a higher temperature, or as in Figure 2.4b, by choosing the gas with a rotating paddle.

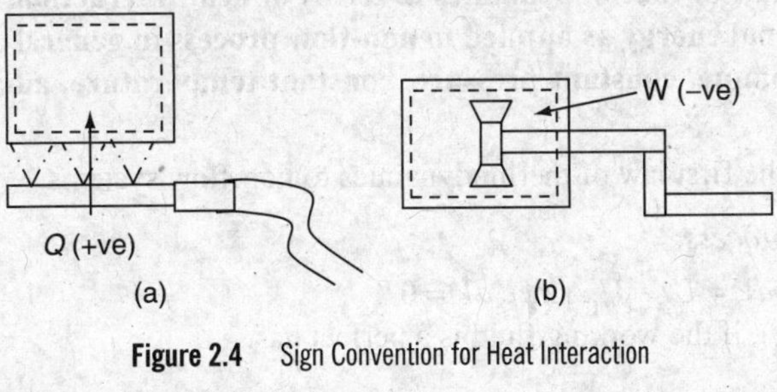

Figure 2.4 Sign Convention for Heat Interaction

6. Define a cyclic process.

Ans: A process or a combination of processes as a result of which the final state of the system is identical with the initial state is called a cycle or cyclic process.

7. State the first law of thermodynamics for a cyclic process and hence, prove that internal energy is a property of the system.

Ans: If a system goes through a cycle, the energy transferred to it as heat equals the work done by it. Mathematically

$$(Q_{\text{to the system}} - W_{\text{by the system}})_{\text{cycle}} = 0.$$

This law may be stated in various other forms but they are all equivalent and they all imply the conservation of energy.

Let us apply the first law to two cycles as shown in Figure 2.5: 1-*a*-2-*b*-1 and 1-*a*-2-*c*-1. Then

$$(Q_{1\text{-}a\text{-}2} + Q_{2\text{-}b\text{-}1})_{\text{to the system}} - (W_{1\text{-}a\text{-}2} + W_{2\text{-}b\text{-}1})_{\text{by the system}} = 0 \qquad (1)$$

and

$$(Q_{1\text{-}a\text{-}2} + Q_{2\text{-}c\text{-}1})_{\text{to the system}} - (W_{1\text{-}a\text{-}2} + W_{2\text{-}c\text{-}1})_{\text{by the system}} = 0 \qquad (2)$$

Equation (1) – (2)

$$(Q_{2\text{-}b\text{-}1} - Q_{2\text{-}c\text{-}1})_{\text{to the system}} - (W_{2\text{-}b\text{-}1} - W_{2\text{-}c\text{-}1})_{\text{by the system}} = 0. \qquad (3)$$

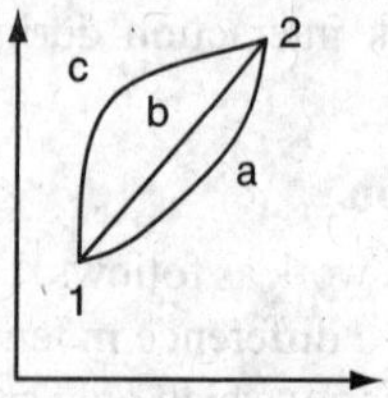

Figure 2.5 Path of Cyclic Processes

We know that $(W_{2\text{-}b\text{-}1} - W_{2\text{-}c\text{-}1}) \neq 0$. Hence, in general $(Q_{2\text{-}b\text{-}1} - Q_{2\text{-}c\text{-}1}) \neq 0$, that is, both heat and work are not point functions.

$$\int \delta Q - \int \delta W = 0$$

$$_{2\text{-}b\text{-}1}\int (\delta Q - \delta W) = {}_{2\text{-}c\text{-}1}\int (\delta Q - \delta W)$$

Though $\int \delta Q$ and $\int \delta W$ depend on the path, the combination $_{2\text{-}1}\int (\delta Q - \delta W)$ is independent of the path. Hence, the quantity $(\delta Q - \delta W)$ is an exact differential, say of a function U.

Then U is a point function.

$$dU = \delta Q - \delta W$$

The function U is defined as the internal energy of the system.

8. State the first law of thermodynamics in terms of heat interaction, work interaction and change in internal energy as applied to non-flow process in general and simplify the same for constant volume, constant pressure, constant temperature, adiabatic and polytropic processes.

Ans: Application of the first law of thermodynamics to non-flow systems

(i) *Constant volume process:*

$_1Q_2 = U_2 - U_1 + \int pdV = U_2 - U_1$, since $dV = 0$
$_1Q_2 = m\,C_v\,(T_2 - T_1)$, if the working fluid is a perfect gas.

(ii) *Constant pressure process:*

$_1Q_2 = U_2 - U_1 + \int pdV = U_2 - U_1 + p\,(V_2 - V_1)$
$_1Q_2 = m\,C_p\,(T_2 - T_1)$, if the working fluid is a perfect gas.

(iii) *Constant temperature process:*

$$ {}_1Q_2 = U_2 - U_1 + p_1V_1 \,\ell n\frac{V_2}{V_1} $$

$$ {}_1Q_2 = m R T_1 \,\ell n\frac{V_2}{V_1} $$, if the working fluid is a perfect gas.

(iv) *Adiabatic process:* It can be shown that pV^{γ} = constant for this process and γ is the ratio of specific heats of the gas. The heat interaction is zero for this process.

$$ 0 = U_2 - U_1 + \frac{p_1 V_1 - p_2 V_2}{(\gamma-1)} $$

$$ = m\, C_v (T_2 - T_1) + \frac{p_1 V_1 - p_2 V_2}{(\gamma-1)} $$, if the working fluid is a perfect gas.

(v) *Polytropic process:* It is the process for which pV^n = constant, where n is index of expansion or compression.

$$ {}_1Q_2 = U_2 - U_1 + \frac{p_1 V_1 - p_2 V_2}{(n-1)} $$

$$ = m\, C_v (T_2 - T_1) + \frac{p_1 V_1 - p_2 V_2}{(n-1)} $$, if the working fluid is a perfect gas.

9. State the differences between a closed system and an open system.

Ans: Differences between a closed system and an open system are listed in Table 2.1.

Table 2.1 Differences Between a Closed System and an Open System

Closed System	Open System
System approach defined by boundary interactions at the boundaries.	Control volume approach defined by boundary interactions at the boundary.
Boundary usually does change its shape.	Boundary does not change its shape.
Material does not flow across the boundary.	Material flows across the control volume.

10. State the steady flow energy equation (SFEE).

Ans: A control volume is any volume of fixed shape and of fixed position and orientation to the observer. We shall derive an expression for the flow work or displacement work. (see Figure 2.6 or 2.7)

$$ \text{Flow work per unit mass} = \frac{p\,A\,dx}{dm} = \frac{p\,A\frac{dV}{A}}{dm} = p\frac{dV}{dm} = pv, \quad \frac{\text{N.m}}{\text{kg}} $$

We know $\frac{d\dot{m}}{dt} = (\dot{m}_1 - \dot{m}_2)$ and for steady flow process $\frac{d\dot{m}}{dt} = 0$. Hence, $\dot{m} = \dot{m}_1 = \dot{m}_2$.

The steady flow energy equation (SFEE) can be written using the energy conservation principle (see Figure 2.8)

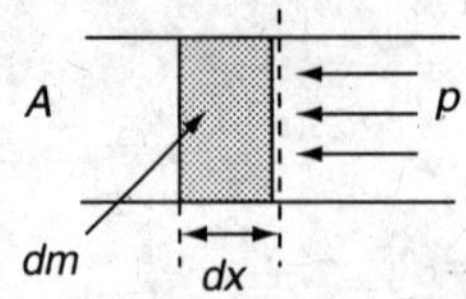

Figure 2.6 Schematic for Flow Work

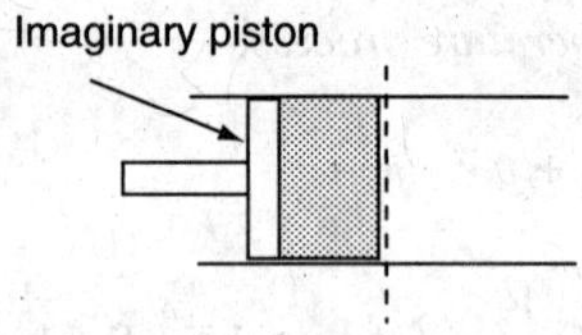

Figure 2.7 Schematic for Displacement Work

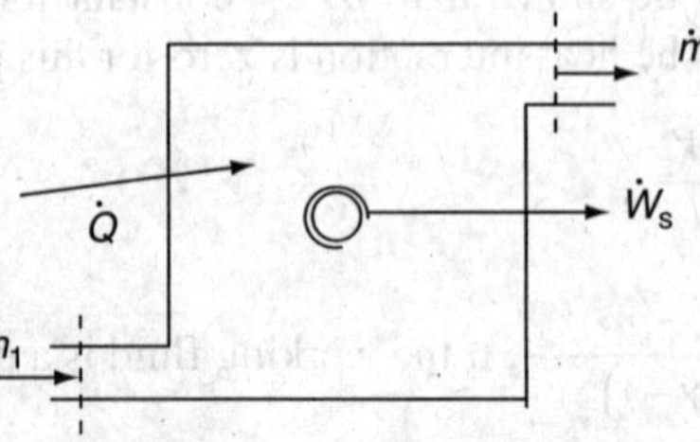

Figure 2.8 Diagram Used in the derivation of the Steady Flow Energy Equation

$$\dot{Q} + \dot{m}\left(\frac{V_1^2}{2} + u_1 + p_1 v_1 + gZ_1\right) = \dot{W}_s + \dot{m}\left(\frac{V_2^2}{2} + u_2 + p_2 v_2 + gZ_2\right)$$

The aforesaid equation can also be rewritten as

$$\frac{\dot{Q} - \dot{W}_s}{\dot{m}} = \left[\frac{\left(V_2^2 - V_1^2\right)}{2} + \left(p_2 v_2 - p_1 v_1\right) + \left(u_2 - u_1\right) + g\left(Z_2 - Z_1\right)\right]$$

or $\frac{\dot{Q} - \dot{W}_s}{\dot{m}} = \int V\,dV + \int dh + \int g dZ$ where V is the velocity of the fluid.

11. Show how the steady flow energy equation can be simplified to the forms applicable to turbine, compressor, nozzle, diffuser and heat exchanger.

Ans: A turbine, whether the fluid is a gas or liquid, is a device as shown in Figure 2.9 in which the fluid does work against some type of blade attached to a rotating shaft. As a result, the device produces work which may be used for some purpose in the surroundings. Pumps, compressors and fans are devices as shown in Figure 2.10 in which work is done on the fluid which results in an increase in pressure of the fluid. Pumps are usually associated with liquids and compressors, and fans are employed for gases. The ratio of outlet-to-inlet pressure across a fan will probably be just slightly above 1, whereas for a compressor the ratio will probably be between 3 and 10. For steady flow through any of these devices, the energy equation reduces to

$$\dot{q} - \dot{w}_s = h_2 - h_1 + \frac{V_2^2 - V_1^2}{2}$$

for unit mass of the fluid flowing through these devices. The potential energy change is normally negligible. Two other terms in this expression require comment. As we have

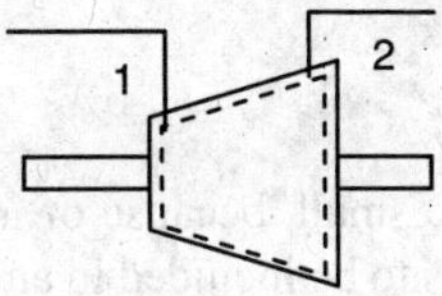

Figure 2.9 Turbine

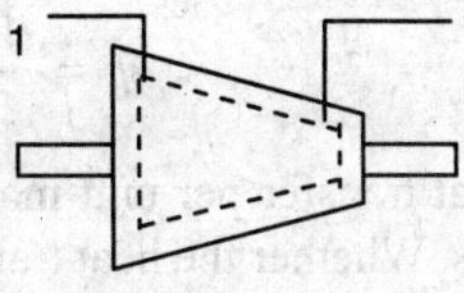

Figure 2.10 Compressor

noted already, the inclusion of heat quantity depends upon the mode of operation. If the device is not insulated, the heat gained or lost by the fluid depends upon whether or not a large temperature difference exists between the fluid and the surroundings. Only experience or experimental specifications will determine the proper approach. As a second point, it is found that the change in kinetic energy is usually quite small in these devices. Velocities may be kept small by a suitable choice of flow area. In many cases, then, the steady flow energy equation for these devices becomes

$$-\dot{w}_s = h_2 - h_1.$$

In this approximate solution, it is seen that the enthalpy decreases for a turbine and increases in the direction of flow for compressors and pumps. The enthalpy change is a composite of the changes in internal energy and the pv product.

Nozzle and Diffuser: In many steady flow processes, there is a need to either increase or decrease the velocity of a flow system. A device that increases the velocity (and hence the kinetic energy) of a fluid at expense of a pressure drop in the direction of flow is called a nozzle. A diffuser is a device for increasing the pressure of a flow stream at the expense of a decrease in velocity. Figure 2.11 shows the sketches of the general shapes of a nozzle and a diffuser. Note that a nozzle is a converging passage. The opposite conditions hold for a diffuser. We are not in a position to explain the reason for these differences at this point, theoretically; so it should be accepted as a matter of experience. One outgrowth of this that is extremely important in rocket design is that a converging-diverging nozzle must be used if a fluid is to be accelerated from a low (subsonic) velocity to a high (supersonic) velocity.

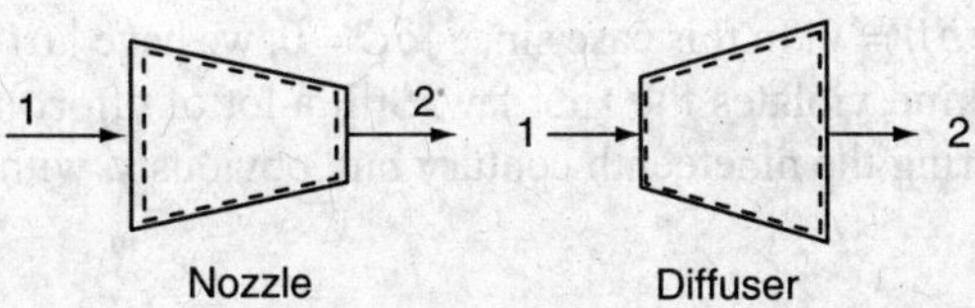

Figure 2.11 Flow Devices with No Work and Heat Interactions

The actual shapes of these devices in practice are not as shown, but will probably have curved surfaces in the direction of flow. The true shape is unimportant for the overall energy balance that we wish to make. Since both the devices are merely ducts, it is apparent that no shaft work is involved, and the change in potential energy, if any, would be negligible under most conditions. For a simple substance, the open, steady flow energy equation per unit mass reduces to

$$\dot{q} = \frac{V_2^2 - V_1^2}{2} + h_2 - h_1.$$

In many cases the heat transfer per unit mass may be quite small, because of insulation around the duct, or for other reasons. Whether the heat transfer term needs to be included in an analysis depends on its magnitude relative to other terms in the energy balance. Thus, nozzles and diffusers are variable area ducts employed to accelerate or decelerate a fluid. Specification of the inlet conditions is insufficient to determine the outlet state of a fluid in these devices. One outlet property or a specification of the type of process is required to fix the final state.

12. What is a throttling process? Explain the significance of this process.

Ans: Steady flow systems such as turbines and nozzles produce useful work or increase the kinetic energy of the fluid as it passes through these devices. Associated with these effects is a drop in pressure. In some instances, it is useful to produce a significant pressure drop without any work interactions or changes in kinetic or potential energy. This type of process is called throttling. Flow through a restriction such as a valve or a porous plug as shown in Figure 2.12 fulfils the necessary conditions. Although the velocity may be quite high in the region of the restriction, measurements upstream and downstream from the actual valve area will indicate that the change in velocity, and hence the change in kinetic energy, across the restriction is very small. Since the control volume is rigid and no rotating shafts are present, no work interactions are involved. In most applications, either the throttling device is insulated or heat transfer is insignificant. Thus, for a throttling process, the enthalpy change is zero, or

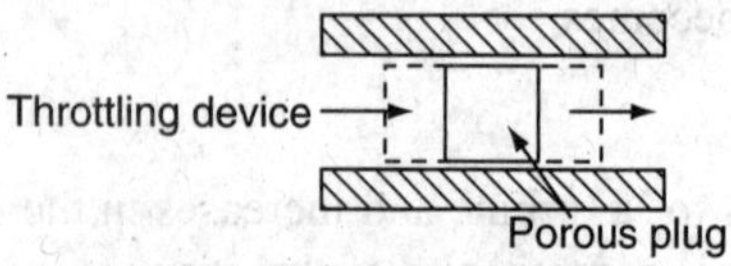

Figure 2.12 Throttle Valve

$$h_1 = h_2.$$

This statement does not say that the enthalpy is constant for the process, but merely requires that the initial and final enthalpies be the same.

13. Explain the concept of Perpetual Motion Machine of the First Kind (PMMI).

Ans: A machine that operating in a cycle produces work continuously or cyclically without absorption of energy at the same time is known as the perpetual motion machine of the first kind. According to the first law, we have $\oint \delta Q - \oint \delta W = 0$. In this case since $\oint \delta Q = 0$, we have $\oint \delta W = 0$. This means perpetual motion machine of the first kind violates the first law. Still a lot of effort has been spent in inventing such machines, especially during the nineteenth century but, obviously, without success.

Solved Problems

1. During an expansion process for a closed system, the volume of a gas changes from 1 m^3 to 2 m^3. The pressure changes during this quasi-static process according to the relation $p = 20\,V + 5$, where p is in N/cm^2 and V is in m^3. Find the work done by the gas in joules.

$$W = \int p\,dV = \int_1^2 (20V + 5) \times 10^4\,dV = [10\,(2^2 - 1^2) + 5\,(2 - 1)] \times 10^4$$

$$= 35 \times 10^4 \text{ J or N.m.}$$

2. The Clausius equation of state for a gas is given by the expression $p\ (v - b) = RT$, where b and R are constants. Derive an expression for the work done during a quasi-static isothermal expansion process, in terms of R, T, v and b.

For isothermal process, T = constant and hence, $p\ (v - b)$ = constant.

$$W = \int p\ dv = \int_{v_1}^{v_2} \frac{RT}{(v-b)} dv = RT\ell n \frac{(v_2 - b)}{(v_1 - b)}$$

3. Two kilogram of a gas is contained in a piston-cylinder assembly at initial conditions of 0.06 m³ and 6.8 bar. The gas is allowed to expand to a final volume of 0.14 m³. Determine the amount of work done for the following processes:
   (i) The pressure remains constant,
   (ii) The product pV is a constant and
   (iii) The product pV^2 is a constant.

(i) $p = C;\ W = \int p\,dV = p\int_{V_1}^{V_2} dV = 6.8 \times 10^5\,(0.14 - 0.06) = 6.8 \times 10^5 \times 0.08 = 54400$ N.m.

(ii) $pV = C;\ W = p_1 V_1 \ell n \frac{V_2}{V_1} = 6.8 \times 10^5 \times 0.06 \times \ell n \frac{0.14}{0.06} = 34569.75$ N.m .

(iii) $pV^2 = C;\ pV^n = \text{Constant},\ W = \frac{(p_1 V_1 - p_2 V_2)}{(n-1)},\ p_2 = (p_1\ V_1^2)/(V_2^2) = 1.25 \times 10^5$ bar.

$$W = \frac{(6.8 \times 0.06 - 1.25 \times 0.14) \times 10^5}{(2-1)} = \frac{(0.408 - 0.175) \times 10^5}{(1)} = 23300 \text{ N.m.}$$

4. An ideal gas is initially in a piston-cylinder assembly at 1.4 bar and 0.03 m³. The gas is first heated at constant pressure until the volume is doubled. It is then allowed to expand isothermally until the volume is again doubled. Determine the total work done by the gas and plot the processes on a p-V diagram (Figure 2.13). The initial gas temperature is 280°C.

Given data: $p_1 = 1.4$ bar, $V_1 = 0.03$ m³, $V_2 = 2\ V_1 = 0.06$ m³, $p_2 = p_1 = 1.4$ bar, $V_3 = 2\ V_2 = 0.12$ m³, $p_2\ V_2 = p_3\ V_3$

$${}_1W_2 = p\,(V_2 - V_1)$$
$$= 1.4 \times 10^5 \times 0.03 = 4200 \text{ N.m.}$$

$$\frac{V_1}{T_1} = \frac{V_2}{T_2},\ \text{hence, } T_2 = \frac{V_2}{V_1} T_1$$

$$T_2 = 2 \times (273 + 280) = 1106 \text{ K.}$$

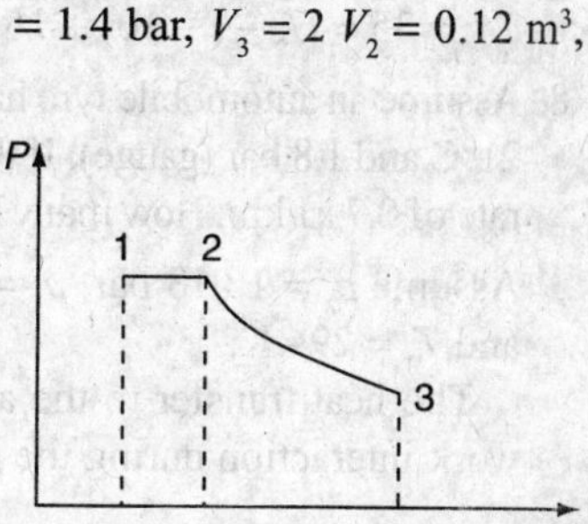

Figure 2.13 p-V Diagram

$${}_2W_3 = p_2 V_2 \ell n \frac{V_3}{V_2} = 1.4 \times 10^5 \times 0.06 \times \ell n 2 = 5822.436 \text{ N.m.}$$

Total work done = 4200 + 5822.436 = 10022.436 N.m.

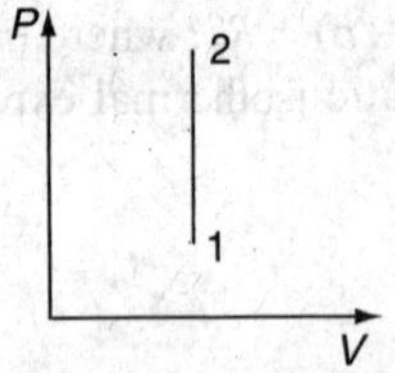
Figure 2.14 p-V Diagram

5. Five kg of a perfect gas is heated by supplying 166 kJ of heat. During the process the volume is held constant at 3.7 m^3 but the pressure increases from 1.02 bar to 1.2 bar. Compute the work done and the change in temperature. Sketch the process on a p-V diagram (Figure 2.14).

   Work done during a constant volume process is zero $p\,V = m\,R\,T$,
   $1.02 \times 10^5 \times 3.7 = 5 \times 287 \times T_1$, hence, $T_1 = 262.9$ K
   $1.2 \times 10^5 \times 3.7 = 5 \times 287 \times T_2$, hence, $T_2 = 309.4$ K
   The change in temperature = 46.5°C.

6. A closed system undergoes a process during which 8 kJ of heat is removed and 25 kJ of work is done by the system. The system is then restored to its initial state by adding 7 kJ of heat to it. There is also a work interaction in addition to the heat interaction. What are the magnitude and direction of work interaction during the second process?

   We know,
   $$_1Q_2 = U_2 - U_1 + {_1W_2} \quad (1)$$
   and
   $$_2Q_1 = U_1 - U_2 + {_2W_1} \quad (2)$$
   On substitution, $-8 = U_2 - U_1 + 25$; hence, $U_2 - U_1 = -33$ kJ and using this $U_2 - U_1$ in the equation (2), we get $7 = 33 + {_2W_1}$ or ${_2W_1} = -26$ kJ.
   Work is done on the system = 26 kJ.

7. For each of the following cases of a process involving a closed system, fill in the blank spaces (quantities are expressed in kJ):

	Q	W	U_1	U_2	dU
(i)	25	−10		−10	
(ii)	−10			65	−20
(iii)		15	−10		20

   (i) ${_1Q_2} = U_2 - U_1 + {_1W_2}$, $25 = -10 - U_1 - 10$, hence, $U_1 = -45$ and $dU = U_2 - U_1 = 35$ kJ.
   (ii) ${_1Q_2} = U_2 - U_1 + {_1W_2}$, $-10 = -20 + {_1W_2}$, hence, ${_1W_2} = 10$, $dU = U_2 - U_1 = 65 - U_1 = -20$; hence, $U_1 = 85$ kJ.
   (iii) ${_1Q_2} = U_2 - U_1 + {_1W_2}$, ${_1Q_2} = 20 + 15$, hence, ${_1Q_2} = 35$ kJ,
   $35 = U_2 - (-10) + 15 = U_2 + 25$, hence, $U_2 = 10$ kJ.

8. Assume an automobile tyre has a constant volume of 0.057 m^3 and an initial temperature and pressure of 21°C and 1.8 bar (gauge). If this tyre is driven on a hot road, heat is transferred to the air in the tyre at the rate of 0.7 kJ/km. How many kms of driving will be required to raise the tyre pressure to 2.4 bar (gauge)?

   Assume $p_a = 1.013$ bar, $p_1 = 1.8 + 1.013 = 2.813$ bar, $p_2 = 2.4 + 1.013 = 3.413$ bar, $V_1 = 0.057$ m^3 and $T_1 = 294$ K.

   The heat transfer to the air in the tyre is one of constant volume process and hence there is no work interaction during the process.

   $$\text{Mass of air in the tyre, } m = \frac{2.813 \times 10^5 \times 0.057}{287 \times 294} = 0.19 \text{ kg and}$$

   $$T_2 = \frac{3.413}{2.813} \times 294 = 356.7 \text{ K.}$$

$_1Q_2 = m \times C_V \times (T_2 - T_1) = 0.19 \times 0.718 \times (62.7) = 8.55$ kJ.

Hence, the distance covered to raise the tyre pressure $= \dfrac{8.55}{0.7} = 12.2$ km.

9. 0.085 m³ of carbon dioxide at 1.38 bar and 4.5°C is contained in a piston cylinder assembly. Electric current flows through a resistor within the cylinder until 35 kJ of electrical energy has entered the gas. If the process is constant pressure and the final volume is 0.142 m³, determine the magnitude and direction of any heat transfer in kJ.

   Gas constant = (8314.4)/(44) = 188.96 J/kg K and C_V for CO_2 = 655 J/kg K.

   For constant pressure process, $\dfrac{V_1}{T_1} = \dfrac{V_2}{T_2}$ and $T_2 = \dfrac{V_2}{V_1} T_1 = \dfrac{0.142}{0.085} \times 277.5 = 463.5$ K.

   Mass of the gas $= \dfrac{1.38 \times 10^5 \times 0.085}{188.96 \times 277.5} = 0.2237$ kg.

   Applying the first law of thermodynamics,

   $$_1Q_2 + 35 = 0.655 \times 0.2237 \times (463.5 - 277.5) + 1.38 \times 100 \times (0.142 - 0.085),$$

   $_1Q_2 = 0.119$ kJ, that is, 0.119 kJ of heat enters the system.

10. A closed system consists of a vertical piston-cylinder assembly containing a fluid which is compressed by a frictionless piston weighing 3044 N. During an interval of time, paddle wheel within the cylinder does 6781 N.m of work on the fluid. If the heat transferred during this time is –10.55 kJ and the change in internal energy is –1.05 kJ, calculate the distance the piston moves. The area of the piston is 51.6 cm² and the atmospheric pressure is 1.01 bar.

    Applying the first law of thermodynamics,

    $$_1Q_2 = U_2 - U_1 + W_p + {_1W_2},$$

    $-10.55 = -1.05 - 6.781 + {_1W_2}$, hence, $_1W_2 = -2.719$ kJ or kN.m

    Force acting on the piston = 3044 + 1.01 × 10 × 51.6 = 3565 N
    The distance moved by the piston = (– 2719)/(3565) = –0.762 m (the piston moves downwards).

11. Air initially at 1.4 bar and 60°C flows through a cross-sectional area of 0.0093 m² at the rate of 45 kg/min. Downstream from this point the pressure is 2.8 bar, the temperature is 60°C and the velocity is 15.24 m/s. Determine (i) the inlet velocity and (ii) the outlet area in m².

    Given data: p_1 = 1.4 bar, T_1 = 333 K, A_1 = 0.0093 m², mass flow rate = 45 kg/min, p_2 = 2.8 bar, T_2 = 333 K, Velocity, V_2 = 15.24 m/s.

    Density of air, $\rho_1 = \dfrac{1.4 \times 10^5}{287 \times 333} = 1.464 \text{ kg/m}^3$ and $\rho_2 = \dfrac{2.8 \times 10^5}{287 \times 333} = 2.928 \text{ kg/m}^3$

    Velocity of air, $V_1 = \dfrac{m}{\rho_1 A_1} = \dfrac{45}{60 \times 1.464 \times 0.0093} = 55 \text{ m/s}$

    Outlet area, $A_2 = \dfrac{45}{60 \times 2.928 \times 15.24} = 0.0168 \text{ m}^2.$

12. An ideal gas enters a steady flow system at 1.013 bar and 38°C. The inlet to the system has a diameter of 20.32 cm, and the velocity at that point is 9.14 m/s. At the outlet, the diameter of the

duct is 7.62 cm and the density of the fluid at that section is 4.0 kg/m³. If the molar mass of the gas is 30, determine the velocity at the outlet and the mass flow rate in kg/min.

Gas constant, $R = 8314.4/30 = 277$ J/kg K, and density, $\rho_1 = \dfrac{1.013 \times 10^5}{277 \times 311} = 1.175\,\text{kg/m}^3$

Mass flow rate, $m = \rho_1 A_1 V_1 = 1.175 \times (\pi/4) \times (0.2032)^2 \times 9.14 = 0.348$ kg/s or 20.88 kg/min

$$\text{Outlet velocity, } V_2 = \frac{m}{\rho_2 A_2} = \frac{0.348}{4.0 \times (\pi/4) \times (0.0762)^2} = 19\,\text{m/s}.$$

13. Calculate the power required by a compressor if air flowing at the rate of 0.9 kg/s enters at 1.014 bar, 5°C with a velocity of 61 m/s and leaves at 2.06 bar, 71°C with a velocity of 122 m/s. Heat transferred from the air to the cooling water circulating around the compressor casing amounts to 18.6 kJ/kg of air.

The steady flow energy equation is

$$\dot{Q} + \dot{m}\left[h_1 + \frac{V_1^2}{2} + g\,Z_1\right] = \dot{W}_S + \dot{m}\left[h_2 + \frac{V_2^2}{2} + g\,Z_2\right].$$

$$\text{Hence, } \dot{W}_S = 0.9\left[1.005 \times 278 + \frac{61^2}{2000} - 1.005 \times 344 - \frac{122^2}{2000}\right] - 0.9 \times 18.6 = -81.46\ \text{kJ/s}.$$

The power required by the compressor is 81.46 kW (Note: the negative sign indicates work enters the system).

14. Air is compressed isothermally from 1.01 bar at 5°C to 4.81 bar. Flow through the compressor is steady at 0.9 kg/s. Kinetic and potential energies are negligible. Assuming that air is an ideal gas, calculate the power input to the compressor and the heat removed from the air, both in kJ/s.

For isothermal compression, $W = \int p\,dv$ or $-\int v\,dp$ provided the changes in kinetic and potential energies are negligible.

$$W = p_1 v_1 \ell n \frac{v_2}{v_1} \quad \text{or} \quad mRT_1 \ell n \frac{p_1}{p_2}$$

$$W = 0.9 \times 0.287 \times 278\,\ell n \frac{1.01}{4.81} = -112.07\ \text{kJ/s}.$$

Work is done on the system and heat is rejected by the system.

15. An air compressor takes in air from the atmosphere at 1.013 bar and 18°C. The pressure ratio is 7.0 and the final specific volume is 0.234 m³/kg. For every kg of air that passes through the compressor, 19.8 kJ of heat is removed. If the changes in the kinetic and potential energies are negligible, determine the work input in N.m/kg.

Given data: $p_1 = 1.013$ bar; $T_1 = 291$ K; $p_2 = 7 \times 1.013 = 7.091$ bar and $v_2 = 0.234$ m³/kg.

$$\text{Final temperature, } T_2 = \frac{7.091 \times 10^5 \times 0.234}{287} = 578\,\text{K}.$$

The steady flow energy equation for unit mass with $\Delta K.E$ and $\Delta P.E \approx 0$ reduces to

$\dot{q}+(h_1+......)=\dot{w}_S+(h_2+......)$, that is, $\dot{w}_S=-19.8+[1.005(291-578)]=-308.2\,\text{kJ/kg}$

Work input = 308200 N.m/kg.

Practice Problems

1. In many real processes, it is found that gases at low pressures fulfil the relationship $pV^n = c$, where n and c are constants. If n is taken as 1.3, determine the work done by air in a piston-cylinder assembly when it expands from 0.0566 m³ and 4.16 bar to a final pressure of 1.34 bar.

   (18055.8 N.m)

2. A piston-cylinder assembly contains 0.085 m³ of an ideal gas at 1.38 bar and 15°C. The gas is compressed to a final pressure of 4.14 bar and a temperature of 15°C.
   (i) Determine the minimum work of compression, if the process is isothermal.
   (ii) Consider now that a process occurs between the same two end states as in (i) such that it follows a straight line on p-V co-ordinates. Determine the work of compression for this process and compare the two processes on a p-V diagram.

   (−12886.7 N.m; −15640.9 N.m)

3. A gas system executes a process such that at any state the product of its pressure and volume is a constant. If the initial state is defined by the pressure 5.2 bar and volume 0.28 m³ and 33907 N.m of work flow from the system, determine the final volume and pressure of the system.

   (0.3532 m³, 4.119 bar)

4. A certain system undergoes a process such that at all states pV^n = constant, where the pressure p is in N/m² and the volume V is in m³, n is a constant. Show that the work flow in such a case between two states 1 and 2 is

$$W=\frac{(p_1V_1-p_2V_2)}{(n-1)}.$$

   If the initial pressure and volume are 13.8 bar and 0.085 m³, respectively, the final pressure is 2.76 bar and $n = 1.3$, then determine the work flow.

   (121.3 kN.m)

5. A gas in a closed system expands in a quasi-equilibrium process from a volume of 0.2 m³ at 200 kPa, 80°C to a volume of 0.6 m³ in such a manner that $p = (225 - 125\,V)$, with p in kPa and V in m³. How much work is done?

   (70 kN.m)

6. Ten grams of an ideal gas (molecular mass = 32) expands quasi-statically at constant pressure in a closed system from 1.38 bar, 20°C to 80°C. During the process, 0.548 kJ of heat is added to the closed system. Compute the average value C_v for the gas in kJ/kg K.

   (0.654 kJ/kg K)

7. One-tenth of a kilogram of air initially at 2.76 bar and 60°C expands isothermally and quasi-statically until the volume is doubled. Evaluate the quantity of heat added or removed in kJ/kg and the change in internal energy of the air.

   (6.62 kJ, 0 kJ)

8. Carbon dioxide is expanded isothermally in a non-flow process from initial conditions of 1.38 bar, 171°C and 0.057 m^3 to a final volume of 0.114 m^3. Calculate the work done in kN.m and the heat transfer in kJ.

(5.45 kN.m, 5.45 kJ)

9. Nitrogen at 3.1 bar and 116°C is allowed to expand adiabatically and without friction in a closed system to a final pressure of 1.7 bar. Compute the work done by each kg of nitrogen in kN.m. Assume $\gamma = 1.4$ for nitrogen.

(45.58 kN.m)

10. 2.5 kg of air is compressed adiabatically and without friction from an initial pressure of 1.38 bar and 2.83 m^3 to a final volume of 0.566 m^3. Calculate the change in internal energy of the air.

(882.25 kJ)

11. Air enters the nozzle at 11 bar and 43°C. For a frictionless adiabatic process, what is the velocity of air at the exit of the nozzle when the pressure is 3.4 bar? Neglect the entrance velocity.

(425.5 m/s)

12. Air expands through a nozzle from 3.5 bar, 150°C to 1.04 bar, 27°C. Neglect the inlet velocity. Compute the exit velocity if $q_{out} = 4.5$ kJ/kg. State all necessary assumptions.

(488 m/s)

13. Air at 0.7 bar and 42°C enters a compressor of a jet plane with a velocity of 213 m/s and leaves the compressor at a pressure of 2.95 bar, a temperature of 200°C, and a velocity of 243 m/s. The cross-sectional area of the inlet to the compressor is 0.0836 m^2. Determine the power required to drive the compressor if the process is adiabatic.

(2282 kW)

14. An air turbine forms part of an aircraft refrigerating plant. Air at a pressure of 2.95 bar and a temperature of 58°C flows steadily into the turbine with a velocity of 45 m/s. The air leaves the turbine at a pressure of 1.15 bar, a temperature of 2°C and a velocity of 150 m/s. The shaft work delivered by the turbine is 54 kN.m/kg of air. Neglecting changes in elevation, determine the magnitude and sign of the heat transfer per unit mass of air flowing (Assume $C_P = 1005$ J/kg K and enthalpy, $h = F(T)$ only for air).

(+7.96 kJ/kg)

15. Air enters a converging nozzle at 1.72 bar, 60°C and 30.5 m/s. At the outlet of the adiabatic passage the pressure is 1.03 bar and the velocity is six times its initial value. If the inlet area is 0.0093 m^2, determine the outlet temperature of the gas and the exit area of the nozzle in m^2.

(43.8°C; 0.00246 m^2)

Objective Type Questions

1. The thermodynamic cycle is a concept of

(i) any system (ii) an open system
(iii) a closed system (iv) an isolated system

2. An insulated tank initially contains 0.25 kg of a gas with an internal energy of 200 kJ/kg. Additional gas with an internal energy of 300 kJ/kg and an enthalpy of 400 kJ/kg enters the tank until the total mass of the gas contained is 1 kg. The internal energy of the gas finally in the tank, in kJ/kg, is
   (i) 250 (ii) 275
   (iii) 300 (iv) 350
3. Which equation is valid for a thermodynamic system undergoing a quasi-equilibrium process?
   (i) $\delta q = du + \delta w$ (ii) $\delta q = du + pdv$
   (iii) $Tds = du + \delta w$ (iv) $Tds = du + pdv$
4. If the work interaction by a closed system containing a perfect gas ($\gamma = 1.4$) during a polytropic expansion ($pv^{1.3}$ = constant) is 40 kN.m, then the heat interaction during the process, in kJ, is
   (i) 10 (ii) 20
   (iii) 30 (iv) 40
5. Identify the extensive property that when multiplied by the intensive property pressure gives flow energy to an open system.
   (i) Area (ii) Entropy
   (iii) Mass (iv) Volume
6. Which integral represents the work interaction for a steady flow device with negligible changes in kinetic and potential energies (symbols have their usual meaning)?
   (i) $\int p\,dv$ (ii) $-\int v\,dp$
   (iii) $\int F\,dx$ (iv) $\int VI\,dt$
7. If the work output of the ideal cycle represented as circle in the following *p-V* diagram (Figure 2.15) is 9π kN.m, then the work output of the cycle ABCD, in kN.m, is

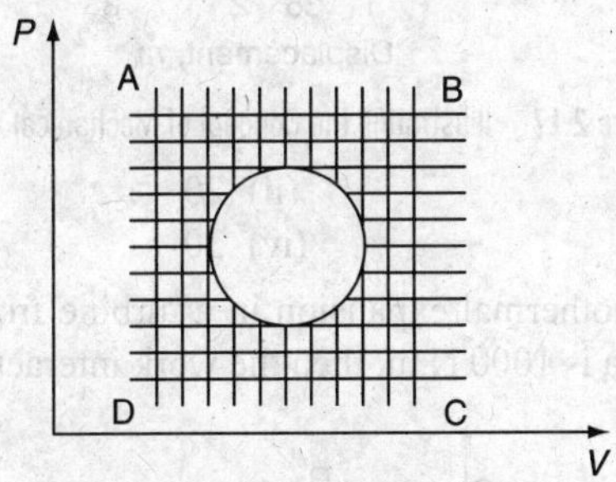

Figure 2.15 *p-V* Diagram

   (i) 64 (ii) 72
   (iii) 81 (iv) 100
8. The rectangle 1234 and the square 4536 in the following pressure-volume diagram (Figure 2.16) represent two cyclic processes for a system (5 is the mid point of the side 12). If the work interaction for the rectangular cycle is 24 Nm, then the work interaction of the square cycle, in Nm, is
   (i) 12 (ii) 18
   (iii) 24 (iv) 30

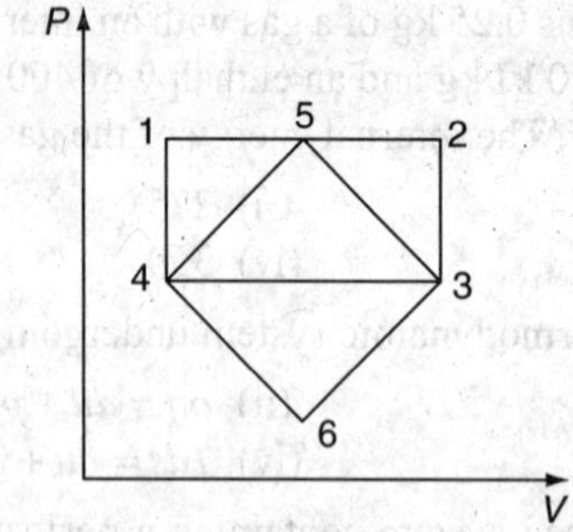

Figure 2.16 p-V Diagram

9. The balloon inflating operation is an example of

(i) isentropic process (ii) isobaric process
(iii) isometric process (iv) isothermal process

10. The following Figure 2.17 depicts the variations of force in an elementary system that undergoes a process during which the plunger position changes from 0 m to 3 m. If the change in internal energy of the system is 2.5 J, then the heat absorbed by the system during the process, in J, is

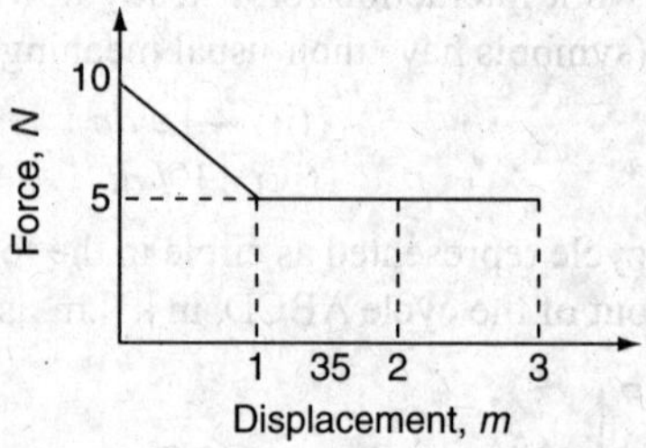

Figure 2.17 Illustrating the Concept of Mechanical Work

(i) 15 (ii) 20
(iii) 25 (iv) 30

11. An ideal gas undergoes an isothermal expansion in a turbine from state R to state S as shown in Figure 2.18. If the shaded area is 1000 N.m, then the work interaction during the process, in N.m, is

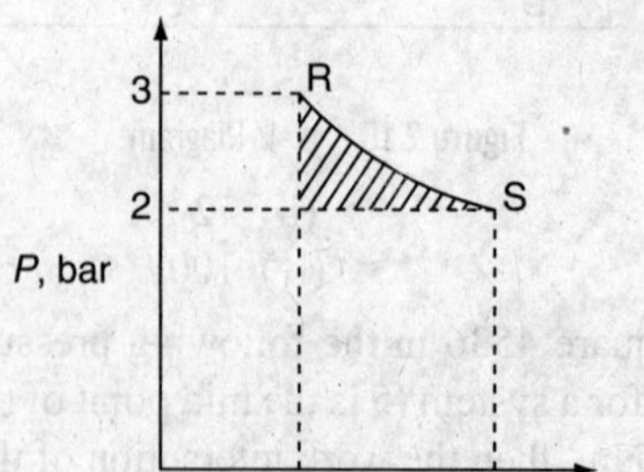

Figure 2.18 p-V Diagram for Flow System

(i) 14000 (ii) 12000
(iii) 11000 (iv) 10000

12. Which one of the following processes is called wire drawing in thermodynamics?
    (i) Isentropic (ii) Isenthalpic
    (iii) Isothermal (iv) Isobaric
13. The work interaction during an isothermal process is represented by a line integral when the process is outlined on
    (i) temperature-volume co-ordinates
    (ii) enthalpy-entropy co-ordinates
    (iii) pressure-volume co-ordinates
    (iv) temperature-entropy co-ordinates
14. The work interaction during the ideal process 1–2 shown in Figure 2.19 that occurs in a rotary machine, in kNm, is

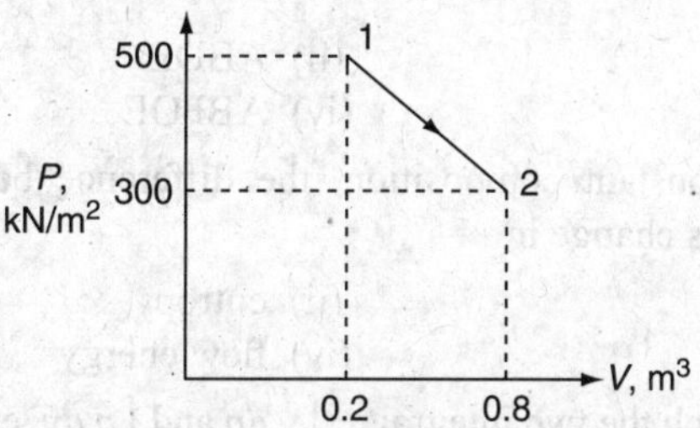

Figure 2.19 *p*-V Diagram

    (i) 100 (ii) 200
    (iii) 300 (iv) 400
15. The work interaction for the cycle shown in Figure 2.20, in kNm, is

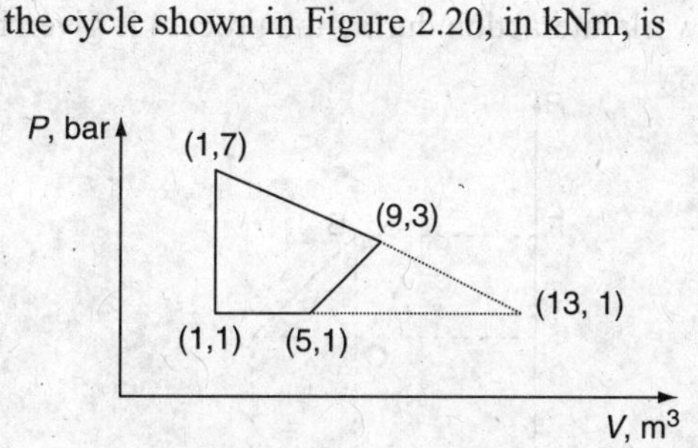

Figure 2.20 *p*-V Diagram

    (i) 1200 (ii) 2800
    (iii) 5600 (iv) 7200
16. Assuming the heat interaction (δQ), work interaction (δW), changes in internal energy (dU), enthalpy (dH) and entropy (dS) for a system undergoing a thermodynamic process, the first law of thermodynamics relates
    (i) δQ, δW and dS (ii) δQ, δW and dU
    (iii) δW, dH and dS (iv) δQ, dU and dS

17. The heat rejection in a reciprocating air compressor during the reversible compression process AB, shown in the temperature-entropy diagram (Figure 2.21), is represented by the area

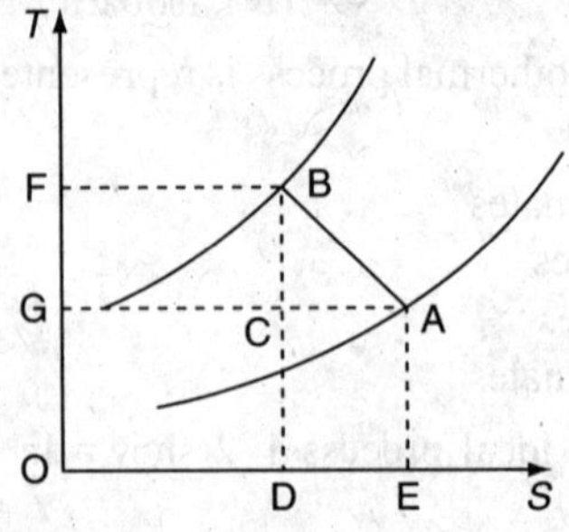

Figure 2.21 Temperature-entropy Diagram

(i) ABC (ii) ABDE
(iii) ABFG (iv) ABFOE

18. For a closed system of constant composition, the difference between the net heat and work interactions is identifiable as change in

(i) enthalpy (ii) entropy
(iii) internal energy (iv) flow energy

19. Identify the process for which the two integrals $-\int v\,dp$ and $\int p\,dv$, evaluated between any two given states, give the same value.

(i) Isenthalpic (ii) Isothermal
(iii) Isentropic (iv) Polytropic

20. The following pressure-volume diagram (Figure 2.22) shows the reversible compression process in a rotary compressor. The work interaction during the process is given by the area

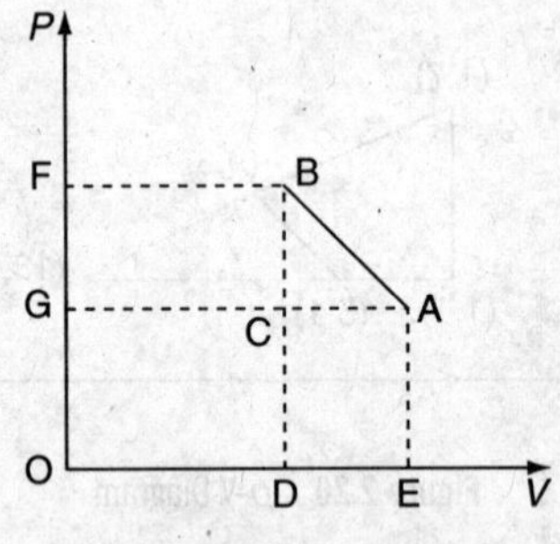

Figure 2.22 *p-V* Diagram

(i) ABC (ii) ABDE
(iii) ABFG (iv) ABFOE

21. Identify the areas on the *p-V* diagram (Figure 2.23) where the workable processes of a system are encountered with reference to the initial state 1.

(i) A2 and A3 (ii) A1 and A4
(iii) A1 and A2 (iv) A3 and A4

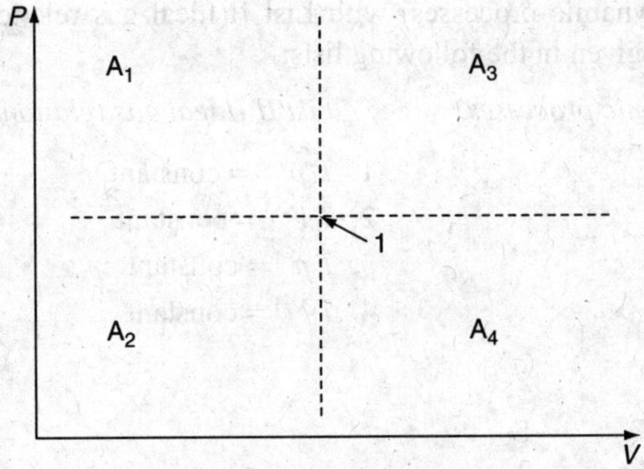

Figure 2.23 p-V Diagram

22. The chemical reaction that occurs in fuel cell is

(i) isentropic
(ii) isobaric
(iii) isometric
(iv) isothermal

23. Match List I with List II and select the correct answer using the codes given in the following lists:

List I (Apparatus)	*List II (Thermodynamic process)*
A. Separating calorimeter	1. Adiabatic process
B. Throttling calorimeter	2. Isobaric process
C. Sling psychrometer	3. Isometric process
D. Gas thermometer	4. Isenthalpic process

Codes:

	A	B	C	D
(i)	4	3	2	1
(ii)	2	4	1	3
(iii)	3	4	1	2
(iv)	2	1	4	3

24. Which terms in the steady flow energy equation represent flow energy of the system?

$$\dot{Q}+\dot{m}\left[C_vT_1+p_1v_1+\frac{V_1^2}{2}+g\,Z_1\right] = \dot{W}+\dot{m}\left[C_vT_2+p_2v_2+\frac{V_2^2}{2}+g\,Z_2\right]$$

(i) C_vT_1, C_vT_2
(ii) $g\,Z_1, g\,Z_2$
(iii) $V_1^2/2, V_2^2/2$
(iv) p_1v_1, p_2v_2

25. Match List I(Thermodynamic processes) with List II(Ideal gas relations) and select the correct answer using the codes given in the following lists:

List I (Thermodynamic processes)	*List II (Ideal gas relations)*
A. Isometric process	1. $T^{\gamma} p^{\gamma-1}$ = constant
B. Isobaric process	2. $T V^{n-1}$ = constant
C. Adiabatic process	3. $T p^{-1}$ = constant
D. Polytropic process	4. $T V^{-1}$ = constant

Codes:

	A	B	C	D
(i)	4	3	1	2
(ii)	2	1	4	3
(iii)	3	4	2	1
(iv)	3	4	1	2

26. Four compression processes undergone by an ideal gas, whose specific heats ratio $\gamma = 1.4$, are shown in the following *T-s* diagram (Figure 2.24). The index of compression (n) for the process 1–3 is

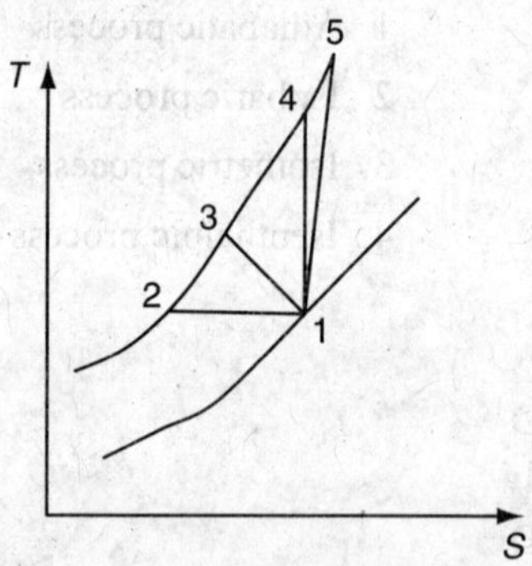

Figure 2.24 *T-s* Diagram

(i) 1.0
(ii) 1.3
(iii) 1.4
(iv) 1.6

27. The enthalpy change per unit mass of a thermodynamic system during an isentropic process is identical to the integral of

(i) $v\,dp$
(ii) $p\,dv$
(iii) $c_v\,dT$
(iv) $R\,dT$

28. Which compression law requires minimum work for a reciprocating compressor?

(i) pV^n = constant ($n > \gamma$)
(ii) pV^{γ} = constant
(iii) pV^n = constant ($1 < n < \gamma$)
(iv) pV = constant

29. Which one of the following processes most nearly approaches reversibility?
    (i) Air escaping from a tyre
    (ii) Compressing a spring
    (iii) Steam expanding through a nozzle
    (iv) Flue gases leaving a chimney

30. Match List I (Gadgets encountered in thermodynamics) with List II (Ideal thermodynamic processes) and select the correct answer using the following codes:

List I (Gadgets)	*List II (Thermodynamic processes)*
A. Porous plug	1. Isobaric
B. Nozzle	2. Isometric
C. Gas calorimeter	3. Throttling
D. Bomb calorimeter	4. Adiabatic

Codes:

	A	B	C	D
(i)	3	4	1	2
(ii)	3	2	4	1
(iii)	4	1	2	3
(iv)	4	3	1	2

31. A mixture of gases expands from 0.03 m^3 to 0.06 m^3 at constant pressure of 1 MPa and absorbs 84 kJ of heat during the process. The change in internal energy for the mixture, in kJ, is
    (i) 114 (ii) 84
    (iii) 54 (iv) 30

32. Which thermodynamic property remains invariant when a fluid flows through a porous plug of infinite fineness?
    (i) Pressure (ii) Enthalpy
    (iii) Entropy (iv) Volume

33. The pressure exerted on 10 kg of a gas contained in a cylinder by a frictionless piston in its upper end is 2 bar. If the volume of the gas changes from 3 m^3 to 1 m^3 with a corresponding change in its internal energy from 400 kJ to 200 kJ, then the heat interaction during the process, in kJ, is
    (i) 600 (ii) 200
    (iii) −200 (iv) −600

34. Select the multiplying factor that completes the following equation that gives the amount of heat interaction for a closed system during a reversible polytropic process, pv^n = constant.

$$q = \{\quad\} \times \frac{R(T_1 - T_2)}{(n-1)}$$

(i) $\frac{\gamma}{(\gamma-1)}$ (ii) $\frac{n}{(\gamma-1)}$

(iii) $\frac{(\gamma-n)}{(\gamma-1)}$ (iv) $\frac{(\gamma-1)}{(\gamma-n)}$

35. The index n in the relation pV^n = constant for the four expansion processes $a-b$, $a-c$, $a-d$ and $a-e$ undergone by the contents of a closed system and shown in the following p-V diagram (Figure 2.25) are respectively.

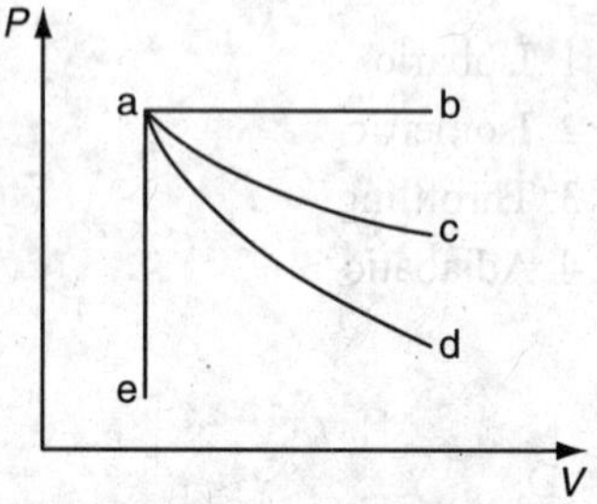

Figure 2.25 *p-V* Diagram

(i) 0, 1, γ, ∞ (ii) ∞, γ, 1, 0
(iii) 0, γ, 1, ∞ (iv) ∞, 1, γ, 0

36. Which one of the following events is reversible?
    (i) Flow of river over a water fall
    (ii) Flow of electric current through inductors and capacitors
    (iii) Hydraulic jump in water channel
    (iv) Magnetization of material exhibiting hysteresis
37. Which one of the following fluid (air) motions is isentropic?
    (i) Coutte flow
    (ii) Fanno flow
    (iii) Nozzle (converging) flow
    (iv) Rayleigh flow
38. In steady adiabatic flow of incompressible fluids, energy dissipation due to friction and turbulence results in
    (i) an increase in the internal energy
    (ii) a decrease in the internal energy
    (iii) an increase in the total pressure
    (iv) a decrease in the entropy
39. The enthalpy change is sufficient to determine the heat interaction for a closed system (containing a perfect gas) undergoing
    (i) any workable process
    (ii) an isothermal process
    (iii) an isobaric process
    (iv) an isometric process

40. Select the four temperatures that are equivalent.

(i) 35°C, 95°F, 368 K, 555 R (ii) 35°C, 95°F, 308 K, 495 R
(iii) 35°C, 63°F, 308 K, 523 R (iv) 35°C, 95°F, 308 K, 555 R

Answers

1. (iii)	2. (iv)	3. (iv)	4. (i)	5. (iv)	6. (ii)	7. (iv)	8. (iii)	9. (ii)	10. (ii)
11. (iii)	12. (ii)	13. (i)	14. (i)	15. (ii)	16. (ii)	17. (ii)	18. (iii)	19. (ii)	20. (iii)
21. (ii)	22. (iv)	23. (ii)	24. (iv)	25. (iv)	26. (ii)	27. (i)	28. (iv)	29. (ii)	30. (i)
31. (iii)	32. (ii)	33. (iv)	34. (iii)	35. (i)	36. (ii)	37. (iii)	38. (i)	39. (iii)	40. (iv)

3

Second Law of Thermodynamics and Entropy

1. State the importance of the second law of thermodynamics.

Ans: The first law of thermodynamics is simply a statement of the conservation of energy principle to a process undergone by the thermodynamic system. The first law places no restriction on the direction of the process, but satisfying the first law does not ensure that the process will actually occur. This inadequacy of the first law to identify whether a process can take place is solved by introducing another general principle, the second law of thermodynamics. A process will not occur unless it satisfies both the first and the second laws of thermodynamics. The use of the second law of thermodynamics is not limited to identifying the direction of process alone, but also asserts that energy has quantity as well as quality. The first law is concerned with the quantity of energy and the transformation of energy from one form to another with no regard to its quality. Preserving the quality of energy is a major concern to engineers and the second law provides the necessary means to determine the quality as well as the degree of degradation of energy during a process. More of high temperature energy can be converted to work than the same amount of energy at a lower temperature. The second law is also used in determining the theoretical limits for the performance of commonly used engineering systems such as heat engines and refrigerators as well as predicting the degree of completion of chemical reactions.

2. Define a reversible process.

Ans: A process that can be exactly reversed, that is, the system can be made to go through various states in the order opposite to the one in the original process, by an infinitesimal change in the surroundings is called a reversible process.

Another definition of a reversible process is that it is one that brings about such changes in the system and the surroundings that both the system and surroundings can be restored to their original conditions by reversing the direction of interaction. In general, any process that involves work interaction equivalent to infinitesimally slow motion of a frictionless piston due to infinitesimal difference of pressure across the piston and heat interaction due to infinitesimal difference of temperature is reversible.

$W = \int pdv$ is valid for a reversible process only. If the surroundings do an amount of work dw on the system, some of it is lost or dissipated and only the amount = $(-pdv)$, where dv is change (in this case decrease) in the volume of system, is transferred to the system.

3. What are the factors that render a process irreversible?

Ans: Motion involving friction, sudden expansion of a gas and transfer of heat from surroundings to the system or vice versa due to finite temperature difference are the factors that will render the process irreversible.

4. Define and explain heat engine. Define also its efficiency.

Ans: We have learnt that work can be converted to heat directly and completely, but converting heat into work requires the use of some special devices. These devices are called heat engines. These engines receive heat from a high-temperature source. These engines convert a part of this heat into work and reject the remaining waste heat to a low temperature sink (see Figure 3.1(a)). They operate on a cycle. Heat engines and other devices usually involve a fluid to and from which heat is transferred while undergoing a cycle. This fluid is called the working fluid. The term heat engine is often used in a broader sense to include work-producing devices that do not operate in a thermodynamic cycle. Engines that involve internal combustion such as gas turbine and car engines fall into this category. These devices operate in a mechanical cycle but not in a thermodynamic cycle. The working fluid (combustion gases) does not undergo a complete cycle. Instead of being cooled to the initial temperature, the exhaust gases are purged out and replaced by fresh air-and-fuel mixture at the end of cycle. The work producing device that best fits into the definition of a heat engine is the steam power plant, which is an external combustion engine. The schematic of a basic steam power plant is shown in Figure 3.1(b).

The efficiency of the heat engine is defined as the ratio of net work output to the heat supplied.

$$\text{Efficiency, } \eta = \frac{W}{Q_1}$$

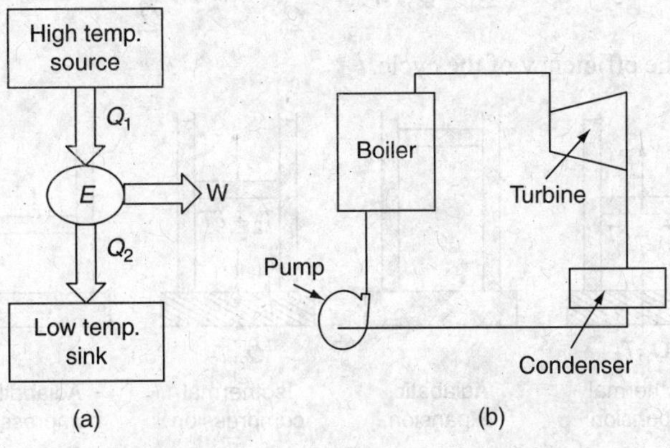

Figure 3.1 (a) Heat Engine, (b) Steam Power Plant

5. State the second law of thermodynamics due to Planck and Clausius and show that violation of either statement will result in violation of the other.

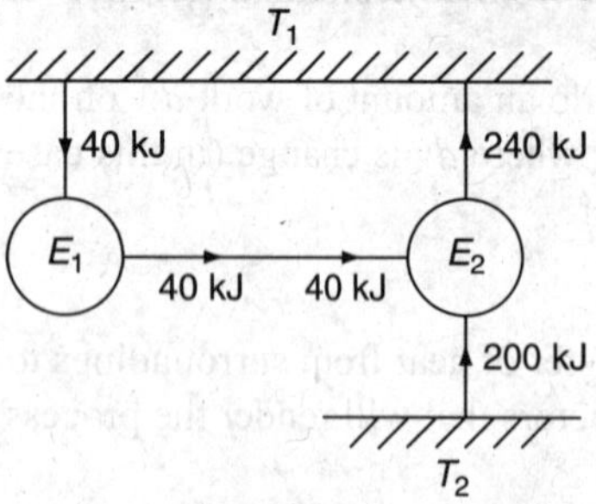

Figure 3.2 Violation of Planck Statement

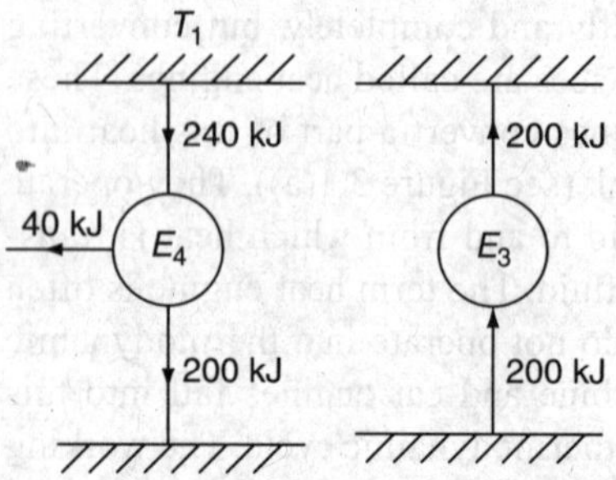

Figure 3.3 Violation of Clausius Statement

Ans: Planck's statement

'It is impossible to construct a device which will produce no effect other than cooling one body and delivering work during a complete cycle'.

1-T engine cannot operate continuously; it does not violate first law of thermodynamics.

Clausius statement: 'It is impossible to construct a device which will produce no other effect than transferring heat from one body at lower temperature to one at higher temperature.'

Let us suppose that Planck's statement is incorrect:

(i) E_1 violates Planck's statement (Figure 3.2) and E_2 does not violate Clausius statement.If we couple both devices, we see that 200 kJ of heat are transferred from T_2 toT_1 ($T_2 < T_1$). This contradicts the statement of Clausius.

Let us suppose that Clausius statement is incorrect:

(ii) E_3 violates Clausius statement (Figure 3.3) and E_4 does not violate any statement. If we combine E_3 and E_4, we see that 40 kJ of heat are converted into work without any change in surroundings. This contradicts Planck's statement. Hence both statements are equivalent.

6. Sketch the Carnot cycle on pressure volume diagram indicating the process, work done and heat transferred. Also derive the efficiency of this cycle in terms of temperature of the different process of the cycle.

Ans: Carnot cycle

A device working on Carnot's cycle is called Carnot's engine. In one cycle, the working substance absorbs heat Q_1 from the reservoir at a higher temperature called the source, does some work, say W, and rejects heat Q_2 to the reservoir at a lower temperature called the sink (see Figure 3.4).

$$W = Q_1 - Q_2;\ \eta = \frac{Q_1 - Q_2}{Q_1} = 1 - \frac{Q_2}{Q_1}$$

where η is defined as the efficiency of the cycle.

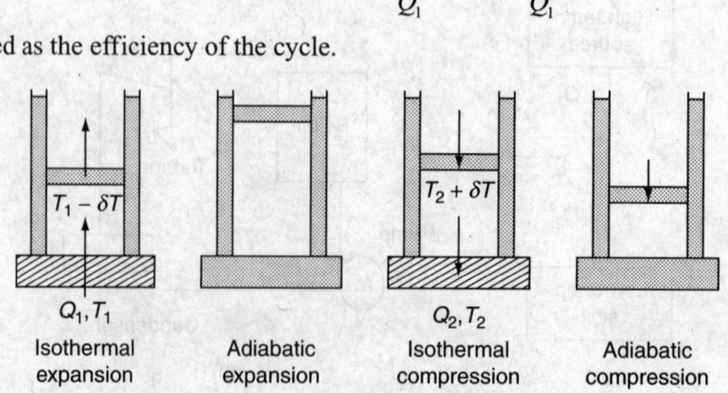

Figure 3.4 Working of Carnot Cycle

Carnot's engine is a hypothetical device. Among other practical difficulties, this engine will work at infinitely slow rate. But the concept is very important, as it serves as a standard for comparison of the efficiencies of practical devices.

An expression for η: we shall derive an expression for the efficiency of a reversible engine working between T_1 and T_2 and using perfect gas as the working substance (see Figure 3.5).

$${}_Aw_B = \int p\,dv = p_A v_A \ell n \frac{v_B}{v_A} \text{ (assume unit mass of working substance)}$$

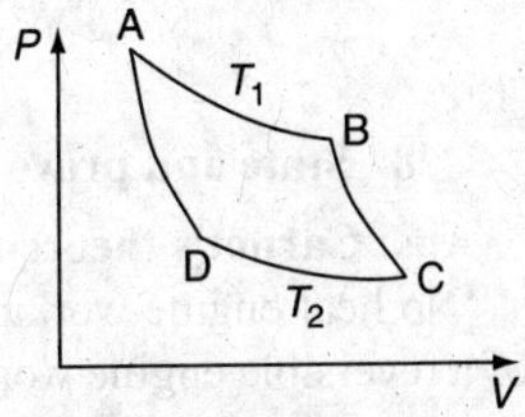

Figure 3.5 The Carnot Cycle on p-v Diagram

Also, $T_A = T_B$ and hence $u_A = u_B$ therefore $q_1 = p_A v_A \ell n \frac{v_B}{v_A} = RT_1 \ell n \frac{v_B}{v_A}$ (1)

Since BC is adiabatic

$${}_Bw_C = \int pdv = \frac{p_B v_B - p_C v_C}{\gamma - 1} = \frac{R(T_1 - T_2)}{\gamma - 1}, \text{ since } PV^\gamma = \text{constant} \quad (2)$$

For CD and DA, we get: (following equation (1) and (2))

$${}_Cw_D = RT_2 \; \ell n \frac{v_D}{v_C} \text{ and } {}_Dw_A = \frac{R(T_2 - T_1)}{\gamma - 1}$$

$$(w)_{\text{cycle}} = RT_1 \; \ell n \frac{v_B}{v_A} + \frac{R(T_1 - T_2)}{\gamma - 1} - RT_2 \ell n \frac{v_C}{v_D} + \frac{R(T_2 - T_1)}{\gamma - 1}$$

For the adiabatics BC and DA,

$$\frac{v_C}{v_B} = \left(\frac{T_1}{T_2}\right)^{1/(\gamma-1)} \quad \frac{v_D}{v_A} = \left(\frac{T_1}{T_2}\right)^{1/(\gamma-1)} \quad \frac{v_C}{v_B} = \frac{v_D}{v_A} \text{ or } \frac{v_B}{v_A} = \frac{v_C}{v_D}.$$

$$W_{\text{cycle}} = R\,(T_1 - T_2)\,\ell n \frac{V_B}{V_A}$$

$$\eta = \frac{W_{cycle}}{q_1} = \frac{R(T_1 - T_2)\ell n \frac{V_B}{V_A}}{RT_1 \ell n \frac{V_B}{V_A}} = \frac{T_1 - T_2}{T_1}.$$

The efficiency does not depend on the nature of working substance; hence, for all reversible engines, irrespective of the working substance, the efficiency

$$\eta = \frac{T_1 - T_2}{T_1}, \text{ where } T_1 \text{ and } T_2 \text{ are measured on perfect gas scale.}$$

7. Define the term coefficient of performance (COP) as applied to a Carnot refrigerator and a Carnot heat pump.

Ans: The performance of a refrigerating machine or a heat pump is expressed as the ratio of useful heat to work.

$$(\text{COP})_{\text{Refrigerator}} = \frac{Q_2}{W} = \frac{Q_2}{Q_1 - Q_2} \quad \text{or} \quad = \frac{T_2}{T_1 - T_2}.$$

$$(\text{COP})_{\text{Heat pump}} = \frac{Q_1}{W} = \frac{Q_1}{Q_1 - Q_2} \quad \text{or} \quad = \frac{T_1}{T_1 - T_2}.$$

8. State and prove the Carnot's theorem.

Ans: Carnot's theorem

'No heat engine working between given temperature T_1 and T_2 can have efficiency greater than that of a reversible engine working between the same temperatures.'

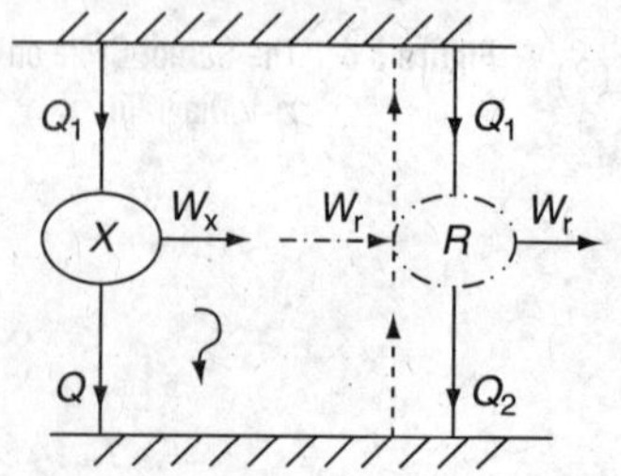

Figure 3.6 Illustration to Prove Carnot's Theorem

$$W_x = Q_1 - Q \qquad W_r = Q_1 - Q_2$$

$$\eta_x = \frac{Q_1 - Q}{Q_1}, \qquad \eta_r = \frac{Q_1 - Q_2}{Q_1}$$

Let us assume that $\eta_x > \eta_r$

$$(Q_1 - Q) > (Q_1 - Q_2) \quad \text{or} \quad W_x > W_r$$

Let us look at their combination (Figure 3.6) as it shows that $(W_x - W_r)$ amount of work is coming out from the sink without any effort. This contradicts the statement of second law of thermodynamics.

$$\eta_x > \eta_r \text{ is incorrect} \quad \text{or} \quad \eta_r \geq \eta_x.$$

A corollary to this theorem is that all reversible engines working between temperatures T_1 and T_2 have the same efficiency. We can conclude that for a reversible engine the efficiency does not depend on the working substance but only on the temperatures of the source and the sink.

$$\eta_r = f(T_1, T_2).$$

9. Establish the inequality of Clausius and show that entropy is a property.

Ans: Clausius inequality First, we observe that for a reversible engine, absorbing heat Q_1 from the source at T_1 and rejecting heat Q_2 to the sink at T_2,

$$\frac{Q_1}{T_1} - \frac{Q_2}{T_2} = 0.$$

As along the adiabatics, $\Delta Q = 0$, the above may (with due regard to the fact that Q_2 is the heat rejected) be written in the form

$$\oint \frac{\delta Q}{T} = 0.$$

Now, if we consider an irreversible engine working between the same temperatures, absorbing heat Q_1 at T_1, then because its efficiency will be less than that of the reversible one, the work done by it will be less than the corresponding work done by the reversible engine, that is, if Q'_2 is the heat rejected by the irreversible engine, then

$$(Q_1 - Q'_2) < (Q_1 - Q_2)$$

Or, in this case, $\oint \frac{\delta Q}{T} = \frac{Q_1}{T_1} - \frac{Q'_2}{T_2} < 0$,

since the efficiency of the irreversible engine $\left(1 - \frac{Q'_2}{Q_1}\right) < \left(1 - \frac{T_2}{T_1}\right)$

In general, we have $\oint \frac{\delta Q}{T} = 0$ for a reversible cycle and $\oint \frac{\delta Q}{T} < 0$ for an irreversible cycle. Hence, we can write $\oint \frac{\delta Q}{T} \leq 0$.

This is known as Clausius inequality; the equality sign holds if the cycle is reversible. Here, we have derived it by using Carnot's theorem which, itself, is a consequence of Planck's statement of the second law. Some times, Clausius inequality itself is regarded as a statement of the second law.

Entropy: Let a system be taken from state 1 to state 2 along a reversible path 1A2 and brought back to 1 along another reversible path 2B1.

Then 1A2B1 being a reversible cycle, we have

$$\oint \frac{\delta Q}{T} = 0,$$

or

$$\int_{1A2} \frac{\delta Q}{T} = -\int_{2B1} \frac{\delta Q}{T}$$

and since 2B1 is reversible

$$\int_{1A2} \frac{\delta Q}{T} = \int_{1B2} \frac{\delta Q}{T}$$

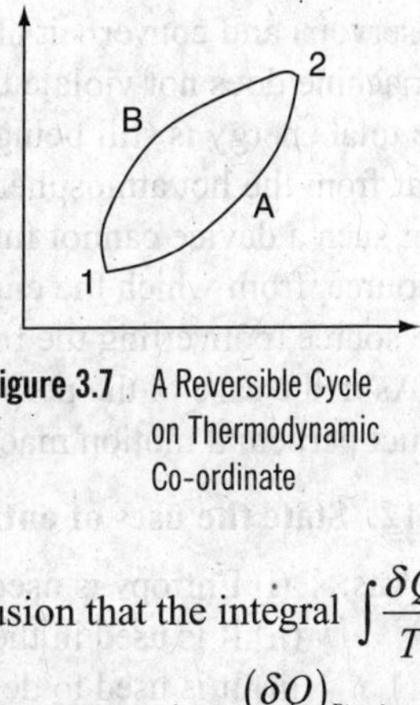

Figure 3.7 A Reversible Cycle on Thermodynamic Co-ordinate

1A2 and 1B2 are any two reversible paths joining 1 and 2. Hence the conclusion that the integral $\int \frac{\delta Q}{T}$ has the same value for all reversible paths joining 1 and 2. In other words, the expression $\frac{(\delta Q)_R}{T}$ is an exact differential of some function, say S. The function S whose differential is $\frac{(\delta Q)_R}{T}$ is called entropy. By definition, it is not a path function. We may write the following expression (using the first law)

$$dS = \frac{(\delta Q)_R}{T} = \frac{(dU + pdV)}{T}$$

Note: To calculate the change in entropy when a system is taken from state 1 to state 2 along any path 1×2 we should not evaluate the integral $\int \frac{\delta Q}{T}$; but join 1 and 2 by a reversible path 1R2 and then evaluate the integral $\int \frac{\delta Q}{T}$.

10. Derive a general expression for the change of entropy and there from obtain expression for entropy change in terms of temperature, pressure and volume when a perfect gas changes its state from p_1, V_1, T_1 to p_2, V_2, T_2.

Ans: The following expression is derived for the calculation of entropy change for a perfect gas:

$$dS = \frac{dU}{T} + \frac{pdV}{T} = mC_V\frac{dT}{T} + mR\frac{dV}{V}$$

The above equation can be integrated and expressed in any of the following forms:

$$S_2 - S_1 = mC_V \ell n\frac{T_2}{T_1} + mR\ \ell n\frac{V_2}{V_1}$$

$$S_2 - S_1 = mC_p \ell n\frac{T_2}{T_1} - m\,R\,\ell n\frac{p_2}{p_1}$$

$$S_2 - S_1 = mC_V \ell n\frac{p_2}{p_1} + m\,C_P\ \ell n\frac{V_2}{V_1}.$$

11. What is a perpetual motion machine of the second kind (PMM II)?

Ans: A cyclic or continuously working device that draws (heat) energy from a single source (reservoir) and converts it all into work is called a perpetual motion machine of the second kind. Such a machine does not violate the first law as it is converting one form of energy into another and as such the total energy is still being conserved. If such a machine could be constructed, an engine could draw heat from the hot atmosphere and deliver work without any expense. However, the second law implies that such a device cannot function; at least two reservoirs are needed for a heat engine to deliver work: a source, from which the engine draws heat and a sink to which it rejects a part of the heat drawn from the source (converting the rest into work).

As in the case of the perpetual motion machine of the first kind (PMM I), efforts were put in to construct perpetual motion machines of the second kind and, of course, they have all been unsuccessful.

12. State the uses of entropy.

Ans: (i) Entropy is used as a convenient coordinate.
(ii) It is used in the calculations of heat transfer in reversible processes.
(iii) It is used to determine whether a process is reversible, irreversible or impossible.
(iv) It is used in the correlation of physical property data.

Solved Problems

1. In a steam power plant, the temperature of the walls of the boiler tubes is 510°C. The cooling water circulating through the condenser is supplied from a river at a temperature of 12°C. What is the maximum possible efficiency of the plant?

$$T_1 = 783\text{ K};\ T_2 = 285\text{ K};\ \eta = \frac{T_1 - T_2}{T_1} = \frac{783-285}{783} = 0.636 \quad \text{or} \quad 63.6\%$$

2. A heat engine is used to drive a heat pump. The heat transfers from the heat engine and from the heat pump are used to heat the water circulating through the radiators of a building. The efficiency of the heat engine is 27 percent and the COP of the heat pump is 4. Evaluate the ratio of the heat transfer to the circulating water to the heat transfer to the heat engine.

$$\eta = \frac{Q_1 - Q_2}{Q_1} = 0.27 \qquad (1)$$

$$(COP)_{heat\ pump} = \frac{Q^*}{W} = \frac{Q^*}{Q_1 - Q_2} = 4$$

Hence, $Q^* = 4(Q_1 - Q_2)$. Heat transfer to the circulating water $= Q_2 + Q^*$ (2)

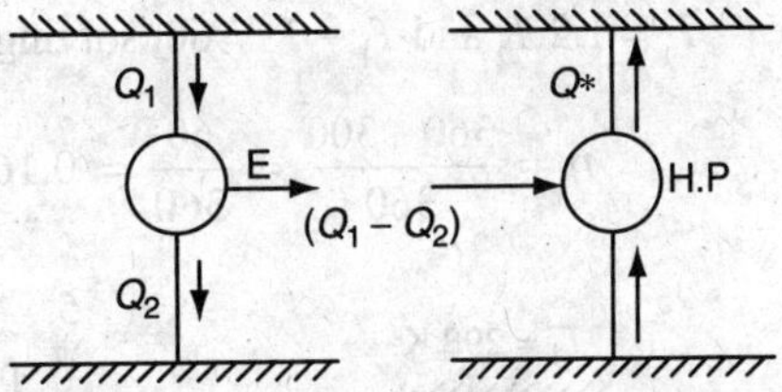

Figure 3.8 Schematic Diagram of a Heat Engine and a Heat Pump

The ratio $\frac{Q_2 + Q^*}{Q_1} = ?$ or $\frac{Q_2}{Q_1} + \frac{Q^*}{Q_1} = ?$ From equation (1), we get $\frac{Q_2}{Q_1} = 0.73$ and Equation (2) gives

$$\frac{Q^*}{Q_1} = 4\left(1 - \frac{Q_2}{Q_1}\right) = 4(1 - 0.73) = 1.08.$$

Therefore the required ratio $\frac{Q_2}{Q_1} + \frac{Q^*}{Q_1} = 0.73 + 1.08 = 1.81.$

3. (i) A reversible heat engine operates between system at constant temperatures of 150°C and 10°C. Evaluate the efficiency of the engine. (ii) The work output from the engine is 2000 Nm. Evaluate the heat transfer from the system at 150°C and the heat transfer to the system at 10°C. (iii) The engine in (a) is reversed and operates as a heat pump. Evaluate its COP and the power input required when the heat transfer rate from the system at 10°C is 40 kW.

(i) $\eta = \frac{T_1 - T_2}{T_1} = \frac{(273+150)-(273+10)}{(273+150)} = \frac{140}{423} = 0.33096 \quad \text{or} \quad 33.096\%$

(ii) $\eta = \frac{2000}{Q_1}; \quad Q_1 = \frac{2000}{0.33096} = 6043 \text{ J}; \quad Q_2 = Q_1 - 2000 = 4043 \text{ J}$

(iii) $(COP)_{\text{heat pump}} = \frac{T_1}{T_1 - T_2} = \frac{423}{140} = 3.02;$

$$(COP) = 3.02 = \frac{Q_2 + W}{W}; \quad 3.02\,W = Q_2 + W; \quad W = \frac{40}{2.02} = 19.801 \text{ kW}$$

Power input required = 19.801 kW.

4. For an engine operating on the Carnot cycle, the work output is 0.2 times the heat transfer to the cold body. The difference in the working fluid temperature when in contact with the hot and the cold body is 60 K. Evaluate the efficiency of the engine and the maximum and minimum temperatures of the working fluid.

Given that: $W = 0.2\,Q_2$ (1)

$T_1 - T_2 = 60$ (2)

$$\eta = \frac{T_1 - T_2}{T_1} = \frac{W}{Q_1} = \frac{Q_1 - Q_2}{Q_1}; \quad \text{hence, } Q_1 - Q_2 = 0.2\,Q_2 \quad \text{or} \quad Q_1 = 1.2\,Q_2$$

From the above relation, we get, $\frac{Q_1}{T_1} = \frac{Q_2}{T_2}$ and $\frac{T_1}{T_2} = 1.2\,; T_1 - T_2 = 60$

$T_1 = 1.2\,T_2$ and $T_1 - T_2 = 60$, solving the equations we get $T_2 = 300$ K and $T_1 = 360$ K

$$\eta = \frac{360-300}{360} = \frac{60}{360} = 0.1667 \text{ or } 16.67\%; \text{ temperatures } t_1 = 87°\text{C and } t_2 = 27°\text{C}.$$

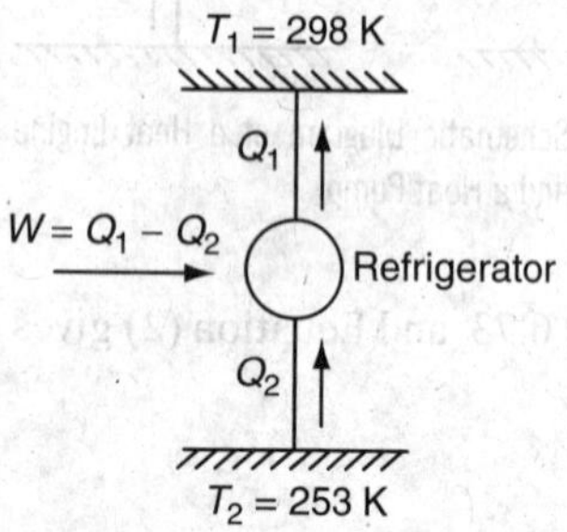

Figure 3.9 Schematic Diagram of a Refrigerator

5. A refrigerator operates on the reversed Carnot cycle. The temperature of the working fluid as it evaporates in the evaporator is −20°C while its temperature in the condenser is 25°C. The heat transfer rate from the cold region to the working fluid (refrigerating effect) is 80 kJ/sec. Evaluate the power required to drive the refrigerator.

$$(COP)_R = \frac{T_2}{T_1 - T_2} = \frac{(273-20)}{(273+25)-(273-20)} = \frac{253}{45} = 5.622$$

$$5.622 = \frac{Q_2}{W}; \quad W = \frac{80}{5.622} = 14.229 \text{ kJ/sec}$$

Power required = 14.23 kW.

6. A perfect gas is cooled reversibly at constant pressure in a cylinder behind a piston from a temperature of 200°C to a temperature of 30°C. Calculate the change of entropy per kg of gas. $C_p = 1.005$ kJ/kg K.

$$s_2 - s_1 = C_p \ell n\left(\frac{T_2}{T_1}\right) - R\ell n\left(\frac{p_2}{p_1}\right) = 1.005\ell n\left(\frac{303}{473}\right) = -0.4476 \text{ kJ/kg K}.$$

7. Show from first principles the entropy change per unit mass of an ideal gas undergoing a polytropic process (pV^n = constant)

$$s_2 - s_1 = \frac{\gamma - n}{\gamma - 1} R\ell n\left(\frac{V_2}{V_1}\right),$$ where γ is the ratio of specific heats; V_1 and V_2 are the initial and final volumes and R gas constant.

For polytropic process, we know $\frac{T_2}{T_1} = \left(\frac{V_1}{V_2}\right)^{(n-1)}$

$$dS = \frac{dQ}{T} = \frac{dU}{T} + \frac{p\,dV}{T} = m\,C_v \frac{dT}{T} + m\,R\frac{dV}{V}$$

Hence, for unit mass of gas $s_2 - s_1 = C_v \ell n\left(\frac{T_2}{T_1}\right) + R\,\ell n\left(\frac{V_2}{V_1}\right)$

$$= \frac{R}{(\gamma-1)} \ell n\left(\frac{V_2}{V_1}\right)^{(1-n)} + R\,\ell n\left(\frac{V_2}{V_1}\right)$$

$$= \left(\frac{1-n}{\gamma-1} + 1\right) R\,\ell n\left(\frac{V_2}{V_1}\right) = \frac{(\gamma-n)}{(\gamma-1)} R\,\ell n\left(\frac{V_2}{V_1}\right)$$

8. An ideal gas undergoes a constant volume heat addition in which 420 kJ/kg of heat is added. The initial temperature is 30°C. Assume the specific heat at constant volume is 0.718 kJ/kg K. Calculate the change of entropy for the gas.

$$420 = 0.718(T_2 - 303);\ T_2 = 887.9 \text{ K}$$

$$s_2 - s_1 = C_V \ell n\frac{T_2}{T_1} \text{ for constant volume process}$$

$$s_2 - s_1 = 0.718\ell n\frac{887.9}{303} = 0.7719 \text{ kJ/kg K.}$$

9. An ideal gas undergoes a reversible polytropic expansion process in a closed system from a temperature T_1 to a temperature T_2 according to the law $pV^n = C$. If the magnitude of change in entropy of the same gas during the heating process at constant volume between the same temperature limits is the same as that of the polytropic process; find the value of the exponent, n in terms of the ratios of the specific heats of the gas.

For polytropic process, the entropy change

$$\Delta s = C_v \ell n\left(\frac{T_2}{T_1}\right) + R\,\ell n\left(\frac{v_2}{v_1}\right) = C_v \ell n\left(\frac{T_2}{T_1}\right) + R\,\ell n\left(\frac{T_2}{T_1}\right)^{1/(1-n)} = \frac{R}{\gamma-1}\ell n\left(\frac{T_2}{T_1}\right) - \frac{R}{n-1}\ell n\left(\frac{T_2}{T_1}\right)$$

$$= \frac{(n-\gamma)}{(\gamma-1)(n-1)}R\ell n\left(\frac{T_2}{T_1}\right)$$

For heating process at constant volume between the same temperatures limits

$$\Delta s = C_v \ell n\left(\frac{T_1}{T_2}\right) = \frac{R}{(\gamma-1)}\ell n\left(\frac{T_1}{T_2}\right)$$

Equating

$$-\frac{R}{(\gamma-1)}\ell n\left(\frac{T_2}{T_1}\right) = \frac{(n-\gamma)}{(\gamma-1)(n-1)}R\ell n\left(\frac{T_2}{T_1}\right);$$

$$-1 = \frac{(n-\gamma)}{(n-1)};\ \text{Hence, } n = \frac{(\gamma+1)}{2}.$$

10. During a process the pressure of one kg of air increases from 1.8 bar to 10 bar. Find the change of entropy if the process is (i) at constant volume and (ii) at constant temperature. Assume $C_v = 0.718$ kJ/kg K and $C_p = 1.005$ kJ/kg K.

(i) $s_2 - s_1 = C_v\ \ell n\left(\frac{p_2}{p_1}\right) = 0.718\,\ell n\left(\frac{10}{1.8}\right) = 1.231 \text{ kJ/K}$

(ii) $s_2 - s_1 = -R\,\ell n\left(\frac{p_2}{p_1}\right) = -0.287\,\ell n\left(\frac{10}{1.8}\right) = -0.492 \text{ kJ/K}$

Practice Problems

1. 2000 kJ of heat per minute are transferred from a reservoir at 320°C to an engine that operates on the Carnot cycle. The engine rejects heat at 30°C. Calculate the thermal efficiency and power developed by the engine in kW.

   (48.9%; 16.3 kW)

2. Exhaust steam from a process plant is to be used as a hot system for a heat engine. The steam is available at a pressure of 140×10^3 N/m^2 with a dryness fraction of 0.6. Heat transfer from the steam to the engine causes the steam to condense to saturated liquid at 140×10^3 N/m^2. River water is available at a mean temperature of 14°C. Evaluate the maximum possible power which could be developed by the engine when the exhaust steam flow rate is 1800 kg/hr.

   (166.94 kW)

3. A Carnot engine operates between two reservoirs at temperatures of T_1 and T_2 K. The work output of the engine is 0.6 times that of heat rejected. Given that the difference in temperature between the source and sink is 200°C, calculate the temperature of source, sink and thermal efficiency.

   (533.3 K; 333.3 K; 37.5%)

4. In a reversed heat engine, the work done on the engine is 75×10^3 Nm and the heat transfer to the engine from the low temperature region is 220×10^3 J. Evaluate the heat transfer to the high temperature region and the coefficient of performance as a refrigerator and as a heat pump.

   (295 kJ; 2.93; 3.93)

5. An ice-making plant produces ice at atmospheric pressure and at 0°C from water at 0°C. The mean temperature of the cooling water circulating through the condenser of the refrigerating machine is 18°C. Evaluate the minimum electrical work in kWh required to produce one tonne of ice. (The enthalpy of fusion of ice is 333.5 kJ/kg at atmospheric pressure.)

   (6.108 kWh)

6. The coefficient of performance of a heat pump is 5 when the power supplied to it is 50 kW. (i) Evaluate the magnitude of heat transfer rates to and from the working fluid. (ii) The heat transfer from the heat pump is used to heat the water flowing through the radiators of a building. Evaluate the mass flow rate of the heated water given that its temperature increases from 50°C to 70°C in flowing through the heat pump. Assume the water velocity to be negligible.

   (200 kJ/sec, 250 kJ/sec; 2.986 kg/sec)

7. An office block is heated by means of a heat pump. The mean air temperature within the building is 20°C when the outside mean air temperature is –4°C. The heat transfer rate to the heat pump is 1,08,000 kJ/hr and the power required to drive the pump is 10 kW. (i) Evaluate the heat transfer rate to the building and the coefficient of performance of the heat pump. (ii) Evaluate the maximum possible coefficient of performance of a heat pump for these conditions, and the minimum power input to the heat pump to satisfy the given heating requirement.

   (40 kJ/sec, 4; 12.208, 3.276 kW)

8. Air initially at 30°C and 1 bar pressure is compressed reversibly and isothermally to a state where the specific volume is 0.2 m^3 by a steady flow process. Find the change of entropy.

   (–0.4218 kJ/kg K)

9. 5.2 kg of air is compressed in a cylinder by the law $pV^{1.3}$ = constant. If the compression ratio is 15 and the initial temperature is 200°C, find the change of entropy for air.

   (–1.0083 kJ/K)

10. An ideal gas at 27°C is heated reversibly at constant pressure of one atmosphere in a steady flow system to a temperature of 115°C. Calculate the change of entropy per kg of the gas if the specific heat at constant pressure for this gas is 1.005 kJ/kg K.

(0.2585 kJ/kg K)

Objective Type Questions

1. The assertion that more of high temperature energy can be converted to work than the same amount of energy at a lower temperature is due to
   (i) the zeroth law of thermodynamics
   (ii) the first law of thermodynamics
   (iii) the second law of thermodynamics
   (iv) the third law of thermodynamics
2. Which of the following laws of thermodynamics should be satisfied to ensure whether a system will undergo a process?
   (i) The zeroth law and first law
   (ii) The first law and second law
   (iii) The second law and third law
   (iv) The zeroth law and second law
3. The heat engines Q, R and S with the efficiencies shown below are respectively

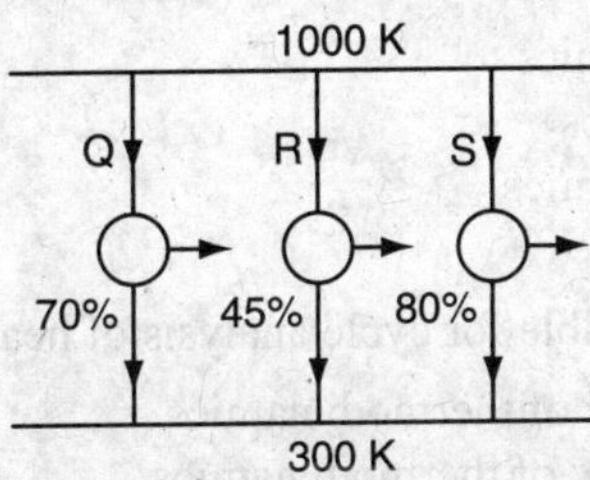

Figure 3.10 Schematic Diagram of Hypothetical, Reversible and Irreversible Heat Engines

   (i) reversible, irreversible and hypothetical
   (ii) irreversible, hypothetical and reversible
   (iii) hypothetical, reversible and irreversible
   (iv) reversible, hypothetical and irreversible
4. Which gadget is in compliance with the Clausius statement of the second law of thermodynamics?
   (i) Closed cycle gas turbine
   (ii) Steam power plant
   (iii) Internal combustion engine
   (iv) House hold refrigerator
5. Which of the following thermodynamics laws/relations are directly involved in the evaluation of the final properties for an adiabatic mixing process?
   (i) First law and second law of thermodynamics
   (ii) First law of thermodynamics and perfect gas relationship
   (iii) Second law of thermodynamics and steady flow energy equation
   (iv) Steady flow energy equation and isentropic relationship

6. The derivation of steady flow energy equation (SFEE) is based on the Newton's second law of motion and
   (i) the first law of thermodynamics
   (ii) the second law of thermodynamics
   (iii) the perfect gas relationship
   (iv) the isentropic relationship
7. Match List I with List II and select the correct answer using the codes given below the lists:

List I (Thermodynamic laws)	*List II (Property of the system)*
A. Zeroth law	1. Absolute temperature zero
B. First law	2. Absolute temperature scale
C. Second law	3. Temperature
D. Third law	4. Energy

Codes:

	A	B	C	D		A	B	C	D
(i)	3	1	4	2	(ii)	2	3	1	4
(iii)	1	4	2	3	(iv)	3	4	2	1

8. Two blocks that are at different temperatures are brought into contact and allowed to reach a final state of equilibrium. The final temperature attained is specified by
   (i) the zeroth law of thermodynamics
   (ii) the first law of thermodynamics
   (iii) the second law of thermodynamics
   (iv) the third law of thermodynamics
9. The fundamental axioms indispensable for cycle analysis of heat engines are
   (i) the zeroth law and the first law of thermodynamics
   (ii) the first law and the second law of thermodynamics
   (iii) the second law and the third law of thermodynamics
   (iv) the zeroth law and the second law of thermodynamics
10. The statement that only those processes which would result in an increase in entropy of the universe is due to
    (i) the zeroth law of thermodynamics
    (ii) the first law of thermodynamics
    (iii) the second law of thermodynamics
    (iv) the third law of thermodynamics
11. Which law of thermodynamics asserts on the direction of a spontaneous process?
    (i) The zeroth law (ii) The first law
    (iii) The second law (iv) The third law
12. The actions of entropy is characterized by
    (i) the zeroth law of thermodynamics
    (ii) the first law of thermodynamics
    (iii) the second law of thermodynamics
    (iv) the third law of thermodynamics

13. A spinning flywheel mounted in frictionless bearing is an example of
    (i) machine working on Carnot principle
    (ii) perpetual-motion machine of the first kind
    (iii) perpetual-motion machine of the second kind
    (iv) perpetual-motion machine of the third kind

14. The property relation $\left(\frac{\partial h}{\partial S}\right) = T$ which forms the basis for the construction of Mollier diagram (h-S diagram) of pure substance is obtained from the study of
    (i) the zeroth law of thermodynamics
    (ii) the first law of thermodynamics
    (iii) the second law of thermodynamics
    (iv) the first and second laws of thermodynamics

15. Which statement is referred to as the Caratheodory's principle?
    (i) It is impossible to operate in any manner a cyclic device that will produce work while exchanging heat with a single heat reservoir
    (ii) It is impossible to operate a cyclic device in such a manner that the sole effect external to the device is the transfer of heat from one energy reservoir to another at a higher temperature
    (iii) In the vicinity of an arbitrary initial state of a system, there will exist adjacent states which cannot be reached from the initial state along adiabatic paths
    (iv) The entropy of a pure crystalline substance may be taken to be zero at the absolute zero thermodynamic temperature

16. Which thermal property change will definitely provide a criterion for the occurrence of a workable process?
    (i) Enthalpy (ii) Entropy
    (iii) Internal energy (iv) Temperature

17. Identify the process during which entropy decreases.
    (i) Breaking (ii) Burning
    (iii) Crystallizing (iv) Falling

18. The application of statistical methods in thermodynamics was developed by
    (i) Ludwig Boltzmann (ii) Satyendra Bose
    (iii) Albert Einstein (iv) Enrico Fermi

19. An ideal gas contained in a cylinder with movable piston undergoes a compression process from state p_1, V_1 to p_2 according to the law pV^n = constant. If the final entropy of the gas has its minimum value, then the compression follows
    (i) isometric process ($n = -\infty$) (ii) isothermal process ($n = 1$)
    (iii) polytropic process ($1 < n < \gamma$) (iv) adiabatic process ($n = \gamma$)

20. Assertion (A): The heat interaction for a closed system can be computed during a reversible process by $Q = \int TdS$.
    Reason (R): The work interaction for a closed system can be computed during a reversible process by $W = \int pdV$.
    (i) Both A and R are true and R is the correct explanation of A
    (ii) Both A and R are true and R is not the correct explanation of A

(iii) A is true but R is false
(iv) A is false but R is true

21. Assigning the basic dimensions to mass, length, time and temperature, respectively, as M, L, T and θ, the dimensions of entropy is

(i) $M L T^{-2} \theta$
(ii) $M L^2 T^{-1} \theta^{-1}$
(iii) $M L^2 T^{-2} \theta^{-1}$
(iv) $M L^3 T^{-2} \theta^{-1}$

22. The trend of entropy change is unique during

(i) adiabatic process
(ii) isobaric process
(iii) isometric process
(iv) isothermal process

23. Select the multiplying factor that completes the following equation which gives the entropy change during a reversible polytropic process

$$s_2 - s_1 = \{ \qquad \} Ln\left(\frac{T_2}{T_1}\right)$$

(i) $\dfrac{C_v(n-\gamma)}{(n-1)}$
(ii) $\dfrac{C_p(n-\gamma)}{(n-1)}$
(iii) $\dfrac{C_v(n-\gamma)}{(\gamma-1)}$
(iv) $\dfrac{C_p(n-\gamma)}{(\gamma-1)}$

24. Which one of the following correlations of physical/thermal properties facilitates the calculation of entropy?

(i) $Tds = dh - pdv$
(ii) $Tds = dh - vdp$
(iii) $Tds = du - pdv$
(iv) $Tds = du - vdp$

25. Assertion (A): One army that consists of ten divisions(10 × 100 soldiers) is more powerful than ten armies (10 numbers) each consisting of a single division(100 soldiers).
Reason (R): The second category belongs to high entropy group than the first category.

(i) Both A and R are true and R is the correct explanation of A
(ii) Both A and R are true and R is not the correct explanation of A
(iii) A is true but R is false
(iv) A is false but R is true

26. Which thermodynamic property is evaluated with help of Maxwell equations from the data on other measurable properties of a system?

(i) Enthalpy
(ii) Entropy
(iii) Latent heat
(iv) Specific heats

27. The entropy change for an ideal gas which undergoes an isothermal compression is

(i) zero
(ii) always positive
(iii) always negative
(iv) indeterminate

28. Consider the following statements for an isolated system undergoing a process.

1. The entropy either increases or in the limit remains constant.
2. The entropy either decreases or in the limit remains constant.
3. The entropy can never decrease.

Of these statements

(i) 1, 2 are correct
(ii) 2, 3 are correct
(iii) 1, 3 are correct
(iv) 3 alone is correct

29. The entropy change for any closed system undergoing an irreversible adiabatic process

(i) may be positive, negative or zero
(ii) may be positive or zero, but not negative
(iii) is negative
(iv) is positive

30. Identify the correct statement as regards entropy.

(i) Entropy is an intensive property
(ii) Entropy of a substance cannot decrease
(iii) The more probable states are states of higher entropy
(iv) The more probable states are states of lower entropy

31. Which property aids and strengthens progress when used in the fields of physics, thermodynamics, statistical mechanics and information theory?

(i) Enthalpy
(ii) Entropy
(iii) Pressure
(iv) Temperature

32. Two constituents contained in separate compartments of a rigid tank are allowed to mix adiabatically. If the entropy change for the constituents is zero, then they should be

(i) real gases
(ii) ideal gases
(iii) any gases
(iv) same gas

33. Who introduced the integrating factor $\frac{1}{T}$ which renders dQ an exact differential in providing a basis for the concept of entropy?

(i) Caratheodory
(ii) Clausius
(iii) Kelvin
(iv) Planck

34. The heat added to a closed system during a reversible process is given by $Q = \alpha T + \beta T^2$, where α and β are constants. The entropy change of the system as its temperature changes from T_1 to T_2 is

(i) $\alpha + \beta (T_2 - T_1)$
(ii) $\alpha (T_2 - T_1) + \beta (T_2^2 - T_1^2)/2$
(iii) $\alpha (T_2^2 - T_1^2)/2 + \beta (T_2^3 - T_1^3)/3$
(iv) $\alpha ln (T_2/T_1) + 2\beta (T_2 - T_1)$

35. Identify the personnel whose function in a firm/establishment can be equated to the role of entropy in the thermodynamic study of thermal power plants.

(i) Boss
(ii) Manager
(iii) Security
(iv) Worker

36. The process undergone by gases in the Joule-Thomson experiment is

(i) isentropic
(ii) isobaric
(iii) isenthalpic
(iv) isometric

37. Which gas acquires more internal energy on heating?

(i) Gas flowing through a nozzle
(iii) Gas in an enclosure of constant volume
(ii) Gas in an expandable enclosure
(iv) Gas flowing through a pipe

38. The expression (symbols have their usual meaning) that furnishes the quantity 'energy in transit' is
    (i) $\frac{V^2}{2}$ (ii) gZ
    (iii) $C_v T$ (iv) pv
39. The ordinate and abscissa of the following diagram (Figure 3.11) that shows an isochoric process of an ideal gas are
    (i) temperature and entropy (ii) enthalpy and entropy
    (iii) pressure and enthalpy (iv) temperature and pressure

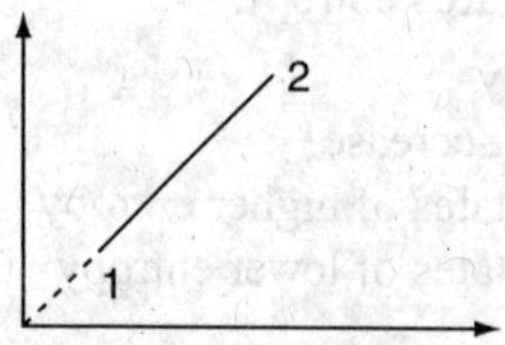

Figure 3.11 Isochoric Process of an Ideal Gas

40. The total energy of fluid at entry or exit point of an open system is the sum of its kinetic energy, potential energy and
    (i) enthalpy (ii) flow work
    (iii) internal energy (iv) shaft work

Answers

1. (iii)	2. (ii)	3. (i)	4. (iv)	5. (ii)	6. (i)	7. (iv)	8. (i)	9. (ii)	10. (iii)
11. (iii)	12. (iii)	13. (iv)	14. (iv)	15. (iii)	16. (ii)	17. (iii)	18. (ii)	19. (ii)	20. (i)
21. (iii)	22. (i)	23. (i)	24. (ii)	25. (i)	26. (ii)	27. (iii)	28. (iv)	29. (iv)	30. (iii)
31. (ii)	32. (iv)	33. (i)	34. (iv)	35. (ii)	36. (iii)	37. (iii)	38. (iv)	39. (iv)	40. (i)

4

Properties of Pure Substance

1. What are phases of a pure substance?

Ans: A phase is any homogeneous part of a system that is physically distinct. Each phase in a system is separated from other phases by interfaces called phase boundaries. A system consisting of an ice cube in liquid water has two phases, solid and liquid, and the phase boundary is the surface of ice cube. A gas mixture is always a single phase. A system composed of liquid water with a layer of oil would involve two phases even if the liquids are thoroughly agitated so that the oil is distributed in tiny globules. A mixture of water and alcohol would be a single phase as the liquids are miscible and form a homogeneous mixture.

2. Explain the concept of a pure substance and show how the change of state of pure substance can be represented on a diagram with a pair of independent properties as coordinates.

Ans: A system that is homogeneous in composition and homogeneous and invariable in chemical aggregation is called a pure substance. A substance cannot be treated as pure if it undergoes a chemical reaction. A system consisting of pure oxygen existing as vapour, a liquid, a solid or any combination of these is a pure substance. The same is true for ordinary water. Air may be considered a pure substance as long as it is all vapour or all liquid. A pure substance in the absence of motion, gravity, capillarity, electricity and magnetism has only two independent properties. Any change of state that consists of a continuous series of homogeneous states assumed by a pure substance can be represented on a diagram having any pair of independent properties as coordinates. Figure 4.1 shows a constant-volume change and a constant-pressure change undergone by a pure substance by lines *Ov* and *Op* on the pressure-volume plane, the temperature-pressure plane and on the plane of internal energy and volume.

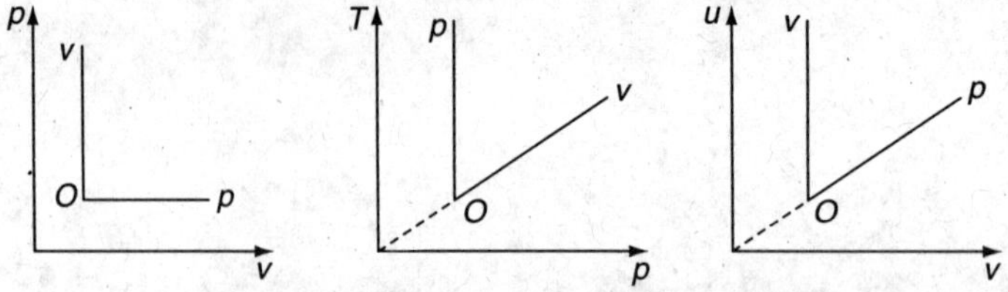

Figure 4.1 Two Dimensional Property Diagrams

3. Explain what do you understand by saturation condition of a pure substance and also saturation pressure and saturation temperature.

Ans: For any one phase of a substance, the pressure and temperature can be varied independently over wide ranges and hence are independent properties. However, when two phases of a pure substance coexist in equilibrium, there is a fixcd relationship between their pressure and temperature. At any given pressure there is one temperature at which two specific phases will exist together in equilibrium. The two phases can coexist in equilibrium at a given temperature only at a particular pressure. The conditions under which two or more phases of a pure substance can exist together in equilibrium are called saturation conditions. The pressure and temperature are called the saturation pressure and the saturation temperature. Liquid water and water vapour in equilibrium with each other are termed as saturated liquid and saturated vapour. Ice in equilibrium with liquid or vapour or both is termed as a saturated solid. A phase is saturated, even if it exists alone, if it is at a pressure and a temperature under which two or more phases could exist together in equilibrium. Liquid water at its boiling temperature is a saturated phase.

4. Sketch the liquid-vapour saturation curves for a few substances and show clearly the critical state and triple-phase state of the pure substance.

Ans: The relationship between liquid-vapour saturation pressures and temperatures is shown for several substances in Figure 4.2. These curves are called liquid-vapour saturation curves, vapourization curves or condensation curves. The upper and lower ends of these curves are called respectively as the critical states and the triple-phase states of the substance.

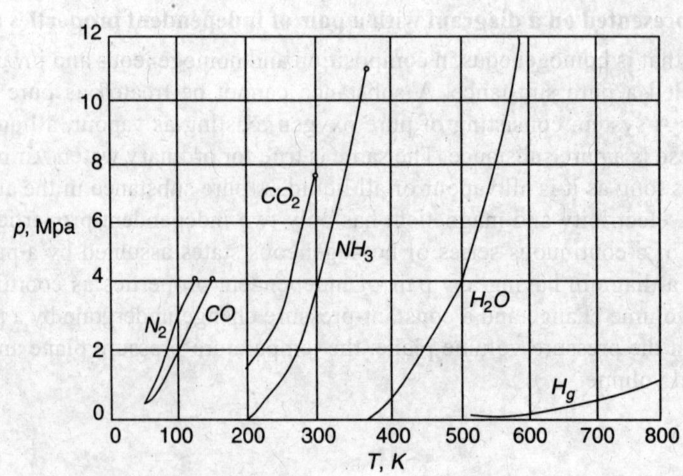

Figure 4.2 Liquid-vapour Saturation Lines for Several Pure Substances

5. Explain the term sublimation and also the sublimation line.

Ans: For most substances, there is some pressure below which liquid cannot exist. Below this pressure, the solid and vapour phases can coexist in equilibrium and the relation between pressure and temperature for these solid-vapour saturation states is fixed for each substance. The transformation from a solid to vapour is known as sublimation; so the solid-vapour saturation line on a *p-T* diagram is often called a sublimation line.

6. Sketch the solid-liquid saturation curve for water.

Ans: The melting or freezing temperature of a substance also depends on the pressure, although the variation of this temperature with the pressure is usually small. For water that expands on freezing, increasing the pressure lowers the freezing temperature so the melting (or freezing) line appears as in Figure 4.3.

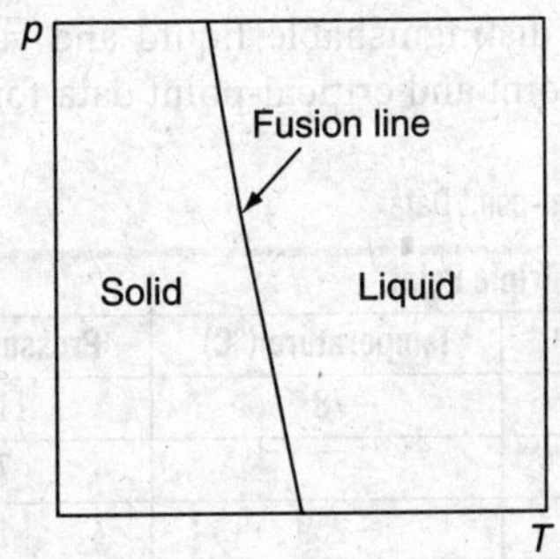

Figure 4.3 Solid-liquid Saturation Line (Melting or Freezing Line) for a Substance that Expands on Freezing

7. Sketch the *p-T* diagram (phase diagram) for water showing all three phases. Indicate the vapourization line, melting line and sublimation line. Show the triple point. Furnish triple-point and critical-point data for a few substances.

Ans: The *p-T* diagram showing all three phases is called a phase diagram of a pure substance. The general form of the phase diagram for water that expands on freezing is shown in Figure 4.4. In this diagram each area represents the pressure and temperature ranges in which a single phase can exist. Each line represents the combinations of pressure and temperature under which two phases may exist together. The intersection of the vapourization line, melting line and sublimation line of a phase (*p-T*)

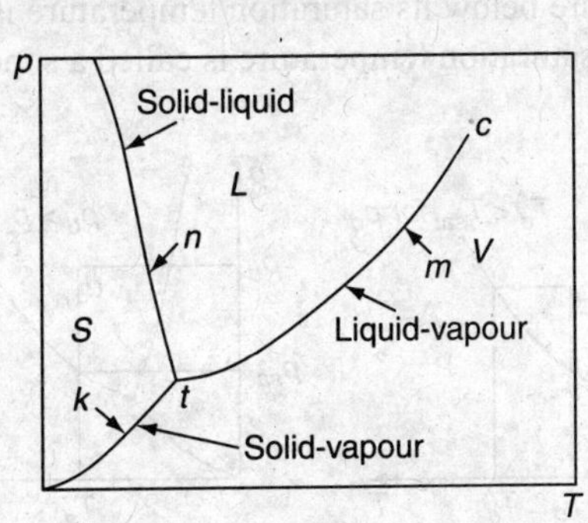

Figure 4.4 Phase Diagram for a Substance that Expands on Freezing

diagram represents the condition under which three phases can exist in equilibrium and is called the triple-phase state, triple-phase point or triple point. These conditions are represented by a point only on a *p-T* diagram; on other property diagrams they are represented by a line or an area. In Figure 4.4 each of the points *L*, *V* and *S* represents a single state. Each of these states is a single phase and so pressure and temperature are independent variables and fully specify the state. Each of the points *m*, *n* and *k* represent an infinite number of states: one phase alone, the other phase alone or any ratio of the two phases in a mixture. This illustrates that whenever two phases exist in equilibrium, *p* and *T* are not independent; so are insufficient to fully specify the state. The state of each phase within the mixture, however, is specified by either *p* or *T* and the knowledge that another phase is or could be present.

As pressure and temperature increase, the specific volume of saturated liquid increases and that of saturated vapour decreases; so that at the end of the vaporization line, point *c* on Figure 4.4, they are equal. At this point, the liquid and vapour are indistinguishable from each other. Point *c* is called the critical state or critical point. The critical pressure and the critical temperature are the highest pressure and temperature at which distinguishable liquid and vapour phases can exist together in equilibrium. Table 4.1 gives triple-point and critical-point data for a few substances.

Table 4.1 Triple-point and Critical-point Data

Substance	Triple Point		Critical Point	
	Pressure (kPa)	Temperature (°C)	Pressure (MPa)	Temperature (°C)
NH_3	6.1	−78	11.3	132
CO_2	517.0	−57	7.39	31
H_2	7.0	−259	1.3	−213
N_2	12.5	−210	3.4	−147
O_2	0.15	−219	5.0	−119
H_2O	0.61	0.01	22.1	374

8. Explain subcooled, compressed solid and superheated vapour states of a pure substance.

Ans: Liquid existing at a temperature lower than its saturation temperature, state *d* in Figure 4.5(a), or, at a pressure higher than its saturation pressure, as shown in Figure 4.5(b), is called a compressed liquid or a subcooled liquid to distinguish it from a saturated liquid. Thus, all points in the liquid region that is not on the liquid-vapour or vapour-solid saturation line represent states of compressed liquid. In a like manner, a solid at a temperature below its saturation temperature is called a compressed solid and a vapour at a temperature above its saturation temperature is called a superheated vapour.

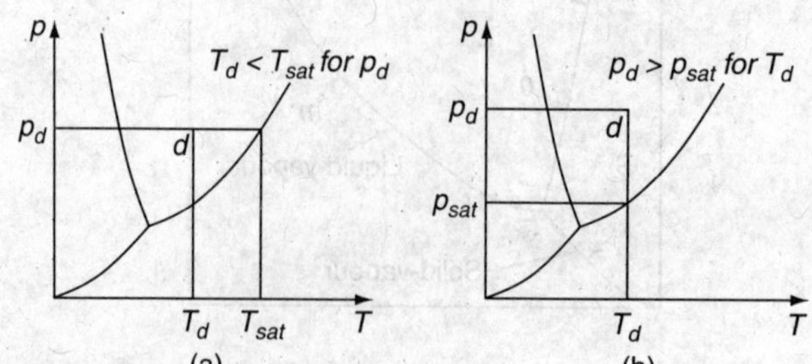

Figure 4.5 Compressed Liquid State Described in Two Ways: (a) Liquid at a Temperature Lower than its Saturation Temperature and (b) Liquid at a Pressure Greater than its Saturation Pressure

If three phases of a pure substance exist together in a system, only one value of pressure and only one value of temperature are possible. The state of each phase present in the system is fixed. However, to determine the state of the system at the triple-point pressure and temperature, more information is required, namely, the proportions of the phases present. For example, if only 1 percent of the mass is in the vapour phase, the specific volume of the system is much less than if 20 percent of the mass is vapour.

In a similar manner, each point on the vapourization (or condensation) line fully defines the state of each phase present. However, the state of the system is not determined because each point represents an infinite number of mixture states ranging from all liquid and no vapour to no liquid and all vapour. Consequently, each point on the vapourization line on a *p-T* diagram represents states with a range of mixture-specific volumes.

The *p-T* diagrams show the conditions under which a single phase, two phases or three phases of a pure substance can exist in equilibrium. Other property diagrams are useful for other purposes.

9. Sketch and explain the enthalpy-pressure diagram for a pure substance (water) showing the three phases. Also show a few isotherms on it.

Ans: Enthalpy changes can be determined from the constant-pressure experiments, if measurement is made of the heat transfer to the system; this may be done by measuring the electrical work to the heater and allowing the heat transfer to the system.

The first law written for unit mass is $\delta Q = du + p\,dv$

where δQ is an elementary heat transfer, du is the elemental change in the internal energy and $p\,dv$ is the displacement work done in the fully resisted expansion.

The previous equation may be written in terms of enthalpy, h as follows:

By definition, $$h = u + pv$$

Differentiating, we have $$dh = du + p\,dv + v\,dp$$

Hence, $$\delta Q = dh - v\,dp$$

Now pressure changes are absent, so $v\,dp$ equals 0, which then runs $\delta Q = dh$.

Thus, the heat transfer equals the change in enthalpy of the substance in the constant-pressure process; so an enthalpy-pressure diagram can be constructed from the experimental observation. In order to construct the *h-p* diagram, it is necessary to calculate enthalpy changes during the shifts from one pressure to the next.

The results of such experiments for a pure substance are indicated in Figure 4.6. It is evident that the *h-p* diagram is more complicated than the *p-T* diagram. This is consequence of the fact, whereas

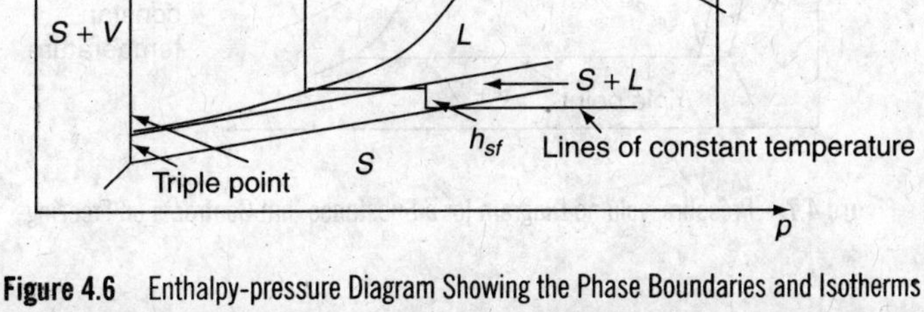

Figure 4.6 Enthalpy-pressure Diagram Showing the Phase Boundaries and Isotherms

the system temperature does not vary during the change of phase, the enthalpy does (the heat transfer is continuous); the result is that now the phase boundaries have been broadened into bands of finite vertical height. These bands are indicated on Figure 4.6 by the letters $S + V$, $S + L$ and $L + V$, signifying the two phases that are present in the corresponding regions. The regions of pure-solid, pure-liquid and pure-vapour phases are marked S, L and V, as before.

The triple point, like all other points (but one) on the *p-t* phase boundaries, extends vertically when represented on the *h-p* diagram, so that now it becomes a line. This is the boundary between the $S + V$, $S + L$ and $L + V$ regions of Figure 4.6; these letters emphasizing that it is in the nature of the triple point to have all three phases in contact.

The critical point is the exception to the rule that phase-boundary points become extended vertically; or rather its extension is zero. The critical point is on the extreme right of the $L + V$ envelope of Figure 4.6; to the right of it there is no such envelope.

The vertical height of a two-phase region ($S + V$, $L + V$ or $S + L$) represents the enthalpy increase of the fluid when it changes from one phase to the other. This enthalpy increase is known as the latent heat of sublimation, vapourization or melting (fusion) as the case may be. The latent heats, being enthalpy changes, are given the symbol h with the subscripts *sg*, *sf* or *fg*, where *s* stands for solid, *f* stands for liquid and *g* stands for vapour (gas). Thus, the enthalpy of vapourization has the symbol h_{fg}.

Isotherms: A few lines corresponding to states of constant temperature are shown in Figure 4.6. Several features may be noted:

(i) The isotherms are vertical in the two-phase regions $S + V$, $S + L$ and $L + V$, signifying that no temperature change occurs during a constant-pressure phase change.

(ii) Outside the two-phase regions, the isotherms have a generally horizontal tendency except in the neighbourhood of the critical point. In fact, the isotherm will have a vertical tangent at the critical point.

(iii) The higher temperatures correspond to the higher enthalpies. Particularly at high temperatures, it is notable that the isotherms become closely horizontal. The isotherms are also practically horizontal in the pure-liquid and pure-solid regions.

10. Sketch the pressure-volume diagram for a pure substance (water) showing clearly the different regions, critical point and triple point.

Ans: If volume measurements are made during the constant-pressure experiments, a pressure-volume diagram can be plotted. Figure 4.7 shows an example for a substance that contracts on freezing.

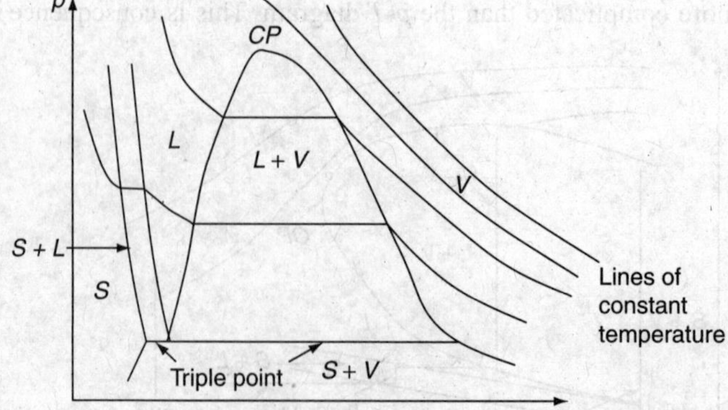

Figure 4.7 Pressure-volume Diagram for a Substance that Contracts on Freezing

The single-phase and two-phase regions are indicated as before. A constant-pressure process is represented by a horizontal line; so the isotherms are horizontal in the two-phase regions far from the critical point, however, the isotherms tend to become rectangular hyperbolae with the p and v axes as asymptotes. The triple point is again a line and the critical point occurs at the apex of the $L + V$ bulge. Inspection of a horizontal line from left to right on Figure 4.7 indicates that when the system undergoes a constant-pressure heating process, the volume increases at each phase change. This is not always the case however. Water, for example, expands as it changes, at constant pressure, from the liquid to the solid state. Such behaviour makes the p-v diagram considerably more complicated in the $S + L$ region. Also the specific volumes of the solid and the liquid are very much smaller than those of the vapour at a given pressure.

11. Sketch a constant volume process on the pressure-volume diagram for water.

Ans: A constant-volume process is represented by a vertical line on a p-v diagram. Figure 4.8 shows three such processes, a, b and c. In case a, we suppose that a mixture of liquid and vapour in a closed container (Figure 4.9) is gradually heated. The initial ratio of liquid to vapour is relatively small. As the temperature rises (and the pressure with it), the state point approaches and then crosses the saturated-vapour line on Figure 4.8 signifying that the meniscus separating liquid and vapour descends to the bottom of the vessel. Finally, only vapour is present; the liquid has boiled away.

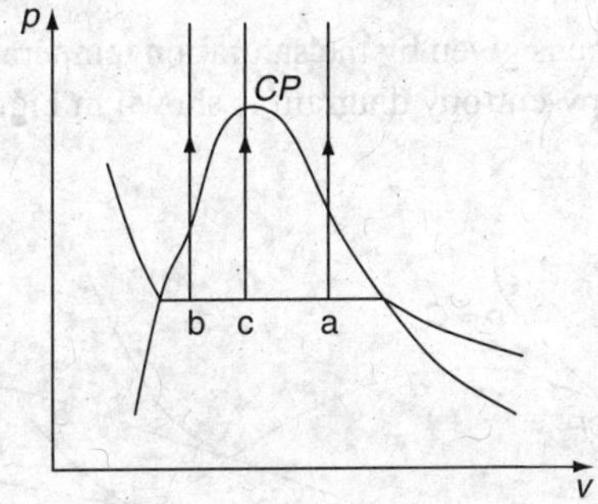

Figure 4.8 Pressure-volume Diagram Showing Constant-volume Processes

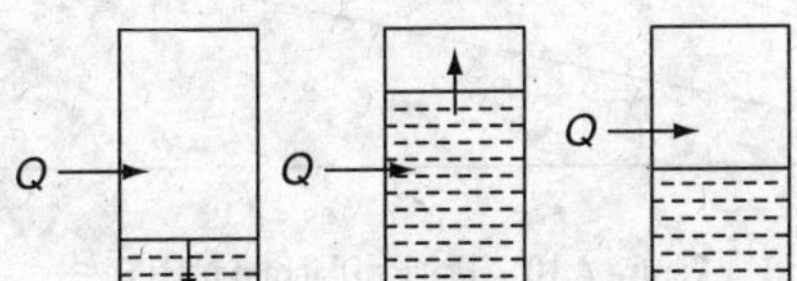

Figure 4.9 Showing the Direction of Motion of the Meniscus When Different Liquid and Vapour Mixtures are Heated at Constant Volume

12. Define the term dryness fraction of steam.

Ans: This is useful in relating the properties of a mixture of two phases to the properties of the pure saturated phases at the same pressure; such a mixture of liquid and vapour phases is often called a wet mixture. To describe this, we introduce first the concept of dryness, or dryness fraction, as a measure of the relative proportions of vapour and liquid phases in a mixture.

The dryness of an equilibrium liquid-vapour mixture, x, is the mass fraction of vapour in the mixture.

If in 1 kg of liquid-vapour mixture, x kg is the mass of vapour and $(1 - x)$ kg is the mass of liquid, then x is known as the quality or dryness fraction of the liquid-vapour mixture. Therefore, quality indicates the mass fraction of vapour in a liquid-vapour mixture, that is, $x = \dfrac{m_v}{m_v + m_l}$, where m_v and m_l are the masses of vapour and liquid, respectively. Saturated vapour has a dryness of unity, or 100 percent; saturated liquid has a dryness of zero.

We can write relations between h, u and v for a mixture of liquid and vapour, and h_g, u_g, v_g, h_f, u_f and v_f, the properties of the saturated phases.

$h = h_f + xh_{fg}$; $v = v_f + xv_{fg}$ or $v = (1 - x)v_f + xv_g$; $u = h - pv$, where $h_{fg} = h_g - h_f$, latent heat of vapourization and $v_{fg} = v_g - v_f$. Since the specific volume of vapour will be more than that of liquid, it is some times convenient to use an approximation $v = xv_g$ instead of $v = (1 - x)v_f + xv_g$.

13. How is enthalpy-entropy diagram constructed? Sketch a typical chart (Mollier chart) and explain the usefulness of this chart and the steam tables.

Ans: We use the relation $T\,ds = dh - v\,dp$ in the construction of this chart as the slope of an isobar line, that is, $\left(\dfrac{dh}{ds}\right)_p = T$ on h-s diagram is its absolute temperature. If the temperature remains constant during the phase transition, the slope is given by the saturation temperature. This diagram is also known as the Mollier diagram. The enthalpy-entropy diagram is shown in Figure 4.10.

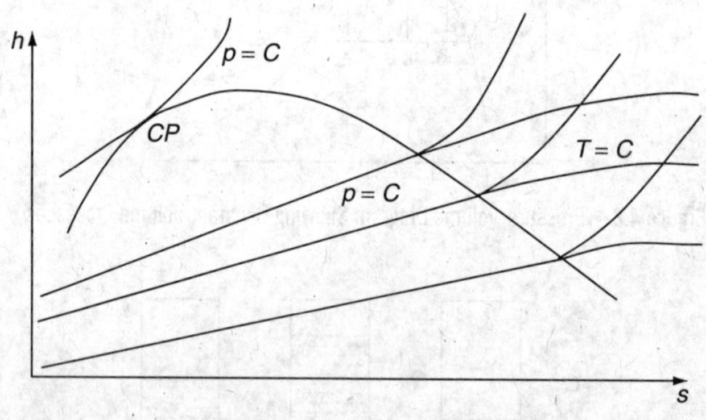

Figure 4.10 Mollier Diagram for H_2O

Steam Tables: The method of reading the thermodynamic properties of a pure substance is the use of tables. Examples are the steam tables and the students are expected to become thoroughly familiar with the use of these tables. The internal energy of saturated water at the triple point ($t = 0.01$°C) is arbitrarily chosen to be zero. The entropy of saturated water is also chosen as zero at 0.01°C. Since the change in properties will be used in all our calculations, it is immaterial where we assign zero values for these properties. It is to be noted that enthalpy is tabulated, but not internal energy. The internal energy can always be deduced from the enthalpy by way of equation $u = h - pv$.

14. Show on a temperature-entropy diagram the heating of water, evaporation and superheating at constant pressure.

Ans: The heating of water at constant pressure is shown in Figure 4.11.

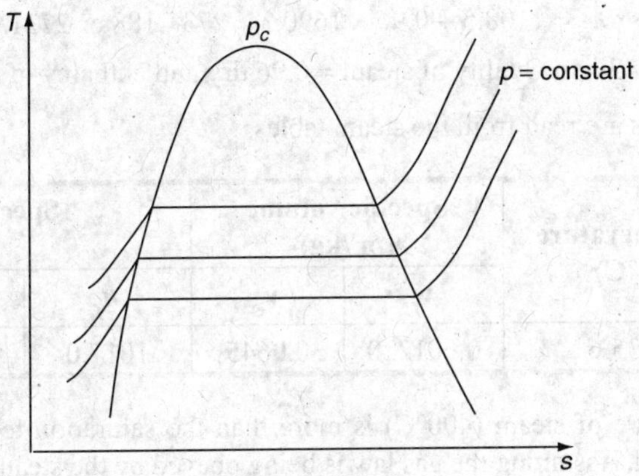

Figure 4.11 *T-S* Diagram for H_2O

Solved Problems

1. Fill in the data omitted in the table using steam tables.

	Pressure (bar)	Temperature (°C)	Specific Volume (m^3/kg)	Enthalpy (kJ/kg)	Quality, Degree of Superheating or Degree of Subcooling
(i)		255	0.0442		
(ii)	31	400			
(iii)		145		2325	
(iv)	69		0.0249		
(v)	69	120			
(vi)	28	290			
(vii)	8.2				$x = 0.9$
(viii)		120		2093	

(i) The following values are read from the steam tables.

Pressure (bar)	Temperature (°C)	Specific Volume (m^3/kg)		Specific Enthalpy (kJ/kg)		
		v_f	v_g	h_f	h_{fg}	h_g
43	255	0.001262	0.0461	1108.5	1690.3	2798.8

Since the given specific volume is less than v_g, the steam is wet. Hence, using the approximate expression $v = x\,v_g$, we can evaluate the dryness fraction of steam.

$$0.0442 = x\,(0.0461);\ x = 0.9587 \text{ or } 0.96;$$

$$h = h_f + x\,h_{fg} = 1108.5 + 0.96 \times 1690.3 = 2731.188 \text{ or } 2731.2 \text{ kJ/kg}.$$

Pressure of steam = 43 bar, Quality of steam = 0.96 dry and enthalpy of steam = 2731.2 kJ/kg.

(ii) The following values are read from the steam tables.

Pressure (bar)	Temperature (°C)	Specific Volume (m^3/kg)		Specific Enthalpy (kJ/kg)		
		v_f	v_g	h_f	h_{fg}	h_g
31	235.6	0.001220	0.0645	1017.0	1785.4	2802.4

Since the temperature of steam (400°C) is more than the saturation temperature (235.6°C), the steam is superheated. Assuming the gas law is being obeyed by the steam in the superheated state, we can use the gas laws.

As the pressure is the same, we get $\dfrac{v_{sup}}{T_{sup}} = \dfrac{v_{sat}}{T_{sat}}$; hence, $\dfrac{v_{sup}}{673} = \dfrac{0.0645}{508.6}$, i.e. $v_{sup} = 0.0853$.

Assuming C_p for the superheated steam in this pressure ranges as 2.436 kJ/kg K, we can evaluate the enthalpy of superheated steam as

$$h_{sup} = h_g + C_p(T_{sup} - T_{sat}) = 2802.4 + 2.436(673 - 508.6) = 3202.87 \text{ kJ/kg}.$$

Specific volume of superheated steam = 0.0853 m^3/kg.

$$\text{Degree of superheat} = T_{sup} - T_{sat} = 673 - 508.6 = 164.4\text{°C}.$$

Enthalpy of superheated steam = 3202.87 kJ/kg.

(iii) The following values are read from the steam tables.

Pressure (bar)	Temperature (°C)	Specific Volume (m^3/kg)		Specific Enthalpy (kJ/kg)		
		v_f	v_g	h_f	h_{fg}	h_g
4.2	145.4	0.001086	0.441	612.3	2127.5	2739.8

Since the enthalpy is less than enthalpy of saturated steam, the condition of steam is wet.

We know the formula $h = h_f + x\,h_{fg}$

Hence, $2325 = 612.3 + x\,(2127.5)$, that is, $x = 0.805$.

Specific volume of the steam $= x\,v_g = 0.805\,(0.441) = 0.355$ m^3/kg.

Pressure of steam is 4.2 bar.

(*Note*: The property values for steam at the temperature ($t = 145$°C) are not available in the steam tables. Hence, the aforesaid properties are assumed for this case.)

(iv) The following values are read from the steam tables.

Pressure (bar)	Temperature (°C)	Specific Volume (m³/kg)		Specific Enthalpy (kJ/kg)		
		v_f	v_g	h_f	h_{fg}	h_g
69	284.8	0.001348	0.0278	1262.2	1512.5	2774.7

Since the given specific volume of steam (0.0249) is less than the specific volume of saturated steam at this pressure (0.0278), the steam is wet. Hence, $v = x(v_g)$.

Therefore, $x = (0.0249)/(0.0278) = 0.8956$

$$h = h_f + x(h_{fg}) = 1262.2 + 0.8956(1512.5) = 2616.8 \text{ kJ/kg.}$$

The temperature of steam is 284.8°C.

(v) The following values are read from the steam tables.

Pressure (bar)	Temperature (°C)	Specific Volume (m³/kg)		Specific Enthalpy (kJ/kg)		
		v_f	v_g	h_f	h_{fg}	h_g
69	284.8	0.001348	0.0278	1262.2	1512.5	2774.7

Since t(120°C) is less than t_{sat}(284.4°C), the water is in subcooled state. Hence, the specific volume of water at the given temperature is 0.001348 m³/kg. The degree of subcooling is (284.8 – 120) = 164.8°C. The enthalpy of water is calculated from the formula

$$h = \frac{h_f}{C_p\left(t_f - t_0\right)} \times C_p\left(t - t_0\right)$$

where t_0 is the reference temperature (i.e. in our case 0.01°C)

$$\text{Therefore, } h = \frac{1262.2}{284.8} \times 120 = 531.82 \text{ kJ/kg.}$$

(vi) The following values are read from the steam tables.

Pressure (bar)	Temperature (°C)	Specific Volume (m³/kg)		Specific Enthalpy (kJ/kg)		
		v_f	v_g	h_f	h_{fg}	h_g
28	230	0.001209	0.0714	990.5	1811.5	2802.0

Since t(290°C) is more than t_{sat}(230°C), the steam is superheated.
The degree of superheat is (290 – 230) = 60°C.
The specific volume is calculated from the formula

$$v_{sup} = \frac{v_{sat}}{T_{sat}} \times T_{sup} = \frac{0.0714}{503} \times 563 = 0.0799$$

or specific volume = 0.0997 m³/kg.

Assuming C_p of superheated steam at this pressure as 2.754, the enthalpy of superheated steam is calculated from the expression

$$h = h_g + C_p\,(\Delta t) = 2802 + 2.754\,(60) = 2967.2,$$

that is, enthalpy of superheated steam = 2967.2 kJ/kg.

(vii) The following values are read from the steam tables.

Pressure (bar)	Temperature (°C)	Specific Volume (m³/kg)		Specific Enthalpy (kJ/kg)		
		v_f	v_g	h_f	h_{fg}	h_g
8.2	171.4	0.001116	0.235	725.4	2043.0	2768.4

The temperature of steam is 171.4°C.

Specific volume of steam $v = 0.9(0.235) = 0.2115$ m³/kg.

Enthalpy of steam $h = h_f + x(h_{fg}) = 725.4 + 0.9(2043.0) = 2564.1$ kJ/kg.

(viii) The following values are read from the steam tables.

Pressure (bar)	Temperature (°C)	Specific Volume (m³/kg)		Specific Enthalpy (kJ/kg)		
		v_f	v_g	h_f	h_{fg}	h_g
2.0	120.2	0.001061	0.885	504.7	2201.6	2706.3

Since the given h(2093) is less than h_g(2706.3), the steam is wet.

$$h = h_f + x(h_{fg});\ 2093 = 504.7 + x(2201.6);\ \text{hence, } x = 0.72.$$

Pressure of steam is 2.0 bar.

Specific volume $v = x(v_g) = 0.72(0.885) = 0.6372$ m³/kg.

The aforesaid results are shown in the following table:

	Pressure (bar)	Temperature (°C)	Specific Volume (m³/kg)	Enthalpy (kJ/kg)	Quality, Degree of Superheating or Degree of Subcooling
(i)	_43_	255	0.0442	_2731.2_	_$x = 0.96$_
(ii)	31	400	_0.0853_	_3202.7_	_$\Delta t = 164.4$°C_
(iii)	_4.2_	145	_0.355_	2325	_$x = 0.805$_
(iv)	69	_284.8_	0.0249	_2616.8_	_$x = 0.8956$_
(v)	69	120	_0.001348_	_531.82_	_$\Delta t = 164.8$°C
(vi)	28	290	0.0799	_2967.2_	_$\Delta t = 60$°C_
(vii)	8.2	_171.4_	_0.2115_	_2564.1_	$x = 0.9$
(viii)	_2.0_	120	_0.6372_	2093	_$x = 0.72$_

2. Using steam tables, determine the value of the enthalpy, the internal energy and the specific volume of a mixture of 0.1 kg of saturated water and 0.9 kg of saturated steam at a pressure of 14 bar.

   The following values are read from the steam tables.

Pressure (bar)	Temperature (°C)	Specific Volume (m^3/kg)		Specific Enthalpy (kJ/kg)		
		v_f	v_g	h_f	h_{fg}	h_g
14	195.04	0.001148	0.14073	830.1	1957.7	2787.8

   Dryness fraction of steam = 0.9/(0.9 + 0.1) = 0.9.

   Specific volume of the steam, $v = x(v_g) = 0.9(0.14073) = 0.126657$ m^3/kg.

   Enthalpy of steam, $h = h_f + x(h_{fg}) = 830.1 + 0.9(1957.7) = 2592.03$ kJ/kg.

   Internal energy, $u = h - pv = 2592.03 - 14(0.126657) \times 100 = 2414.701$ kJ/kg.

3. Using steam tables, determine the value of the enthalpy, the internal energy and the specific volume of a mixture of 0.1 kg of saturated water and 0.9 kg of saturated steam at a temperature of 250°C.

   The following values are read from the steam tables.

Pressure (bar)	Temperature (°C)	Specific Volume (m^3/kg)		Specific Enthalpy (kJ/kg)		
		v_f	v_g	h_f	h_{fg}	h_g
39.776	250	0.0012513	0.050037	1085.8	1714.7	2800.5

   Dryness fraction of steam = 0.9/(0.9 + 0.1) = 0.9.

   Specific volume of the steam, $v = x(v_g) = 0.9(0.050037) = 0.0450333$ m^3/kg.

   Enthalpy of steam, $h = h_f + x(h_{fg}) = 1085.8 + 0.9(1714.7) = 2629.03$ kJ/kg.

   Internal energy, $u = h - pv = 2629.03 - 39.776(0.0450333) \times 100 = 2449.91$ kJ/kg.

4. A vessel of volume 0.04 m^3 contains a mixture of saturated water and saturated steam at a temperature of 240°C; the mass of liquid present is 8 kg. For the mixture, evaluate the pressure, the mass, the specific volume, the enthalpy and the internal energy.

   The following values are read from the steam tables.

Pressure (bar)	Temperature (°C)	Specific Volume (m^3/kg)		Specific Enthalpy (kJ/kg)		
		v_f	v_g	h_f	h_{fg}	h_g
33.478	240	0.001229	0.059645	1037.6	1764.6	2802.2

   The pressure of steam is 33.478 bar. Let the mass of saturated steam be x.

   The volume of steam is calculated from the equation $V = v_f m_f + v_g m_g$.

   Hence, Volume of the mixture, $V = 0.04 = 0.001229 \times 8 + 0.059645 \times (x)$.

Solving this equation, we get $x = 0.5058$ kg.

Mass of the mixture = 8.0 + 0.5058 = 8.5058 kg.

Specific volume of the mixture = 0.04/(8.5058) = 0.0047026 m^3/kg.

Enthalpy of the mixture = 1037.6 × 8 + 2802.2 × 0.5058 = 9718.15 kJ.

Internal energy of the mixture = $H - p\,V$ = 9718.15 − 33.478 × 0.04 × 100 = 9584.2 kJ.

5. Steam at 95°C is found to have a specific volume of 1.685 m^3/kg. What is the specific entropy of the substance and what is the pressure?

The following values are read from the steam tables.

Pressure (bar)	Temperature (°C)	Specific Volume (m^3/kg)		Specific Entropy (kJ/kg k)		
		v_f	v_g	s_f	s_{fg}	s_g
0.8455	95	0.001040	1.982	1.2501	6.1653	7.4154

Since the specific volume of dry steam at 95°C is 1.982 m^3/kg (more than 1.685 m^3/kg), the steam is wet. Let the dryness fraction be x.

Therefore, $v = x(v_g)$, that is, $1.685 = x(1.982)$.

Solving this we get $x = 0.85$.

Pressure of steam is 0.8455 bar.

Specific entropy, $s = s_f + x(s_{fg}) = 1.2501 + 0.85(6.1653) = 6.4906$ kJ/kg K.

6. A wet mixture of water substance is maintained in a closed rigid tank at 1.0 bar. What must be the quality of the mixture initially if, on heating through a series of equilibrium states, the final state will be the critical state? What must be the initial ratio of volume of vapour to liquid?

The following values are read from the steam tables.

Pressure (bar)	Temperature (°C)	Specific Volume (m^3/kg)		Specific Enthalpy (kJ/kg)		
		v_f	v_g	h_f	h_{fg}	h_g
1.0	99.6	0.001043	1.694	417.5	2257.9	2675.4

Let the quality of the mixture or dryness fraction be x. In this case we will be using the exact expression for the specific volume; that is, $v = (1 - x)v_f + x(v_g)$.

Specific volume of the mixture = $(1 - x) \times 0.001043 + 1.694x = 0.001043 + 1.692957x$ (1)

Specific volume of steam at critical state = 0.00317 (read from the steam tables) (2)

Since volume remains the same during the heating process in a closed rigid vessel, we equate the equations (1) and (2).

Hence, $0.001043 + 1.692957x = 0.00317$; that is, $x = 0.001256$

Quality of the mixture = 0.001256.

Volume of vapour = $x(v_g)$ = 0.001256(1.694) = 0.002127 m^3/kg.

Volume of liquid = $(1 - x)(v_f) = (1 - 0.001256)(0.001043) = 0.001042$ m³/kg.

Hence, the ratio of volume of vapour to liquid = (0.002127)/(0.001042) = 2.04.

7. One kilogram steam at a pressure of 10 bar and having a dryness fraction of 0.2 receives 1047 kJ heat at a constant pressure. Calculate the dryness fraction, enthalpy, specific volume and internal energy of steam at the end of constant pressure heat supply.

The following values are read from the steam tables.

Pressure (bar)	Temperature (°C)	Specific Volume (m³/kg)		Specific Enthalpy (kJ/kg)		
		v_f	v_g	h_f	h_{fg}	h_g
10	179.9	0.001127	0.194	762.6	2013.6	2776.2

Before heating,

Enthalpy of steam, $h_1 = h_f + xh_{fg} = 762.6 + 0.2(2013.6) = 1165.32$ kJ/kg.

Specific volume of steam, $v_1 = x(v_g) = 0.2 \times 0.194 = 0.0388$ m³/kg.

Internal energy of steam, $u_1 = h_1 - p_1v_1 = 1165.32 - 10 \times 0.0388 \times 100 = 1126.52$ kJ/kg.

After the heating process at constant pressure,

Enthalpy of steam, $h_2 = h_1 + {}_1q_2 = 1165.32 + 1047 = 2212.32$ kJ/kg.

Since the enthalpy of steam after heating is less than the enthalpy of dry and saturated steam at the pressure (10 bar), the condition of steam is wet (i.e. 2212.32 < 2776.2).

Enthalpy of steam, $h_2 = h_f + x_2h_{fg}$; $2212.32 = 762.6 + x_2(2013.6)$; hence, $x_2 = 0.72$.

Specific volume of steam, $v_2 = x_2(v_g) = 0.72(0.194) = 0.13968$ m³/kg.

Internal energy of steam, $u_2 = h_2 - p_2v_2 = 2212.32 - 10 \times 0.13968 \times 100 = 2072.64$ kJ/kg.

8. One kilogram of steam at 7 bar and 10°C is heated at constant pressure to form steam with 35°C of superheat. It is then cooled at constant volume until the pressure is 1 bar. Calculate the heat required to form the steam at constant pressure, the final dryness fraction of steam and the heat removed in cooling at constant volume.

The following values are read from the steam tables.

Pressure (bar)	Temperature (°C)	Specific Volume (m³/kg)		Specific Enthalpy (kJ/kg)		
		v_f	v_g	h_f	h_{fg}	h_g
1	99.6	0.001043	1.694	417.5	2257.9	2675.4
7	165.0	0.001108	0.273	697.1	2064.9	2762.0

Enthalpy of water at 10°C, $h_1 = 4.188 \times 10 = 41.88$ kJ/kg.

Enthalpy of superheated steam at 7 bar with 35°C of superheat, $h_2 = h_g + C_p(\Delta t)$.

Assuming C_p of superheated steam as 2.198 kJ/kg K, the enthalpy is calculated.

Hence, $h_2 = 2762.0 + 2.198(35) = 2838.93$ kJ/kg.

The heat required to form the steam, $q = h_2 - h_1 = 2838.93 - 41.88 = 2797.05$ kJ/kg.

Specific volume of superheated steam: $\frac{v_{sup}}{T_{sup}} = \frac{v_{sat}}{T_{sat}}$; i.e. $\frac{v_{sup}}{473} = \frac{0.273}{438}$.

$v_{sup} = 0.2948$ m³/kg and internal energy of steam $u_2 = h_2 - p_2 v_2$;

$u_2 = 2838.93 - 7 \times 0.2948 \times 100 = 2632.55$ kJ/kg.

Since the cooling process occurs at constant volume, $v_2 = v_3$, $v_2 = 0.2948$ m³/kg.

As the specific volume of steam is less than the specific volume of saturated steam at pressure 1 bar, the condition of steam is wet. Let us assume the dryness fraction as x_3.

Hence, $0.2948 = x_3(1.694)$; therefore, $x_3 = 0.174$.

Enthalpy of steam after the cooling process, $h_3 = h_f + x_3(h_{fg}) = 417.5 + 0.174(2257.9)$;

$h_3 = 810.37$ kJ/kg. Let the internal energy of steam after the cooling process be u_3.

$u_3 = h_3 - p_3 v_3 = 810.37 - 1 \times 0.2948 \times 100 = 780.89$ kJ/kg.

Heat removed during the cooling process $= u_2 - u_3 = 2632.55 - 780.89 = 1851.6$ kJ/kg.

9. If 0.0566 m³ of dry steam at 8 bar expands to a pressure of 4 bar according to the law pV = constant. Calculate the mass of steam present, the condition after expansion and the work done during expansion.

   The following values are read from the steam tables.

Pressure (bar)	Temperature (°C)	Specific Volume (m³/kg)		Specific Enthalpy (kJ/kg)		
		v_f	v_g	h_f	h_{fg}	h_g
4	143.6	0.001084	0.462	604.7	2133.0	2737.7
8	170.4	0.001115	0.240	720.9	2046.5	2767.5

Mass of steam, $m = \frac{V_1}{v_g}$, i.e. $\frac{0.0566}{0.240} = 0.2358$ kg. The steam expands according to the law $p_1V_1 = p_2V_2$.

Volume of steam after expansion, $V_2 = (8) \times (0.0566)/(4) = 0.1132$ m³. Volume of dry and saturated at 4 bar $= 0.2358 \times 0.462 = 0.1089$ m³. Since the volume of dry and saturated steam is less than the volume of steam after expansion, the condition of steam after expansion is superheated.

Hence, $\frac{V_2}{T_2} = \frac{V_{sat}}{T_{sat}}$; i.e. $\frac{0.1132}{\mathrm{T}_2} = \frac{0.1089}{(273+143.6)}$; $T_2 = \frac{0.1132}{0.1089} \times 416.6 = 433$ K. The condition of steam after expansion is superheated and the temperature of steam is 160°C (i.e. degree of superheat = 16.4°C).

Work done during the process, ${}_1W_2 = p_1V_1 \, ln\,(V_2/V_1) = 8 \times 100 \times 0.0566 \times ln\,(2)$

$${}_1W_2 = 31.385 \text{ kJ}.$$

10. Using steam tables, determine the mass, the enthalpy and the internal energy of a system comprising 0.1 m³ of wet steam at a temperature 160°C and a dryness fraction of 0.94. The system undergoes a fully resisted process at constant pressure to a temperature of 250°C. During the process, there are work and heat interactions between the system and its surroundings. Show the path of this process on a sketch of the p-v diagram and evaluate the increase in enthalpy, the increase in internal energy, the work done and the heat transfer.

The following values are read from the steam tables.

Pressure (bar)	Temperature (°C)	Specific Volume (m^3/kg)		Specific Enthalpy (kJ/kg)		
		v_f	v_g	h_f	h_{fg}	h_g
6.1806	160	0.001102	0.30676	675.4	2081.3	2756.7

Specific volume of steam, $v = x(v_g) = 0.94(0.30676) = 0.28835$ m³/kg.

Specific enthalpy, $h = h_f + x(h_{fg}) = 675.4 + 0.94(2081.3) = 2631.822$ kJ/kg.

Specific internal energy, $u = h - pv = 2631.822 - 6.1806 \times 0.28835 \times 100 = 2453.6$.

Mass of steam = (0.1)/(0.28835) = 0.3468 kg.

Specific volume of steam after expansion, v_2

We know

$$\frac{v_2}{T_2} = \frac{v_g}{T_g},\ v_2 = \frac{523}{433} \times 0.30676$$

$$v_2 = 0.37052 \text{ m}^3/\text{kg}.$$

As the steam is superheated, the specific heat at constant pressure can also be calculated from the formula $C_p = (h_2 - h_1)/(t_2 - t_1)$ where h_1 and h_2 are the enthalpies at temperatures t_1 and t_2 in the property values in the superheated tables at the given pressure.

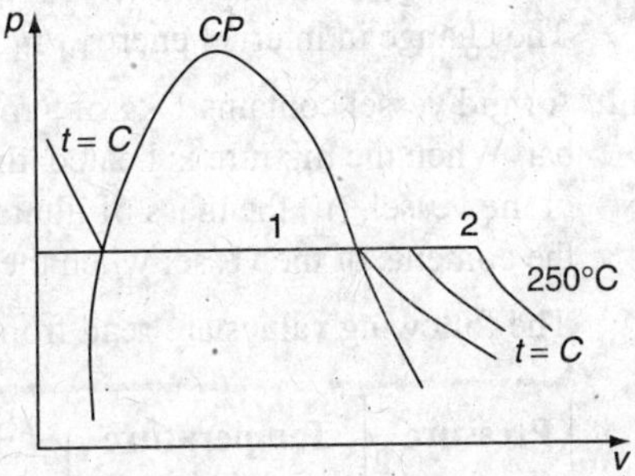

Figure 4.12 Pressure-volume Diagram

The path of process 1-2 is shown in Figure 4.12.

$$h_2 = 2756.7 + C_p(250 - 160) \text{ where } C_p = (2951.6 - 2849.7)/(50) = 2.038$$
$$= 2756.7 + 2.038(90) = 2940.12 \text{ kJ/kg}.$$
$$u_2 = 2940.12 - 6.1806 \times 0.37052 \times 100 = 2711.12 \text{ kJ/kg}.$$

Work done $= p(v_2 - v_1)m = 6.1806 \times (0.37052 - 0.28835) \times 100 \times 0.3468 = 17.612$ kJ.

The increase in enthalpy, $\Delta H = m(h_2 - h_1) = 0.3468 \times (2940.12 - 2631.822) = 106.92$ kJ.

The increase in internal energy, $\Delta U = m(u_2 - u_1) = 0.3468 \times (2711.12 - 2453.6) = 89.31$ kJ.

The heat transfer $_1Q_2 = \Delta U + {_1W_2}$, that is, $_1Q_2 = 89.31 + 17.612 = 106.922$ kJ.

11. One kilogram of steam at 0°C is heated at constant pressure of 7 bar until it is converted into steam at a temperature of 250°C. If this steam loses 250 kJ of heat while the pressure remains constant, determine the final condition of steam, the change in volume and the change in internal energy.

The following values are read from the steam tables.

Pressure (bar)	Temperature (°C)	Specific Volume (m^3/kg)		Specific Enthalpy (kJ/kg)		
		v_f	v_g	h_f	h_{fg}	h_g
7	165	0.001108	0.273	697.1	2064.9	2762.0

The following values are read from the superheated steam tables:

Enthalpy of steam at 7 bar, 250°C = 2953.6 kJ/kg.
Specific volume at 7 bar, 250°C = 0.336 m³/kg.
Internal energy at 7 bar, 250°C = 2718.2 kJ/kg.

Enthalpy of steam after cooling at constant pressure:

$q = h_2 - h_1, -250 = h_2 - h_1$, therefore $h_2 = h_1 - q$, that is, $h_2 = 2953.6 - 250 = 2703.6$.

Since this enthalpy (2703.6) is less than the enthalpy of dry steam (2762), the condition of steam is wet. We know $h_2 = h_f + x(h_{fg})$, $2703.6 = 697.1 + x(2064.9)$.

Hence, the dryness fraction of steam, $x = 0.97$.
Specific volume of steam, $v_2 = x(v_g) = 0.97(0.273) = 0.2648$ m³/kg.
Internal energy of steam, $u_2 = h_2 - pv = 2703.6 - 7 \times 0.2648 \times 100 = 2518.24$ kJ/kg.
The final condition of steam is wet with a dryness fraction of 0.97.
The change in volume, $\Delta v = v_2 - v_1 = 0.2648 - 0.336 = -0.0712$ m³/kg.
The change in internal energy, $\Delta u = u_2 - u_1 = 2518.24 - 2718.2 = -199.96$ kJ/kg.

12. A rigid vessel contains 1 kg of a mixture of saturated water and saturated steam at a pressure of 1.4 bar. When the mixture is heated, the state passes through the critical point. Evaluate (i) the volume of the vessel, (ii) the mass of liquid and of vapour in the vessel initially and (iii) the temperature of the contents of the vessel when the pressure has risen to 300 bar.

The following values are read from the steam tables,

Pressure (bar)	Temperature (°C)	Specific Volume (m³/kg)		Specific Enthalpy (kJ/kg)		
		v_f	v_g	h_f	h_{fg}	h_g
1.4	109.3	0.001051	1.236	458.4	2231.9	2690.3
221.2 (p_c)	374.1 (t_c)	0.00317	0.00317	2107.4	0	2107.4

Since the process is constant volume, the volume of the vessel will be equal to the specific volume of steam at the critical state.

(i) Volume of the vessel = 0.00317 m³.

(ii) For the constant volume heating process v = constant.
Hence, $0.00317 = (1 - x)(0.001051) + x(1.236)$; solving this we get $x = 0.00171$.
Mass of liquid in the vessel initially $= 1 - x = 1 - 0.00171 = 0.99829$ kg.
Mass of vapour in the vessel initially $= x = 0.00171$ kg.

(iii) As the pressure given is 300 bar, the state of steam is supercritical. We can use the supercritical table and evaluate the properties by interpolation only.

(*Note*: we cannot use the gas laws in the supercritical state).

The temperature of steam:

$$t = t_1 + \frac{(v - v_1)}{(v_2 - v_1)}(t_2 - t_1), \text{ that is, } t = 400 + \frac{(0.00317 - 0.00279)}{(0.0053 - 0.00279)}(25) = 403.785°\text{C}.$$

The temperature of steam when the pressure has risen to 300 bar is 403.785°C.

13. A certain quantity of steam is passed in a vessel containing 15 kg of water at 25°C and after the steam is blown the temperature of the mixture rises to 80°C. The pressure of steam blown is 6 bar and dryness fraction is 0.8. The water equivalent of the vessel (tank) is 1 kg. Find out the quantity of steam blown.

The following values are read from the steam tables.

Pressure (bar)	Temperature (°C)	Specific Volume (m^3/kg)		Specific Enthalpy (kJ/kg)		
		v_f	v_g	h_f	h_{fg}	h_g
6	158.8	0.001101	0.315	670.4	2085.0	2755.5

The enthalpy of water at 25°C is 104.9 kJ/kg and at 80°C is 334.9 kJ/kg.

The enthalpy of steam being blown, $h = h_f + x(h_{fg}) = 670.4 + 0.8(2085.0) = 2338.4$

Making an energy balance for the process, we get

$$(m_w + 1)\, h_1 + m_s h_s = (m_w + 1 + m_s)\, h_2.$$

Substituting $(15 + 1)(104.9) + m_s(2338.4) = (15 + 1 + m_s)(334.9)$,

Hence, $m_s = 1.8367$ kg.

The mass of steam blown is 1.8367 kg.

14. Estimate the specific heat at constant pressure for steam at 400°C if the pressure is (i) 0.1 bar, (ii) 1.0 bar and (iii) 300 bar.

Using the superheated and supercritical steam tables, we can evaluate C_p for steam.

(i) $$C_p = \frac{(h_2 - h_1)}{(t_2 - t_1)} = \frac{(3489.1 - 3076.5)}{(500 - 300)} = 2.063\ \frac{\text{kJ}}{\text{kg K}}.$$

(ii) $$C_p = \frac{(h_2 - h_1)}{(t_2 - t_1)} = \frac{(3488.1 - 3074.3)}{(500 - 300)} = 2.069\ \frac{\text{kJ}}{\text{kg K}}.$$

(iii) $$C_p = \frac{(h_2 - h_1)}{(t_2 - t_1)} = \frac{(2614.2 - 1791.5)}{(425 - 375)} = 16.454\ \frac{\text{kJ}}{\text{kg K}}.$$

15. A throttling calorimeter sampling the steam generated by a boiler at a pressure 8 bar gives the readings of pressure 1 bar and temperature 116°C. Evaluate the dryness of the boiler steam.

The functioning of throttling calorimeter is to determine the condition of wet steam provided the dryness is close to unity. Otherwise, a separating calorimeter is required in addition to the throttling calorimeter to measure the dryness of very wet steam. Assuming that the velocities are sufficiently small and using the steady flow energy equation to the throttling process (i.e. adiabatic with no work transfer), we can write

$$\frac{Q - W_s}{m} = 0 = h_2 - h_1 \text{ i.e. } h_2 = h_1.$$

Pressure (bar)	Temperature (°C)	Specific Volume (m^3/kg)		Specific Enthalpy (kJ/kg)		
		v_f	v_g	h_f	h_{fg}	h_g
1	99.6	0.001043	1.694	417.5	2257.9	2675.4
8	170.4	0.001115	0.240	720.9	2046.5	2767.5

The specific heat of superheated steam at 1 bar is calculated from the enthalpy values as shown below:

$$C_p = \frac{(h_2 - h_1)}{(t_2 - t_1)} = \frac{(2776.4 - 2676.2)}{(150 - 100)} = 2.004\ \frac{\text{kJ}}{\text{kg K}}$$

Enthalpy of superheated steam after throttling, $h_2 = 2675.4 + 2.004(116 - 99.6) = 2708.26$.
Hence, $h_1 = 2708.26 = 720.9 + x\,(2046.5)$. Solving this equation we get $x = 0.971$.
The dryness of the boiler steam is 0.971.

Practice Problems

1. State the conditions of steam in the following cases:
   (i) At a pressure of 10 bar, enthalpy is 2500 kJ/kg. (Wet steam)
   (ii) At a pressure of 15 bar, temperature is 198.3°C. (Sat. water or wet or dry steam)
   (iii) At a pressure of 20 bar, the temperature is 220°C. (Superheated steam)
   (iv) At a pressure of 7 bar, specific volume is 0.25 m³/kg. (Wet steam)
   (v) At a pressure of 5 bar, specific volume is 0.43 m³/kg. (Superheated steam)
   (vi) At a pressure of 8 bar, enthalpy is 2800 kJ/kg. (Superheated steam)
2. Five kilogram of steam occupies a volume of 10 m³. Find the quality and the pressure if the temperature is measured at (i) 40°C and (ii) 85°C. (Hint: Use the exact expression for v)
   [(i) 0.1024, 0.0738 bar; (ii) 0.7071, 0.5783 bar]
3. Determine the final volume of a mixture of water and steam if 3 kg of water is heated at constant pressure until the quality is 60 percent if the pressure is (i) 0.25 bar and (ii) 2.7 bar (Hint: Use the exact expression for v).
   [(i) 11.17 m³; (ii) 1.2037 m³]
4. Calculate the mass of 7 m³ of dry saturated steam at a pressure of 10 bar.
   (36.08 kg)
5. Calculate the specific volume, enthalpy, internal energy and entropy of steam at a pressure of 40 bar and dryness fraction 0.8.
   (0.04 m³/kg, 2458 kJ/kg, 2298 kJ/kg, 5.4141 kJ/kg K)
6. A rigid vessel with a volume of 10 m³ contains a water-vapour mixture at 4 bar. If the quality is 60 percent, find the mass. The pressure is lowered to 3 bar by cooling the vessel. Find m_g and m_f.
   (36.02 kg, 16.46 kg, 19.55 kg)
7. 1.5 m³ of steam, at a pressure of 10 bar and dryness fraction 0.95, loses 9420 kJ at constant pressure. Calculate the dryness fraction of steam at the end of the process.
   (0.375)
8. Steam at a pressure of 5 bar and a dryness of 0.95 enters a superheater and emerges without pressure loss at a temperature of 500°C. How much heat has been transferred to the steam per kg flowing.
   (841.7 kJ)
9. A rigid vessel contains saturated water at 260°C. If the correct proportions of liquid and vapour are initially in the vessel, the system will pass through the critical state when heated sufficiently.

Determine the initial proportions of liquid and vapour necessary (i) in percent by volume and (ii) in percent by weight.

[(i) 38.4%, 61.6%; (ii) 95.4%, 4.6%]

10. If 0.0566 m^3 of dry steam at 8 bar expands to a pressure of 4 bar according to the law pV = constant. Calculate (i) the mass of steam present, (ii) the condition after expansion and (iii) the work done during expansion.

[(i) 0.2358 kg, (ii) Superheated at 160°C or degree of superheat = 16.4°C, (iii) 31.38 kJ]

11. Using steam tables, calculate the enthalpy, internal energy and entropy of wet steam at 10 bar and dryness fraction 0.78.

(2333.2 kJ/kg; 2181.9 kJ/kg; 5.605 kJ/kg K)

12. Wet steam at 5 bar and 0.5 dryness is heated at constant pressure to dry and saturated condition. Estimate the amount of heat added.

(1053.7 kJ/kg)

13. A throttling calorimeter is used to measure the dryness fraction of steam in a steam main. The calorimeter readings are: pressure 1 bar; temperature 120°C. The main pressure is 6 bar. Calculate the dryness fraction of the steam in the main.

(0.98)

14. A kilogram of steam at a pressure of 10 bar and a temperature of 250°C is contained in a cylinder closed by a piston. The steam is compressed by the motion of the piston inwards until the pressure is 20 bar. The work done on the steam is 610 kNm and the heat transfer from the steam is 890 kJ. Find the final steam temperature if the steam is superheated, or the dryness if it is wet.

(0.9)

15. 100 kJ of heat are supplied under a constant pressure of 11 bar to 1 kg of dry saturated steam. Find the final state, change in internal energy and the work done by the steam.

(227.7°C, 81.44 kJ/kg, 18.56 kNm/kg)

Objective Type Questions

1. Which one of the following items cannot be considered as a pure substance?
   (i) A mixture of vapour, liquid and ice at triple point
   (ii) A mixture of liquid and vapour
   (iii) A mixture of liquid and air
   (iv) Air

2. The formation of hoar-frost, which is the direct transition from water vapour to ice without intermediate condensation and freezing, is an example of
   (i) the subcooling
   (ii) the supersaturation
   (iii) the dew formation
   (iv) the reverse sublimation

3. Assertion (A): An example of sublimation is the gradual disappearance of fallen snow in prolonged dry, cold weather without intermediate melting.
   Reason (R): The partial pressure of vapour in the atmosphere is below that of triple point.

(i) Both A and R are true and R is the correct explanation of A
(ii) Both A and R are true and R is not the correct explanation of A
(iii) A is true but R is false
(iv) A is false but R is true

4. The ordinate and abscissa of the following diagram (Figure 4.13) that shows the isothermal heating of wet steam till it becomes superheated, represented by '12', are

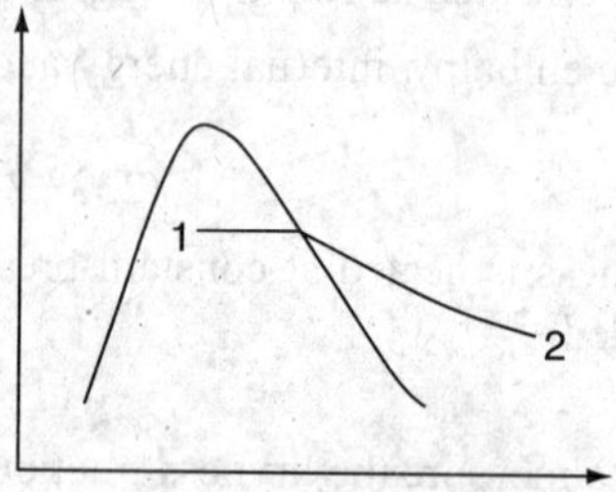

Figure 4.13 Property Diagram for H_2O During Isothermal Heating

(i) pressure-volume
(ii) enthalpy-entropy
(iii) internal energy-volume
(iv) pressure-enthalpy

5. Which diagram of pure substance is easy to construct from the measurable properties?
(i) Internal energy-volume
(ii) Temperature-entropy
(iii) Enthalpy-entropy
(iv) Enthalpy-pressure

6. Assertion (A): On the enthalpy-entropy diagram of a pure substance, quality lines start/originate from the critical point.
Reason (R): All the three phases coexist at the critical point.
(i) Both A and R are true and R is the correct explanation of A
(ii) Both A and R are true and R is not the correct explanation of A
(iii) A is true but R is false
(iv) A is false but R is true

7. The phase transformation lines IJ, IK and IL shown on the temperature-entropy diagram (Figure 4.14) of a pure substance represent respectively

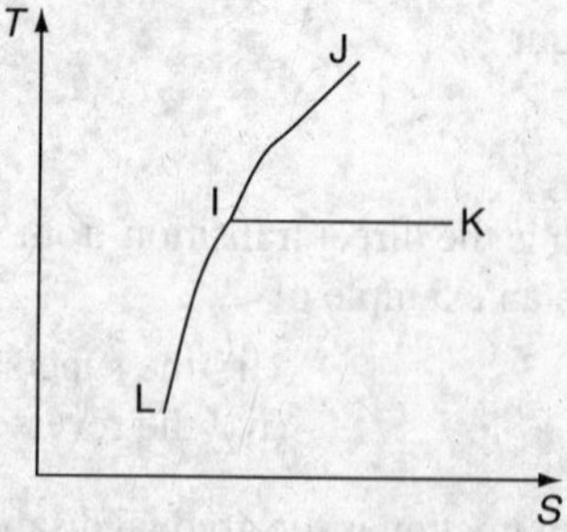

Figure 4.14 Temperature-entropy Diagram for a Pure Substance

(i) vapourization, fusion and sublimation
(ii) sublimation, vapourization and fusion
(iii) fusion, sublimation and vaporization
(iv) vapourization, sublimation and fusion

8. Which form of energy is present during the vapourization of liquid?
   (i) Mechanical energy (ii) Flow energy
   (iii) Potential energy (iv) Kinetic energy

9. When will the ideal gas approximations hold good for a pure substance?
   (i) Saturated vapour state (ii) Saturated liquid state
   (iii) Superheated state (iv) Critical state

10. The wet steam is heated isothermally till it reaches superheated state. The slope of isotherm on the pressure-volume plane
   (i) does not change (ii) changes once
   (iii) changes twice (iv) changes thrice

11. Which equation is always true for a fixed mass of any pure substance (symbols have their usual meanings)?
   (i) $S = \frac{U}{T}$ (ii) $S = \frac{H}{T}$
   (iii) $S = \frac{L}{T}$ (iv) $S = \frac{Q}{T}$

12. The condition of a pure substance will be non-equilibrium in
   (i) superheated vapour state (ii) superheated liquid state
   (iii) saturated liquid state (iv) saturated vapour state

13. The temperature at which the specific volume of a saturated liquid is the same as that of saturated vapour is known as
   (i) boiling point (ii) critical point
   (iii) dew point (iv) triple point

14. The ordinate and abscissa of the following diagram (Figure 4.15) that shows an isometric cooling process '12' of a pure substance are

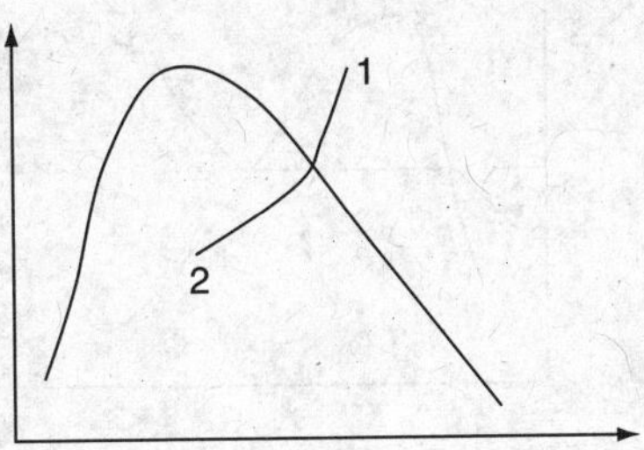

Figure 4.15 Property Diagram for a Pure Substance Showing Isometric Cooling Process

   (i) pressure-volume (ii) enthalpy-entropy
   (iii) temperature-entropy (iv) pressure-enthalpy

15. If saturated vapour is compressed adiabatically, it will
    (i) become superheated
    (ii) liquefy
    (iii) remain dry and saturated
    (iv) become wet
16. Which one of the following triple point representations is correctly matched for a pure substance?
    (i) On *p-V* coordinates – Triangle
    (ii) On *H-S* coordinates – Trapezium
    (iii) On *U-V* coordinates – Straight line
    (iv) On *T-p* coordinates – Point
17. If saturated liquid of a pure substance expands adiabatically, it will become
    (i) subcooled liquid
    (ii) dry and saturated vapour
    (iii) superheated vapour
    (iv) wet vapour
18. An isothermal process undergone by a mixture of saturated liquid and saturated vapour in the two phase region on the enthalpy (ordinate) – entropy (abscissa) diagram is represented by
    (i) horizontal line
    (ii) inclined line
    (iii) vertical line
    (iv) curved line
19. Identify the most common fluid whose properties in all three phases are used in the study of thermodynamics.
    (i) Ammonia
    (ii) Freon 12
    (iii) Helium
    (iv) Water
20. The solid and vapour phases of a pure substance coexist along the
    (i) critical isotherm
    (ii) fusion curve
    (iii) sublimation curve
    (iv) vapourization curve
21. The number of degrees of freedom at the triple point for a pure substance is equal to
    (i) one
    (ii) three
    (iii) five
    (iv) six
22. Figure 4.16 depicts saturation dome for water on the temperature entropy coordinates. The temperature difference shown on a typical isobar line is known as degree of

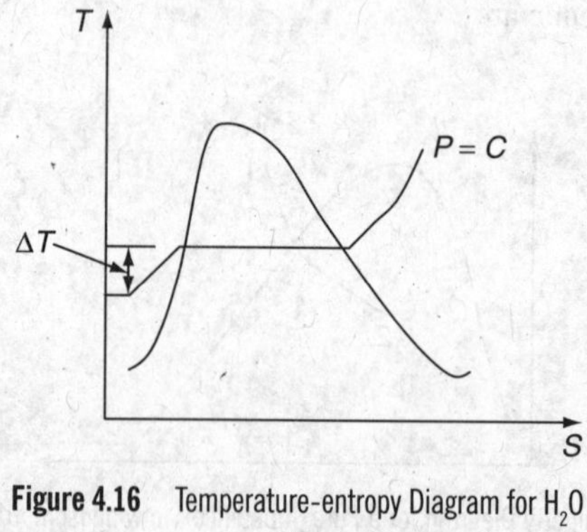

Figure 4.16 Temperature-entropy Diagram for H_2O

    (i) wet-bulb depression
    (ii) saturation
    (iii) subcooling
    (iv) superheating

23. Who introduced the concepts of latent heats of vapourization and fusion during the phase transitions and also proposed the concept of specific heats in thermodynamics?
(i) Joseph Black (ii) Josiah Gibbs
(iii) Benjamin Thomson (iv) James Watt

24. The following table furnishes a portion of properties of a pure substance

P, saturation (MPa)	*T*, saturation (°C)	Specific Volume (m^3/kg) ($V_f \times 10^3$)	Specific Volume (m^3/kg) ($V_g \times 10^3$)
0.10	112	1.05	901.05
0.25	127	1.06	701.06
0.40	142	1.07	501.07

The latent heat of vapourization for this substance at 2.5 bar, in kJ/kg, is
(i) 3200 (ii) 3000
(iii) 2800 (iv) 2600

25. The pressure-volume diagram shown in Figure 4.17 exhibits the phases of a pure substance. The two-phase regions are marked as

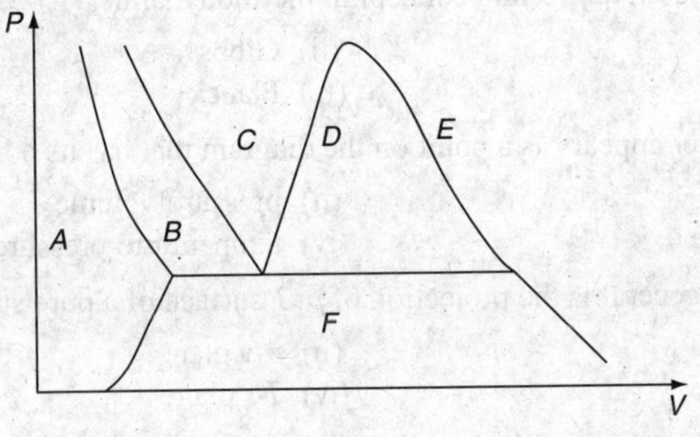

Figure 4.17 Pressure-volume Diagram for a Pure Substance

(i) A, E and F (ii) B, C and D
(iii) B, D and F (iv) A, C and E

26. Assertion (A): All thermodynamic properties of liquid and vapour phases are identical at the critical point of a pure substance.
Reason (R): The vapourization process terminates at the critical point.
(i) Both A and R are true and R is the correct explanation of A
(ii) Both A and R are true and R is not the correct explanation of A
(iii) A is true but R is false
(iv) A is false but R is true

27. The slope of isobar line on the enthalpy-entropy diagram for a pure substance during heating process from ice at –5°C to steam at 100°C at 1 atmosphere
    (i) remains the same
    (ii) changes once
    (iii) changes twice
    (iv) changes thrice
28. The object of using a separating calorimeter in the combined separating and throttling calorimeter is
    (i) to supply dry steam to the throttling calorimeter
    (ii) to increase the quality of steam supplied to the throttling calorimeter
    (iii) to heat the wet steam before being supplied to the throttling calorimeter
    (iv) to reduce the pressure of steam supplied to the throttling calorimeter
29. Which one of the following processes could result in wet vapour at the end?
    (i) Saturated vapour is heated at constant pressure
    (ii) Saturated vapour is compressed reversibly and adiabatically
    (iii) Saturated vapour is expanded reversibly and adiabatically
    (iv) Saturated vapour is throttled adiabatically
30. The heating process undergone by a pure substance below its triple point is
    (i) sublimation
    (ii) vapourization
    (iii) fumigation
    (iv) saturation
31. Who introduced the phase equilibrium concept in thermodynamics?
    (i) Clausius
    (ii) Gibbs
    (iii) Kelvin
    (iv) Planck
32. The triple point of water appears as a point on the diagram that has its coordinates
    (i) enthalpy-entropy
    (ii) pressure-volume
    (iii) enthalpy-pressure
    (iv) temperature-pressure
33. The phase diagram in general is the projection of *pvT* surface of a pure substance on
    (i) *p-T* plane
    (ii) *p-v* plane
    (iii) *T-v* plane
    (iv) *T-s* plane
34. The ordinate and abscissa of the following diagram (Figure 4.18) that shows the throttling process 1-2 of a pure substance are
    (i) pressure and volume
    (ii) enthalpy and entropy
    (iii) temperature and entropy
    (iv) pressure and enthalpy

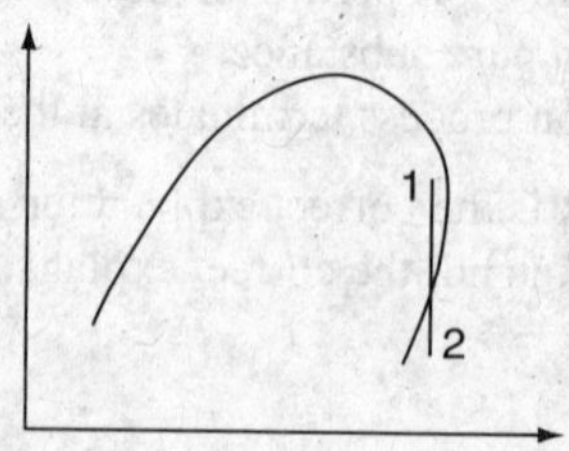

Figure 4.18 Property Diagram Showing Throttling Process for a Pure Substance

35. Identify the fusion curve on the *p*-*T* diagram (Figure 4.19) shown below for a pure substance that expands on freezing.

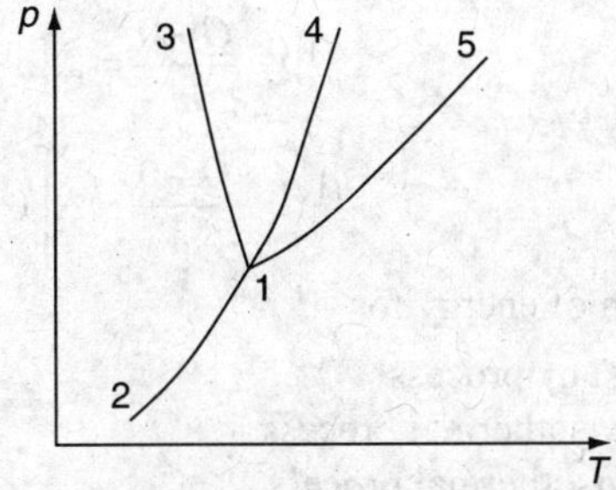

Figure 4.19 Pressure-temperature Diagram for a Pure Substance

(i) Curve 12 (ii) Curve 13
(iii) Curve 14 (iv) Curve 15

36. Which *T*-*s* diagram (Figure 4.20) illustrates correctly the cooling process undergone by the contents of pressure cooker?

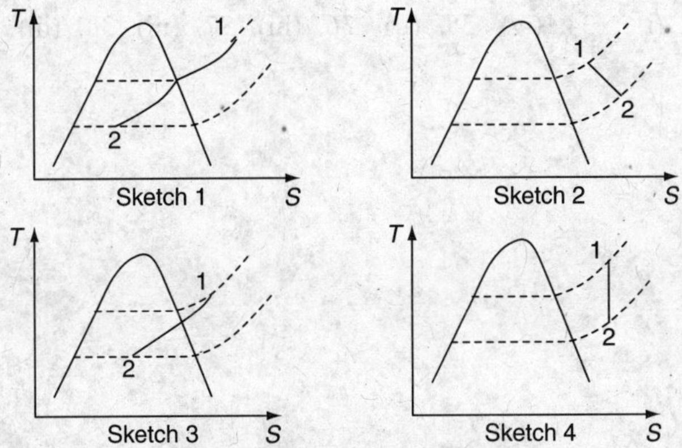

Figure 4.20 Temperature-entropy Diagram for H_2O

(i) Sketch 1 (ii) Sketch 2
(iii) Sketch 3 (iv) Sketch 4

37. Identify the process in which there is liberation of heat.

(i) Diathermic (ii) Endothermic
(iii) Exothermic (iv) Hygroscopic

38. Which process is reversible?

(i) Free expansion of an elastic structure
(ii) Plastic deformation of a metal
(iii) Electrolysis of pure water
(iv) Flow of current through a resister

39. The steady flow energy equation for a pure substance, when simplified and applied to a steam condenser, takes the form

(i) $\frac{0-W_s}{m}=\Delta\left[h+\frac{V^2}{2}+0\right]$ (ii) $\frac{Q-0}{m}=\Delta[h+0+0]$

(iii) $\frac{0-W_s}{m}=\Delta[0+0+gZ]$ (iv) $\frac{Q-0}{m}=\Delta\left[0+\frac{V^2}{2}+gZ\right]$

40. The quantity pv represents a form of energy for

(i) a closed system undergoing any process
(ii) a closed system undergoing isothermal process
(iii) an open system undergoing isothermal process
(iv) an open system undergoing any process

Answers

1. (iii) 2. (iv) 3. (i) 4. (i) 5. (iv) 6. (iii) 7. (iv) 8. (ii) 9. (iii) 10. (ii)
11. (iii) 12. (ii) 13. (ii) 14. (iii) 15. (i) 16. (iv) 17. (iv) 18. (ii) 19. (iv) 20. (iii)
21. (ii) 22. (iii) 23. (iv) 24. (iii) 25. (iii) 26. (i) 27. (iv) 28. (ii) 29. (iii) 30. (i)
31. (ii) 32. (iv) 33. (i) 34. (iv) 35. (ii) 36. (iii) 37. (iii) 38. (iii) 39. (ii) 40. (iv)

5

Properties of Ideal Gas and Real Gases

1. Define an ideal gas and give numerical values of gas constant for oxygen, nitrogen, hydrogen, carbon monoxide, carbon dioxide and steam.

Ans: An ideal gas is defined as a substance for which the equation of state is

$$pv = \frac{\bar{R}}{M} T$$

where $\bar{R}$ is called the universal gas constant and M is the molar mass of the substance. The value of $\bar{R} = 8.31441$ kJ/kmol K. The quantity $\bar{R}/M$ is called the gas constant R for a particular gas, so that the ideal-gas equation becomes

$$pv = RT \tag{1}$$

where R is the gas constant for that gas. R values for several gases are given in Table 5.1.

Table 5.1 Gas Constant R in kJ/kg K

Gas	Air	O_2	N_2 (Atmospheric)	N_2	H_2	CO	CO_2	H_2O
M	28.97	32.0	28.17	28.01	2.016	28.01	44.01	18.015
R	0.287	0.260	0.295	0.297	4.124	0.297	0.189	0.462

2. Explain the validity of perfect gas approximation for real gases.

Ans: Over certain ranges of pressure and temperature, a large number of gases closely approximate perfect gases. In fact, as the pressure approaches zero, equation (1) is found to hold for all gases.

An important characteristic of a perfect gas is that its internal energy is a function of temperature only; i.e.

$$u = f(T) \tag{2}$$

Since enthalpy is defined as

$$h = u + pv \tag{3}$$

Substituting equation (1), $pv = RT$ and equation (2) into equation (3) yields

$$h = f(T) + RT \tag{4}$$

Since R is a constant, enthalpy is also a function of temperature only.

3. Define the two specific heats of a perfect gas.

Ans: The specific heats at constant pressure and constant volume are defined as the temperature derivatives of enthalpy and internal energy at constant pressure and constant volume. For a perfect gas,

$$C_p = \left(\frac{\partial h}{\partial T}\right)_p = \frac{dh}{dT} \tag{5}$$

$$C_v = \left(\frac{\partial u}{\partial T}\right)_v = \frac{du}{dT} \tag{6}$$

Since h and u are functions of temperature (T) only, C_p and C_v are also functions of temperature (T) only for a perfect gas. To evaluate changes in internal energy or enthalpy for a perfect gas, equation (6), $du = C_v\, dT$, and equation (5), $dh = C_p\, dT$, may be integrated,

$$u_2 - u_1 = \int_1^2 C_v\, dT \tag{7}$$

$$h_2 - h_1 = \int_1^2 C_p\, dT \tag{8}$$

A relationship exists among the properties of a system (connecting equilibrium states):

$$T\, ds = du + pdv \tag{9}$$

For a perfect gas,

$$du = C_v\, dT \tag{10}$$

and

$$p = \frac{RT}{v} \tag{11}$$

Substituting equations (10) and (11) into equation (9), we obtain

$$T\, ds = C_v\, dT + \frac{RT}{v} dv$$

or

$$ds = C_v \frac{dT}{T} + R\frac{dv}{v} \tag{12}$$

Integrating equation (12) for a perfect gas

$$s_2 - s_1 = \int_1^2 C_v\, dT + R\, \ell n \frac{v_2}{v_1} \tag{13}$$

In a similar manner,

$$T\, ds = dh - v\, dp \tag{14}$$

For a perfect gas,

$$dh = C_p\, dT \tag{15}$$

and

$$v = \frac{RT}{p} \tag{16}$$

Substituting equations (15) and (16) into equation (14), we obtain

$$T\, ds = C_p\, dT - \frac{RT}{p} dp$$

or

$$ds = C_p \frac{dT}{T} - R\frac{dp}{p} \tag{17}$$

Integrating equation (17) for a perfect gas,

$$s_2 - s_1 = \int_1^2 C_p\, dT - R\,\ell n \frac{p_2}{p_1} \tag{18}$$

4. Derive an expression showing the relation between the two specific heats of a gas and its specific gas constant. Also express the temperature ratio in terms volume ratio or pressure ratio for a reversible adiabatic (isentropic) process.

Ans: The specific heats for many real gases do not vary much with temperature. Assuming that the specific heats are constant, we may proceed to obtain many simplifications.

Equations (7) and (8) can be readily integrated (for constant C_v and C_p) to give

$$u_2 - u_1 = C_v\,(T_2 - T_1) \tag{19}$$

$$h_2 - h_1 = C_p\,(T_2 - T_1) \tag{20}$$

Equations (13) and (18) become

$$s_2 - s_1 = C_v\,\ell n\frac{T_2}{T_1} + R\,\ell n\frac{v_2}{v_1} \tag{21}$$

$$s_2 - s_1 = C_p\,\ell n\frac{T_2}{T_1} - R\,\ell n\frac{p_2}{p_1} \tag{22}$$

For an isentropic process with constant specific heats, $s_2 = s_1$ and from equation (21)

$$C_v\,\ell n\frac{T_2}{T_1} = -R\,\ell n\frac{v_2}{v_1} \tag{23}$$

A relationship among C_p, C_v and R can be derived to put equation (23) into a more useful form. From the definition of h,

$$dh = d\,(u + pv) = du + d\,(pv) \tag{24}$$

Substituting the perfect gas relationship, $dh = C_p\, dT$, $du = C_v\, dT$ and $pv = RT$, we get

$$C_p\, dT = C_v\, dT + R\, dT \tag{25}$$

Simplifying equation (25)

$$C_p - C_v = R \tag{26}$$

Let the ratio of specific heats be denoted by γ,

$$\gamma = \frac{C_p}{C_v} \tag{27}$$

Then $C_p = \gamma C_v$ and substituting in equation (26), $\gamma C_v - C_v = R$

or
$$C_v = \frac{R}{(\gamma - 1)} \tag{28}$$

and
$$C_p = \frac{\gamma}{(\gamma - 1)} R \tag{29}$$

Substituting equation (28) into equation (23)

$$\frac{R}{(\gamma - 1)} \ell n \frac{T_2}{T_1} = -R \ell n \frac{v_2}{v_1}$$

or
$$\ell n \left(\frac{T_2}{T_1}\right)^{\frac{1}{(\gamma-1)}} = \ell n \left(\frac{v_1}{v_2}\right) \tag{30}$$

Taking antilog and rearranging,

$$\left(\frac{T_2}{T_1}\right) = \left(\frac{v_1}{v_2}\right)^{(\gamma-1)} \tag{31}$$

Similarly for an isentropic process, equation (22) becomes

$$C_p \ell n \frac{T_2}{T_1} = R \ell n \frac{p_2}{p_1} \tag{32}$$

Substituting equation (29) into equation (32), we obtain

$$\frac{\gamma}{(\gamma - 1)} R \ell n \frac{T_2}{T_1} = R \ell n \frac{p_2}{p_1}$$

or
$$\left(\frac{T_2}{T_1}\right)^{\frac{\gamma}{(\gamma-1)}} = \left(\frac{p_2}{p_1}\right) \tag{33}$$

From equations (31) and (33), we can write the following relationship for an isentropic process:

$$\frac{T_2}{T_1} = \left(\frac{p_2}{p_1}\right)^{\frac{(\gamma-1)}{\gamma}} = \left(\frac{v_1}{v_2}\right)^{(\gamma-1)} \tag{34}$$

Equation (34) can be written

$$\left(\frac{p_2}{p_1}\right) = \left(\frac{v_1}{v_2}\right)^{\gamma} \quad \text{or} \quad p_1 v_1^{\gamma} = p_2 v_2^{\gamma} \quad \text{or} \quad pv^{\gamma} = \text{constant} \tag{35}$$

The form of equation (35), pv^{γ} = constant, is similar to an equation representing a general class of quasistatic processes for gases. This equation, giving the pressure-volume relation, is

$$pv^{n} = \text{constant} \tag{36}$$

Processes following equation (36) are called polytropic processes. For a polytropic process, pv^n = constant, equations similar to (34) can be derived

$$\frac{T_2}{T_1} = \left(\frac{p_2}{p_1}\right)^{\frac{(n-1)}{n}} = \left(\frac{v_1}{v_2}\right)^{(n-1)} \tag{37}$$

A constant pressure process is a special case of polytropic process and would be represented by equation (36) with $n = 0$; a constant volume process is represented by $n = \infty$; a constant temperature process is represented by $n = 1$; a constant entropy process is represented by $n = \gamma$. Many processes may be approximated by polytropic process with the proper value of n.

5. Define coefficient of volumetric expansion and isothermal compressibility for a perfect gas.

Ans: The isobaric coefficient of volumetric expansion, β, and the isothermal coefficient of compressibility, κ_T, are defined as

$$\beta = \frac{1}{V}\left(\frac{\partial V}{\partial T}\right)_p = \frac{1}{v}\left(\frac{\partial v}{\partial T}\right)_p \tag{38}$$

and

$$\kappa_T = -\frac{1}{V}\left(\frac{\partial V}{\partial p}\right)_T = -\frac{1}{v}\left(\frac{\partial v}{\partial p}\right)_T \tag{39}$$

It is seen that both β *and* κ_T are solely a function of the p-v-T behaviour of the substance. Using the ideal gas equation of state, $pv = RT$, as a valid approximation for low pressure system, we see, for gaseous system that behave essentially like ideal gas,

$$\beta_{\text{ideal gas}} = \frac{1}{T} \quad \text{and} \quad \kappa_{T,\text{ideal gas}} = \frac{1}{p} \tag{40}$$

Hence, for ideal gases, the expressions for β and κ_T are extremely simple. It is easily seen from these equations κ_T for an ideal gas varies rapidly with pressure, whereas β is independent of pressure.

6. State the Helmholtz function and Gibbs function.

Ans: Enthalpy is a derived property defined as the sum of internal energy, u, and the flow energy, pv. Two derived properties that are quite useful in thermodynamics are the Helmholtz function, a, and the Gibbs function, g, which are defined as

$$a = u - Ts \tag{41}$$

$$g = h - Ts \tag{42}$$

7. How accurately does the ideal gas approximation model a real gas?

Ans: The ideal gas equation is accurate for real gases as

1. Molar mass decreases
2. Pressure decreases
3. Temperature increases

For real gases, the ideal gas equation of state can be modified to

$$pv = Z\,RT \tag{43}$$

where Z, the compressibility factor, is an empirically determined function of p and T for each gas. The compressibility factor can be thought of as a 'correction factor' to the ideal gas equation of state when it is applied to real gas. Compressibility factor tables or charts are available for some gases.

8. What are Maxwell equations?

Ans: The Maxwell equations relate entropy to three directly measurable properties p, v and T, for pure simple compressible substances. We shall start with the relation

$$T\,ds = du + p\,dv \tag{44}$$

And then substituting into equation (41) the following equations

$$h = u + pv \tag{45}$$

$$a = u - Ts \tag{46}$$

$$g = h - Ts \tag{47}$$

to obtain

$$du = T\,ds - p\,dv \tag{48}$$

$$dh = T\,ds + v\,dp \tag{49}$$

$$da = -p\,dv - s\,dT \tag{50}$$

$$dg = v\,dp - s\,dT \tag{51}$$

Then, since these are differentials of properties, they are exact differentials, so we can write,

$$\left(\frac{\partial T}{\partial v}\right)_s = -\left(\frac{\partial p}{\partial s}\right)_v \tag{52}$$

$$\left(\frac{\partial T}{\partial p}\right)_s = \left(\frac{\partial v}{\partial s}\right)_p \tag{53}$$

$$\left(\frac{\partial p}{\partial T}\right)_v = \left(\frac{\partial s}{\partial v}\right)_T \tag{54}$$

$$\left(\frac{\partial v}{\partial T}\right)_p = -\left(\frac{\partial s}{\partial p}\right)_T \tag{55}$$

These four equations are called the Maxwell equations. Note that each one involves entropy, s, and all three measurable properties p, v and T.

Some other useful relationships among properties can be derived from equations (41), (47) and (48). For example, taking u as a function of s and v, we have

$$du = \left(\frac{\partial u}{\partial s}\right)_v ds + \left(\frac{\partial u}{\partial v}\right)_s dv \tag{56}$$

Comparing this with equation (48),

$$du = T\,ds - p\,dv \tag{57}$$

and noting that *ds* and *dv* are independent, we can equate the coefficients of *ds* and *dv* to give

$$\left(\frac{\partial u}{\partial s}\right)_v = T \quad \text{and} \quad \left(\frac{\partial u}{\partial v}\right)_s = -p \tag{58}$$

In a similar manner, we can write expressions for *dh*, *da* and *dg* for $h = f(s,p)$, $a = f(v.T)$ and $g = f(p.T)$ and compare them with equations (49), (50) and (51) respectively, to obtain

$$\left(\frac{\partial h}{\partial s}\right)_p = T \quad \text{and} \quad \left(\frac{\partial h}{\partial p}\right)_s = v \tag{59}$$

$$\left(\frac{\partial a}{\partial v}\right)_T = -p \quad \text{and} \quad \left(\frac{\partial a}{\partial T}\right)_v = -s \tag{60}$$

$$\left(\frac{\partial g}{\partial p}\right)_T = v \quad \text{and} \quad \left(\frac{\partial g}{\partial p}\right)_v = -s \tag{61}$$

Solved Problems

1. Determine for ideal gas the slope of (i) a constant pressure line and (ii) a constant volume line on a *T-s* diagram.

   For an ideal gas, the *T-ds* equations $T\,ds = du + p\,dv$ and $T\,ds = dh - v\,dp$ become

   $$T\,ds = C_v\,dT + p\,dv \text{ and } T\,ds = C_p\,dT - v\,dp.$$

   For a constant pressure process, $dp = 0$. For a constant volume process, $dv = 0$. The desired slope can be obtained from the above equations as

   $$\left(\frac{\partial T}{\partial s}\right)_p = \frac{T}{C_p} \quad \text{and} \quad \left(\frac{\partial T}{\partial s}\right)_v = \frac{T}{C_v}$$

   Since $C_p > C_v$, the slope of constant pressure line is less than that of constant volume line on the *T-s* diagram.

2. A mass of 1.5 kmol of an ideal gas of molecular mass 25 is contained in a rigid vessel of volume 4 m^3 at a temperature of 100°C. Evaluate the mass, the pressure, the specific volume and the gas constant of the gas. For this condition, the gas has a value of $\gamma = 1.38$. Evaluate the corresponding values of C_p and C_v. Subsequently the gas cools to atmospheric temperature, 15°C. Evaluate the final pressure of the gas. Assuming the gas to be perfect, evaluate the change in internal energy, the change in enthalpy and the magnitude and sign of heat transfer.

   $$\text{Mass of gas} = \text{No. of moles} \times \text{molecular mass} = \frac{1.5 \times 1000 \times 25}{1000} = 37.5 \text{ kg.}$$

   $$\text{Gas constant} = \frac{8314.3}{25} = 332.57 \frac{\text{J}}{\text{kg K}}.$$

   We know the gas equation, $pV = m\,RT$, i.e. $\text{p} \times 4 = 37.5 \times 332.57 \times (273 + 100)$.

Hence the pressure of the gas, $p_1 = 1.1629 \times 10^6$ N/m² = 1.1629 MN/m² or 11.629 bar.

The specific volume, $v_1 = \dfrac{4}{37.5} = 0.10666 \dfrac{\text{m}^3}{\text{kg}}$

$$C_p - C_v = R = 332.57, \text{ and } \gamma = \frac{C_p}{C_v} = 1.38$$

$$C_p = \gamma\frac{R}{(\gamma-1)} = 1.38 \times \frac{332.57}{0.38} = 1207.7\frac{\text{J}}{\text{kg K}} \text{ and } C_v = \frac{R}{(\gamma-1)} = \frac{332.57}{0.38} = 875.18\frac{J}{kg\,K}.$$

Since the gas is contained in a rigid vessel, it undergoes a constant volume process during cooling. Hence,

$$\frac{p_1}{T_1} = \frac{p_2}{T_2}, \quad p_2 = \frac{p_1}{T_1} \times T_2 = \frac{1.1629}{373} \times 288 = 0.8978\frac{\text{MN}}{\text{m}^2}.$$

The final pressure is 0.8978 MN/m² or 8.978 bar.

$$\Delta U = m\, C_v\, (T_2 - T_1) = 37.5 \times 0.87518 \times (288 - 373) = -2789.63 \text{ kJ.}$$

$$\Delta H = m\, C_p\, (T_2 - T_1) = 37.5 \times 1.2077 \times (288 - 373) = -3849.54 \text{ kJ.}$$

$${}_1Q_2 = \Delta U + {}_1W_2 = -2789.63 + 0 = -2789.63 \text{ kJ.}$$

The heat is transferred from the gas.

3. One kilogram of nitrogen at a temperature of 150°C occupies a volume of 0.2 m³. The gas undergoes a fully-resisted constant pressure expansion, without friction, to a final volume of 0.36 m³. Evaluate the final pressure, the final temperature, the work done and the heat transfer. Assume nitrogen to be a perfect gas for which $C_v = 743$ J/kg K and $R = 297$ J/kg K.

We know the gas equation

$$pv = RT \text{ (for unit mass), hence, } p_1 = \frac{RT}{v} = \frac{297 \times 423}{0.2 \times 10^3} = 628.155\frac{\text{kN}}{\text{m}^2}.$$

For a constant pressure process, $p_2 = p_1 = 628.155$ kN/m² or 6.28155 bar and

$$\frac{v_1}{T_1} = \frac{v_2}{T_2}, T_2 = \frac{0.36}{0.2} \times 423 = 761.4 \text{ K}$$

Work done, ${}_1W_2 = p_1\,(v_2 - v_1) = 628.155 \times (0.36 - 0.2) = 100.5$ kN.m or kJ

Heat transfer, ${}_1Q_2 = \Delta h = C_p\,(T_2 - T_1) = (C_v + R)\,(761.4 - 423)$

$= (743 + 297)\,(338.4) = 351900$ J or 351.9 kJ.

4. In a chemical plant, carbon monoxide flows steadily into a turbine at a pressure of 800 kN/m² and a temperature of 240°C; there it expands reversibly and adiabatically to a final pressure of 100 kN/m². Changes in velocity and elevation are negligible. Calculate the final specific volume, the final temperature and the increase in entropy. Given that the mass flow rate is 10 kg/s, evaluate the power delivered by the turbine. Assume carbon monoxide to be a perfect gas for which $M = 28.01$, $C_v = 20.82$ kJ/kmol K. Take the universal gas constant = 8314 J/kmol K.

Gas constant, $R = (8314)/(28.01) = 296.8$ J/kg K or 0.2968 kJ/kg K

$$C_v = (20.82)/(28.01) = 0.7433 \text{ kJ/kg K}$$

$$C_p = C_v + R = 0.7433 + 0.2968 = 1.0401 \text{ kJ/kg K}$$

$$\gamma = C_p/C_v = 1.0401/0.7433 = 1.3993$$

Final temperature, T_2: $\dfrac{T_2}{T_1} = \left(\dfrac{p_2}{p_1}\right)^{\frac{\gamma-1}{\gamma}}$; $T_2 = 513 \times \left(\dfrac{100}{800}\right)^{\frac{0.3993}{1.3993}} = 283.4\ K$

Hence, $t_2 = 10.4°C$

Final specific volume, $v_2 = (R\,T_2)/p_2 = (296.8 \times 283.4)/(100 \times 10^3) = 0.8411\ m^3/kg$.

For the reversible adiabatic process the change in entropy = 0.

Power delivered = $m\,(h_1 - h_2) = m\,C_p\,(T_1 - T_2) = 10 \times 1.0401 \times (513 - 283.4)$

$= 2388$ kW.

5. One kilogram of air at a pressure of $1.0 \times 10^6\ N/m^2$ and a temperature of 100°C undergoes a reversible polytropic process which may be represented by $pv^{1.1}$ = constant. The final pressure is $200 \times 10^3\ N/m^2$. Evaluate the final specific volume, the final temperature and the increase in entropy. Evaluate also the work done and the heat transfer. Assume air to be a perfect gas for which $R = 287$ J/kg K and $\gamma = 1.4$.

For air $C_v = R/(\gamma - 1) = 287/0.4 = 717.5$ J/kg K and $C_p = C_v + R = 1004.5$ J/kg K

For polytropic process, $\dfrac{T_2}{T_1} = \left(\dfrac{p_2}{p_1}\right)^{\frac{n-1}{n}}$; $T_2 = 373 \times \left(\dfrac{200}{1000}\right)^{\frac{0.1}{1.1}} = 322.23$ K

Final specific volume, $v_2 = (R\,T_2)/p_2 = (287 \times 322.23)/200 \times 10^3 = 0.4624\ m^3/kg$

$$s_2 - s_1 = C_p\ \ell n\left(\frac{T_2}{T_1}\right) + R\ \ell n\left(\frac{p_1}{p_2}\right) = 1.0045\ \ell n\left(\frac{322.23}{373}\right) + 0.287\ \ell n\left(\frac{1000}{200}\right) = 0.3149\frac{kJ}{kg\,K}$$

$$\text{Work done, } {}_1W_2 = \frac{R(T_1 - T_2)}{(n-1)} = \frac{287(373 - 322.23)}{0.1\times 10^3} = 145.71\frac{\text{kJ}}{\text{kg}}$$

Heat transfer, ${}_1Q_2 = C_v\,(T_2 - T_1) + {}_1W_2 = 0.7175 \times (322.23 - 373) + 145.71$

$= -36.427 + 145.71 = 109.283$ kJ/kg.

6. Each of the cylinders of a marine diesel engine has a bore of 400 mm and a clearance volume of $4.5\ dm^3$. The piston stroke is 530 mm. At the end of the induction stroke (i.e. with the piston in its outermost position) the pressure and temperature of the gas in the cylinder are $98 \times 10^3\ N/m^2$ and 40°C, respectively. On the assumption that the gas obeys the ideal-gas rule and that for this gas $R = 286$ J/kg K, calculate the mass and the number of moles of the gas contained at the end of the induction stroke.

Stroke volume, $V_s = (\pi/4) \times 40^2 \times 53 = 66601.8\ cm^3$

$V_{total} = Vs + Vc = 66601.8 + 4500 = 71101.8\ cm^3$ or $0.0711018\ m^3$

We know the relation, $pV = m\,RT$ for the gas contained in the cylinder.

Hence, $98 \times 10^3 \times 0.0711018 = m \times 286 \times (273 + 40)$

Mass of the gas, $m = \dfrac{98 \times 10^3 \times 0.0711018}{286 \times 313} = 0.07784$ kg

Molecular mass $= \dfrac{8314.3}{286} = 29$, hence the number of moles $= \dfrac{0.07784}{29} = 0.002684$ kmol.

7. A volume of 0.36 m^3 of oxygen initially at a temperature of 220°C and a pressure of 400×10^3 N/m^2 is compressed reversibly and isothermally to a final volume of 0.06 m^3. Calculate the mass, the final pressure, the increase in the internal energy, the work done, the heat transfer and the change in entropy. Assume oxygen to be a perfect gas for which $R = 260$ J/kg K.

We know the relation $pV = m\,RT$; $m = \dfrac{400000 \times 0.36}{260 \times 493} = 1.1234$ kg.

For the isothermal process, $p_1 V_1 = p_2 V_2$; $p_2 = \dfrac{400000 \times 0.36}{0.06} = 2.4 \times 10^6$ N/m^2

$\Delta U = 0$ for the isothermal process

Work done, ${}_1W_2 = p_1 V_1 \ell n \dfrac{V_2}{V_1} = 400000 \times 0.36\, \ell n \dfrac{0.06}{0.36} = -258 \times 10^3$ N.m or -258 kN.m

Heat transfer, ${}_1Q_2 = 0 + {}_1W_2 = -258$ kJ

Change in entrotry, $S_2 - S_1 = mR\, \ell n \dfrac{V_2}{V_1} = 1.1234 \times 260\, \ell n \left(\dfrac{0.06}{0.36}\right) = -523 \dfrac{\text{J}}{\text{K}}$ or $0.523 \dfrac{\text{kJ}}{\text{K}}$.

8. The indicator diagram of a reciprocating air compressor shows that during the compression stroke the volume of air in the cylinder is 0.8 dm^3 when the pressure is 140×10^3 N/m^2 gauge. Correspondingly, when the pressure is 520×10^3 N/m^2 gauge the volume is 0.36 dm^3. The mass of air contained in the cylinder at both points is 0.0022 kg. The barometric pressure is 773 mHg. Calculate the temperature of the air at each point. On the assumption that the compression process may be represented by pV^n = constant, where n is a constant, evaluate n. Assuming the process to be reversible and polytropic, evaluate the magnitude and sign of the work and heat transfer. Assume air to be a perfect gas for which $R = 287$ J/kg K and $C_p = 1000$ J/kg K.

The barometric pressure, $p_a = \dfrac{1.01325}{760} \times 773 = 1.0306$ bar or $103.06 \times 10^3 \dfrac{\text{N}}{\text{m}^2}$.

Given data: $p_1 = 140 \times 10^3 + 103.06 \times 10^3 = 243.06 \times 10^3$ N/m^2 (abs); $V_1 = 0.0008$ m^3;

$p_2 = 520 \times 10^3 + 103.06 \times 10^3 = 623.06 \times 10^3$ N/m^2 (abs); $V_2 = 0.00036$ m^3.

For the polytropic process, $p_1 V_1^n = p_2 V_2^n$, i.e. $n = \dfrac{\ell n(p_2/p_1)}{\ell n(V_1/V_2)} = \dfrac{\ell n(623/243)}{\ell n(0.8/0.36)} = 1.179$

$$T_1 = \frac{p_1 V_1}{mR} = \frac{243060 \times 0.0008}{0.0022 \times 287} = 307.9 \text{ K and } T_2 = \frac{p_2 V_2}{mR} = \frac{623060 \times 0.00036}{0.0022 \times 287} = 355.2 \text{ K}$$

The temperatures t_1 and t_2 are, respectively, 34.9°C and 82.2°C.

The work and heat transfer (between the points 1 and 2) are

Work transfer,

$${}_1W_2 = \frac{p_1 V_1 - p_2 V_2}{n-1} = \frac{(243.06 \times 0.0008 - 623.06 \times 0.00036) \times 10^3}{0.179 \times 10^3} = -0.1668 \text{ kJ}$$

$$\Delta U = m\,C_v\,(T_2 - T_1) = 0.0022 \times (1000 - 287) \times (355.2 - 307.9) = 74.19 \text{ J or } 0.07419 \text{ kJ}$$

Heat transfer, ${}_1Q_2 = \Delta U + {}_1W_2 = 0.07419 - 0.1668 = -0.09261$ kJ.

Work is done on the system and heat is leaving the system.

9. Air flows steadily into a compressor at a temperature of 16°C and a pressure of 100×10^3 N/m² and leaves at a temperature of 246°C and a pressure of 600×10^3 N/m². There is no heat transfer to or from the air as it flows through the compressor; changes in velocity and elevation are negligible. Evaluate the external work done per unit mass of air, assuming air to be a perfect gas for which $R = 287$ J/kg K and $\gamma = 1.4$. Evaluate the minimum external work required to compressor the air adiabatically from the same initial state to the same final pressure. Evaluate also the isentropic efficiency of the compressor.

For air, $C_p = \dfrac{\gamma}{\gamma - 1} R = \dfrac{1.4}{0.4} \times 287 = 1004.5 \dfrac{\text{J}}{\text{kg K}}$ or $1.0045 \dfrac{\text{kJ}}{\text{kg K}}$

The steady flow energy equation is

$$\dot{Q} + \dot{m}\left[h_1 + \frac{V_1^2}{2} + g\,Z_1\right] = \dot{W}_S + \dot{m}\left[h_2 + \frac{V_2^2}{2} + g\,Z_2\right]$$, which on simplification gives

$0 + \dot{m}\left[h_1 + 0 + 0\right] = \dot{W}_S + \dot{m}\left[h_2 + 0 + 0\right]$ i.e. $W_s = -C_p(T_2 - T_1)$ for unit mass flow rate

$$W_s = -1.0045(519 - 289) = -231 \text{ kJ/kg.}$$

The temperature of air leaving the compressor for adiabatic compression:

$$T_{2'} = T_1\left(\frac{p_2}{p_1}\right)^{\frac{\gamma-1}{\gamma}} = 289\left(\frac{600}{100}\right)^{\frac{0.4}{1.4}} = 482.2\,K$$

Work required for adiabatic compression, $W_{\text{adia}} = -1.0045(482.2 - 289) = -194$ kJ

Hence, the isentropic efficiency $= \dfrac{(T_{2'} - T_1)}{(T_2 - T_1)} = \dfrac{194}{231} \times 100 = 83.99$ %.

Isentropic Efficiency

Isentropic efficiency of a work producing machine is the ratio of the actual work output to the work output of a reversible adiabatic machine taking fluid at the initial state and exhausting it at the same final pressure.

$$\eta_{isen} = \frac{W_X}{W_{X,R}}$$

In the case of work absorbing machine, for example, rotary compressor, it is the ratio of the work input of a reversible adiabatic machine to the actual work input to the machine.

$$\eta_{isen} = \frac{W_{X,R}}{W_X}$$

10. Helium at a pressure of 90×10^3 N/m² and a temperature of 60°C flows steadily into a rotary compressor, and leaves at a pressure of 210×10^3 N/m². Changes in elevation and velocity are negligible; the compression process may be assumed to be adiabatic. Given that the isentropic efficiency of the compressor is 73 per cent, calculate the temperature of helium leaving the compressor and the power required to compress 6 kg of helium per minute. Assume helium to be perfect gas for which $C_p = 5.19$ kJ/kg K and $M = 4$.

$$\text{Gas constant for helium} = \frac{8314}{4} = 2078.5 \frac{\text{J}}{\text{kg K}} \text{ or } 2.0785 \frac{\text{kJ}}{\text{kg K}},$$

$$C_v = 5.19 - 2.0785 = 3.1115 \text{ kJ/kg K and } \gamma = C_p/C_v = 5.19/3.1115 = 1.668.$$

For adiabatic compression process, $Q = 0$ and the steady flow energy equation simplifies to

$$0 + \dot{m}(h_1 + 0 + 0) = \dot{W}_S + \dot{m}(h_2 + 0 + 0) \text{ i.e. } W_s = -m\, C_p\,(T_2 - T_1).$$

The temperature of air leaving the compressor for adiabatic compression:

$$T_{2'} = T_1\left(\frac{p_2}{p_1}\right)^{\frac{\gamma-1}{\gamma}} = 333\left(\frac{210}{90}\right)^{\frac{0.668}{1.668}} = 467.5\,K$$

$$\text{Isentropic efficiency} = 0.73 = \frac{(T_{2'} - T_1)}{(T_2 - T_1)} = \frac{(467.5 - 333)}{(T_2 - 333)}\text{; hence, } T_2 = 517.246 \text{ K}$$

$$\text{Power required} = -\frac{6}{60} \times 5.19 \times (517.246 - 333) = -95.624 \text{ kW}.$$

Practice Problems

1. A cylinder, fitted with a piston, contains 20 dm³ of methane gas (CH_4) at a pressure of 500×10^3 N/m² and a temperature of 115°C. The methane undergoes a process to a pressure of 100×10^3 N/m² during which the work done by the gas is 10 kNm and the heat transfer to it is 1 kJ. Determine the final temperature of the methane and the increase in entropy. Assume $R = 0.519$ kJ/kg K and $\gamma = 1.3$.

   (282.12 K; 0.006277 kJ/K)

2. A rigid vessel of volume 0.3 m³ contains air at a pressure of 35×10^3 N/m² and a temperature of 15°C. Atmospheric air at a pressure of 100×10^3 N/m² and a temperature of 15°C leaks slowly into the vessel. When the pressure of air in the vessel has risen to 70×10^3 N/m² the leak is stopped; it is then found that the temperature of the air in the vessel is 50°C. Calculate the mass of air which has leaked in.

   (0.0995 kg)

3. A room measuring 4 m × 5 m × 2.5 m is heated electrically from an initial temperature of 0°C to 20°C. The air pressure inside the room is the same as that of surroundings and is 720 torr and also remains constant during the period of heating. The heat capacity of walls and furniture is 30 kJ/K and specific heat of air at constant pressure is 1 kJ/kg K. Find how much air escapes through gaps and windows during this period? How much electrical energy is needed for heating the room?

   (4.179 kg; 0.508 kWh)

4. 1 kg of air is initially at a temperature of 10°C. It goes through the following processes to complete a cycle:

   (i) Reversible adiabatic compression until the pressure is increased to two and half times.
   (ii) Reversible heat addition at constant pressure. The addition of heat is so controlled that initial volume is reached.
   (iii) Constant volume process which returns the system to initial state. Calculate for each process the work done and heat exchange. What would be the efficiency of the cycle?

   (−60.76 kJ, 0 kJ; 97.49 kJ, 341.4 kJ;0 kJ, −304.72 kJ;10.75%)

5. The following processes constitute a cycle:
   (i) Addition of heat at constant volume so that the pressure rises from 6 bar to 15 bar.
   (ii) Polytropic expansion ($n = 1.3$) till the system reaches original temperature of 110°C.
   (iii) Isothermal compression till the system reaches original state.
   Calculate the temperature at the end of constant volume process, pressure at the end of polytropic expansion and change in entropy during each of the three stages per kg of air.
   (957.5 K; 0.2829 bar; (i) 0.6579 kJ/K; (ii) 0.2187 kJ/K; (iii) –0.8766 kJ/K)

6. A nozzle is supplied with a steady stream of a perfect gas ($\gamma = 1.4$) at a pressure of 700×10^3 N/m^2, a temperature of 45°C, a density of 4.5 kg/m^3 and a velocity of 120 m/S.
   (i) Determine the gas constant, R, and specific heat at constant pressure, C_p, in J/kg K.
   (ii) Assuming the flow to be reversible and adiabatic, determine the temperature and velocity of the gas at nozzle exit where the pressure is 400×10^3 N/m^2.
   (iii) If the mass flow rate is 1 kg/s, determine the cross-sectional areas at entry to and exit from the nozzle.
   (489 J/kg K; 1712 J/kg K; –2°C; 418.7 m/S; 18.52 cm^2; 7.91 cm^2)

7. A hot-air engine operates on the Carnot cycle between the temperature limits of 250°C and 50°C; all the processes take place in a piston-cylinder mechanism. The pressure and volume at the start of isothermal expansion are 600×10^3 N/m^2 and 0.1 m^3, respectively. The pressure after isothermal expansion is 300×10^3 N/m^2. Assume air to be a perfect gas with $\gamma = 1.4$ and $R = 287$ J/kg K.
   (i) Evaluate the efficiency of the cycle and the work done per cycle.
   (ii) Evaluate the power developed by the engine when it performs 96 cycles/min.
   (38.24%, 15.92 kN.m; 25.47 kW)

8. Air at 0.8 bar and 5°C enters the intake (i.e. diffuser) of a jet engine with a velocity of 300 m/s. The air decelerates reversibly in the intake to a negligible velocity before entering the compressor. Calculate the pressure of the air at inlet to the compressor (i) if the process can be assumed adiabatic and (ii) if heat is transferred at such a rate that the temperature remains at 5°C throughout the process.
   (1.349 bar; 1.406 bar)

9. A certain gas has $C_p = 1.968$ kJ/kg K and $C_v = 1.507$ kJ/kg K. Find its molecular weight and gas constant. A constant volume chamber of 0.3 m^3 capacity contains 2 kg of this gas at 5°C. Heat is transferred to the gas until the temperature is 100°C. Find the work done, the heat transferred and the changes in internal energy, enthalpy and entropy.
   (18 kg/kmol; 461 J/kg K; 0; 286.33 kJ; 286.33 kJ; 373.92 kJ; 0.8859 kJ/K)

10. 1 kg of an ideal gas is heated from 18°C to 93°C. 160 kJ of heat are added during the process. Find the specific heats, change in internal energy, change in enthalpy and work done. Assume $R = 0.287$ kJ/kg K, $\gamma = 1.18$.
    (1.594 kJ/kg K; 1.881 kJ/kg K; 119.55 kJ; 141.07 kJ; 40.45 kN.m)

Objective Type Questions

1. The statement that the internal energy of an ideal gas is a function of temperature only is due to
   (i) the Avogadro's law (ii) the Boyle's law
   (iii) the Charles's law (iv) the Joule's law

2. Which one of the following expressions (symbols have their usual meanings) is an example of composite property of the system?
   (i) $m\,g\,Z$ (ii) $m\,V^2/2$
   (iii) $m\,u$ (iv) $m\,h$
3. Which thermodynamic property attains a greater value during the throttling process?
   (i) Enthalpy (ii) Pressure
   (iii) Temperature (iv) Volume
4. The dimensions of enthalpy is
   (i) $M\,L\,T^{-2}$ (ii) $M\,L^2\,T^{-1}$
   (iii) $M\,L^2\,T^{-2}$ (iv) $M\,L^3\,T^{-2}$
5. A certain quantity of monatomic gas undergoes quasiequilibrium process in a piston-cylinder gadget at constant pressure. If the work interaction during the process is 200 kNm, then the heat interaction during the process, in kJ, is
   (i) 200 (ii) 300
   (iii) 400 (iv) 500
6. An ideal gas, whose internal energy is given by $u = A + B\,p\,v$, undergoes a reversible adiabatic expansion pv^γ = constant. The index γ for the process is
   (i) $1+\frac{B}{A}$ (ii) $1+\frac{A}{B}$
   (iii) $1+\frac{1}{B}$ (iv) $1+\frac{1}{A}$
7. When a perfect gas escapes from an insulated cylinder, the gas that remains in the cylinder undergoes an
   (i) isothermal process (ii) isochoric process
   (iii) isobaric process (iv) isentropic process
8. The work done expression $RT\,\ell n\left(\frac{v_2-b}{v_1-b}\right)+a\left(\frac{1}{v_2}-\frac{1}{v_1}\right)$ between state 1 and state 2 for the van der Waals gas is applicable for
   (i) any process
   (ii) isobaric process
   (iii) isothermal process
   (iv) isentropic process
9. An insulated, evacuated tank is connected to the atmosphere by a short length of pipe and a valve. When the valve is cracked, air leaks into the tank until the tank pressure rises to the atmospheric value p_0. If the atmospheric temperature is T_0, the final temperature of the air in the tank is
   (i) T_0^{γ} (ii) $T_0\gamma^{1}$
   (iii) γT_0 (iv) $(\gamma-1)\,T_0$
10. The relationship $p=-\frac{\partial E}{\partial V}$, where p, E and V are respectively pressure, internal energy and volume, holds good for a thermodynamic closed system undergoing
   (i) any process (ii) isenthalpic process
   (iii) polytropic process (iv) isentropic process

11. Assertion (A): On the *T-s* plane for the perfect gas, the constant pressure lines are steeper than the constant volume lines.
    Reason (R): The specific heat at constant pressure is more than the specific heat at constant volume for a perfect gas.
    (i) Both A and R are true and R is the correct explanation of A
    (ii) Both A and R are true and R is not the correct explanation of A
    (iii) A is true but R is false
    (iv) A is false but R is true
12. Assertion (A): On an enthalpy-pressure or internal energy-specific volume diagrams for ideal gas, the isotherms are horizontal lines.
    Reason (R): The small dependence of enthalpy and internal energy on pressure or specific volume can be ignored.
    (i) Both A and R are true and R is the correct explanation of A
    (ii) Both A and R are true and R is not the correct explanation of A
    (iii) A is true but R is false
    (iv) A is false but R is true
13. The Gibbs free energy function is a property composed of
    (i) pressure, volume and temperature
    (ii) enthalpy, temperature and entropy
    (iii) temperature, pressure and enthalpy
    (iv) volume, enthalpy and entropy
14. Which property of an ideal gas remains unchanged during Joule-Thomson process?
    (i) Temperature (ii) Enthalpy
    (iii) Entropy (iv) Pressure
15. Which gas has got high mean specific heat at constant pressure at elevated temperatures?
    (i) Argon (ii) Carbon dioxide
    (iii) Helium (iv) Nitrogen
16. The Joule-Thomson coefficient $\mu = \left(\frac{\partial T}{\partial p}\right)_h$ is positive near normal room temperature for the gas
    (i) helium (ii) hydrogen
    (iii) neon (iv) refrigerant-12
17. The adiabatic equation of state $p\,V^{5/3}$ = constant is applicable to
    (i) ideal gases (ii) monatomic gases
    (iii) diatomic gases (iv) polyatomic gases
18. The average energy per molecules of an ideal gas defines the property
    (i) enthalpy (ii) entropy
    (iii) pressure (iv) temperature
19. The process that will give strongest support for the deviation of real gas effects from the ideal gas law is
    (i) isothermal (ii) isentropic
    (iii) polytropic (iv) throttling

20. A gas whose equation of state is $p(V-b)=RT$, has the Joule-Kelvin or Joule-Thomson coefficient
   (i) $-\dfrac{b}{C_p}$
   (ii) $-\dfrac{V}{C_p}$
   (iii) $-\dfrac{RT}{C_p}$
   (iv) $-\dfrac{p}{C_p}$
21. The internal energy of van der Waals gas $\left(p+\dfrac{a}{v^2}\right)(v-b)=RT$ depends on its
   (i) temperature only
   (ii) temperature and pressure
   (iii) temperature and specific volume
   (iv) pressure and specific volume
22. Identify the substance whose triple point pressure is greater than the atmospheric pressure
   (i) Ammonia
   (ii) Carbon dioxide
   (iii) Helium
   (iv) Water
23. Which property of an ideal gas is pressure independent?
   (i) Specific heat at constant volume
   (ii) Specific heat at constant pressure
   (iii) Coefficient of compressibility
   (iv) Coefficient of volumetric expansion
24. Which gas possesses the maximum root mean square speed at a given temperature?
   (i) Carbon dioxide
   (ii) Hydrogen
   (iii) Nitrogen
   (iv) Oxygen
25. Which class of gaseous molecules lend themselves to ease of analysis in statistical thermodynamics?
   (i) Diatomic gas
   (ii) Monatomic gas
   (iii) Polyatomic gas
   (iv) van der Waals gas
26. A thermally and calorically perfect gas
   (i) has zero viscosity
   (ii) has constant viscosity
   (iii) is incompressible
   (iv) satisfies the relation $p=\rho RT$
27. The assumption of constant molecular mass for the working substance in the internal combustion cycle analysis breaks down during
   (i) the suction stroke
   (ii) the compression stroke
   (iii) the power stroke
   (iv) the exhaust stroke
28. The lowering of partial pressure of solvent when a solute is added is due to
   (i) the Dalton's law
   (ii) the Gibb's law
   (iii) the Henry's law
   (iv) the Raoult's law
29. Which approximation for the gas would give unity for the ratio $((RT_c)/(p_c v_c))$ (where T_c, p_c and v_c are, respectively, temperature, pressure and specific volume at the critical state)?
   (i) Ideal gas
   (ii) van der Waals gas
   (iii) Callender gas
   (iv) Clausius gas

30. The molecular formula of dry ice is

(i) H_2O (ii) NH_3
(iii) CO_2 (iv) CO

31. The triple point for carbon dioxide on the pressure-enthalpy diagram (Figure 5.1) shown below is represented by

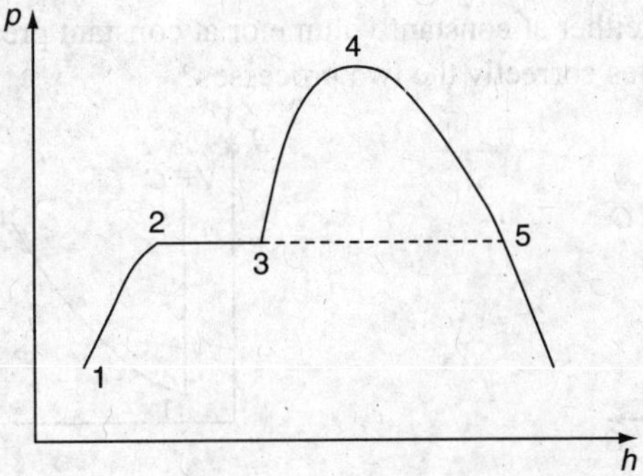

Figure 5.1 Pressure-enthalpy Diagram

(i) Line 12 (ii) Line 23
(iii) Line 34 (iv) Line 45

32. Which approximation for the gaseous equation of state considers the volume occupied by gas molecules alone?

(i) Clausius equation (ii) van der Waals equation
(iii) Redlich-Kwong equation (iv) Virial equation

33. Which slope function is related to the measurable properties of a pure substance during the phase transformations according to the Claperyon equation?

(i) $\dfrac{\partial p}{\partial T}$ (ii) $\dfrac{\partial p}{\partial h}$

(iii) $\dfrac{\partial T}{\partial v}$ (iv) $\dfrac{\partial p}{\partial v}$

34. The ordinate and abscissa of the diagram which shows the isobaric processes of an ideal gas as hyperbolae are

(i) temperature-entropy (ii) pressure-enthalpy
(iii) temperature-density (iv) enthalpy-entropy

35. Which one of the following relationships is correct for the principal heat capacities of a simple substance?

(i) $C_p - C_v = T\left(\dfrac{\partial p}{\partial T}\right)_V\left(\dfrac{\partial V}{\partial T}\right)_p$ (ii) $C_p - C_v = p\left(\dfrac{\partial S}{\partial T}\right)_p\left(\dfrac{\partial T}{\partial p}\right)_V$

(iii) $C_p - C_v = V\left(\dfrac{\partial S}{\partial p}\right)_V\left(\dfrac{\partial p}{\partial V}\right)_T$ (iv) $C_p - C_v = S\left(\dfrac{\partial V}{\partial T}\right)_p\left(\dfrac{\partial p}{\partial V}\right)_T$

36. The slope of isobar line on the temperature (ordinate)-volume (abscissa) diagram for an ideal gas is

(i) $\frac{p}{C_V}$ (ii) $\frac{p}{C_p}$

(iii) $\frac{p}{(C_p - C_v)}$ (iv) $\frac{p}{C_p} - \frac{p}{C_v}$

37. An ideal gas can be heated either at constant volume or at constant pressure from the initial state 1. Which *T-s* diagram illustrates correctly the two processes?

(i) 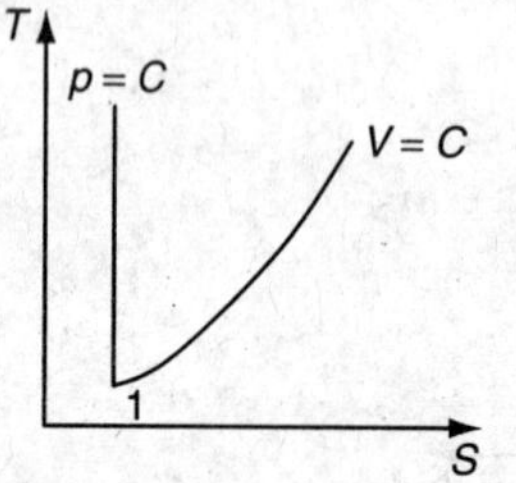

(ii) 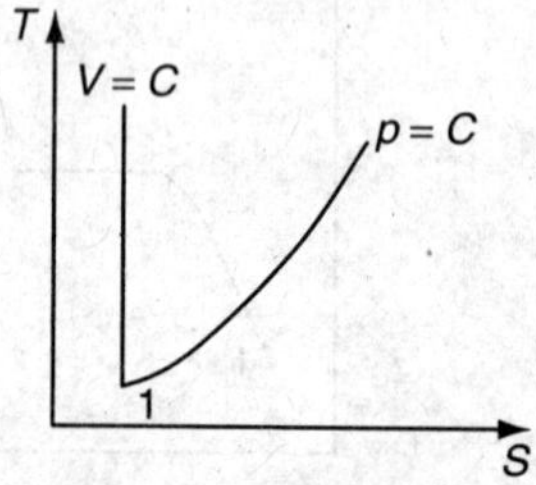

(iii) 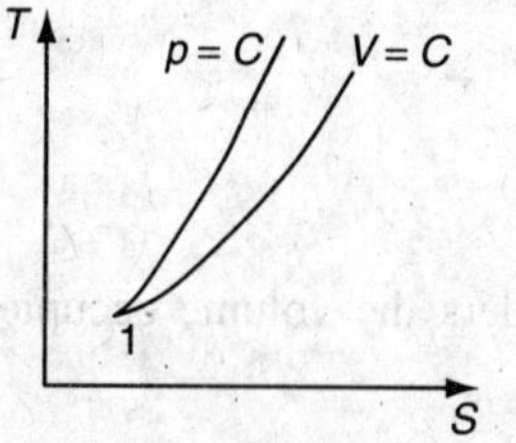

(iv) 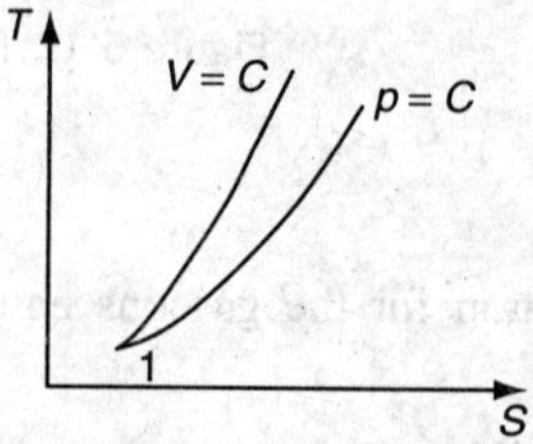

38. The slope of isometric line on the temperature (ordinate)-pressure (abscissa) diagram for an ideal gas is

(i) $\frac{v}{C_V}$ (ii) $\frac{v}{C_p}$

(iii) $\frac{v}{(C_p - C_v)}$ (iv) $\frac{v}{C_p} - \frac{v}{C_v}$

39. Which gas, if used to inflate aircraft tyres to a pressure of 1500 kPa, would have the least mass?

(i) Air (ii) Carbon dioxide

(iii) Helium (iv) Nitrogen

40. Match List I (Properties in differential form) with List II (Properties of ideal gas) and select the correct answer using the codes given in the following lists

List I (Properties in differential form)	*List II (Properties of ideal gas)*
A. *du*	1. $\frac{\gamma}{(\gamma-1)}[d(pv)]$
B. *dh*	2. $T\left[\frac{dp}{p}+\frac{dv}{v}\right]$

C. ds 3. $\frac{1}{(\gamma-1)}[d(pv)]$

D. dT 4. $\frac{1}{T}\left[\frac{\gamma pdv+vdp}{(\gamma-1)}\right]$

Codes:

	A	B	C	D
(i)	4	3	2	1
(ii)	2	1	4	3
(iii)	3	1	4	2
(iv)	4	3	1	2

41. Which one of the following substances has constant specific heats at all pressures and temperatures?
    (i) Monatomic gases (ii) Diatomic gases
    (iii) Triatomic gases (iv) Polyatomic gases

42. Assertion (A): When a gas is forced steadily through the porous plug of an insulated pipe, the enthalpy of gas is the same on both sides of the plug.
    Reason (R): The gas undergoes an isentropic expansion through the porous plug.
    (i) Both A and R are true and R is the correct explanation of A
    (ii) Both A and R are true and R is not the correct explanation of A
    (iii) A is true but R is false
    (iv) A is false but R is true

43. The list that provides properties which are referred to as thermodynamic potentials is
    (i) pressure, internal energy, Helmholtz function
    (ii) temperature, entropy, enthalpy
    (iii) entropy, internal energy, Gibbs function
    (iv) enthalpy, Gibbs function, Helmholtz function

44. Identify the line of constant enthalpy on the temperature-entropy chart (Figure 5.2) shown below for carbon dioxide.

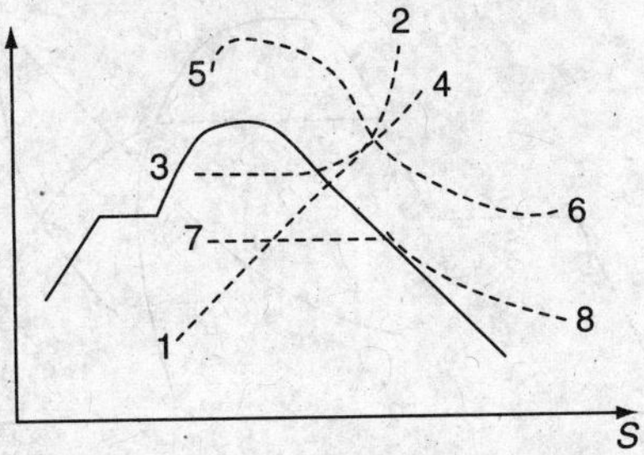

Figure 5.2 Temperature-entropy Diagram

    (i) Line 1–2 (ii) Line 3–4
    (iii) Line 5–6 (iv) Line 7–8

45. The slope of constant internal energy line for a perfect gas is equal to (p/T) when this process is represented on
   (i) the pressure-volume diagram
   (ii) the temperature-entropy diagram
   (iii) the entropy-volume diagram
   (iv) the enthalpy-pressure diagram

46. The abscissa and ordinate of the diagram (Figure 5.3) which shows a family of isenthalpic lines together with the inversion curve for a gas are

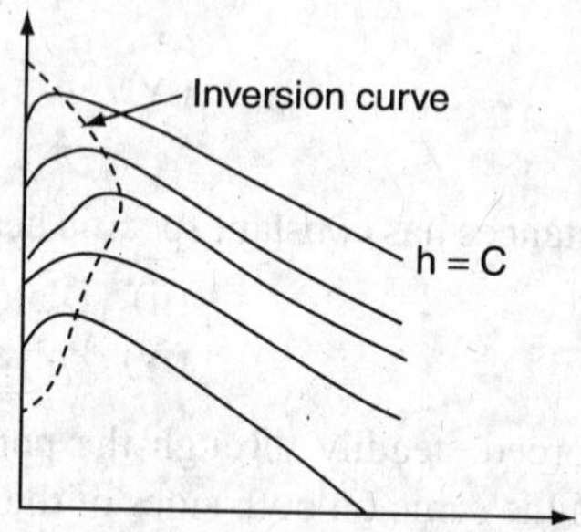

Figure 5.3 Inversion and Isenthalpic Curves of a Real Gas

   (i) pressure-volume
   (ii) volume-temperature
   (iii) temperature-entropy
   (iv) pressure-temperature

47. The following Figure 5.4 shows the pressure-enthalpy diagram for CO_2. With respect to this figure match List I (curves) with List II (properties) and select the correct answer using the codes given in the following lists.

List I	*List II*
A. Curve I	1. Volume
B. Curve II	2. Temperature
C. Curve III	3. Entropy

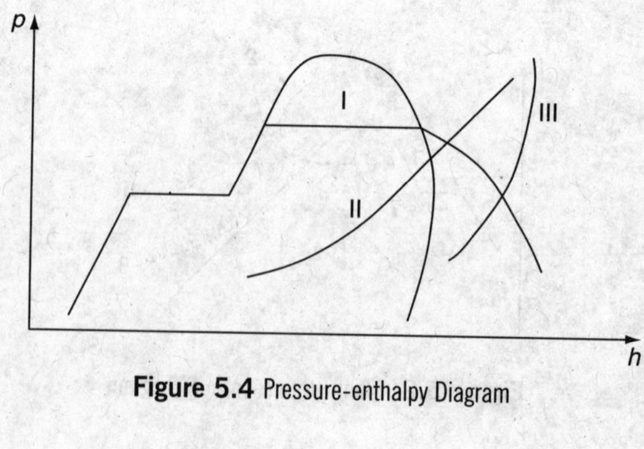

Figure 5.4 Pressure-enthalpy Diagram

Codes:

	A	B	C
(i)	2	1	3
(ii)	1	2	3
(iii)	3	2	1
(iv)	2	3	1

48. Which liquid has the lowest freezing point at atmospheric pressure?
   (i) Alcohol (ii) Mercury
   (iii) Pentane (iv) Water
49. The Gibbs's phase rule, which relates the number of independent properties (F) with the number of components (C) and the number of phases (P) of a thermodynamic system, is expressed mathematically as
   (i) $F = C + 3 - P$ (ii) $F = C - 3 + P$
   (iii) $F = C + 2 - P$ (iv) $F = C - 2 + P$
50. Which one of the following properties is well defined for a thermodynamic system containing only a few molecules?
   (i) Mass (ii) Pressure
   (iii) Temperature (iv) Volume

Answers

1. (iv)	2. (iv)	3. (iv)	4. (iii)	5. (iv)	6. (iii)	7. (iv)	8. (iii)	9. (iii)	10. (iv)
11. (iv)	12. (i)	13. (ii)	14. (ii)	15. (iii)	16. (iv)	17. (ii)	18. (iv)	19. (iv)	20. (i)
21. (i)	22. (ii)	23. (iv)	24. (ii)	25. (ii)	26. (iv)	27. (iii)	28. (iv)	29. (i)	30. (iii)
31. (ii)	32. (i)	33. (i)	34. (iii)	35. (i)	36. (iii)	37. (iv)	38. (iii)	39. (iii)	40. (iii)
41. (i)	42. (iii)	43. (iv)	44. (iii)	45. (iii)	46. (iv)	47. (i)	48. (ii)	49. (iii)	50. (i)

6

Vapour Power Cycles

1. Sketch the Carnot cycle for steam on *p-v* and *T-s* diagrams and derive an expression for thermal efficiency. Why Carnot cycle is not used as a standard of comparison for steam power plants?

Ans: The Carnot cycle consists of two constant pressure operations ($p = c$; $T = c$) and two adiabatic operations ($b - c/b_1 - c_1/b_2 - c_2$ and $d - a$) as shown in the Figure 6.1. The cycle is shown on temperature - entropy diagram in the Figure 6.2.

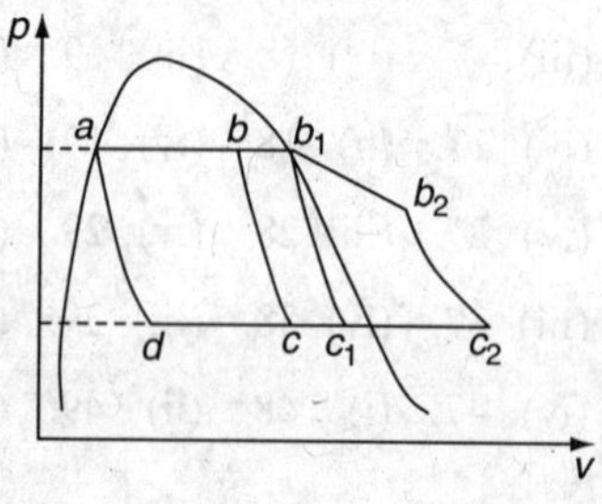

Figure 6.1 *P-v* Diagram of Carnot Vapour Cycle

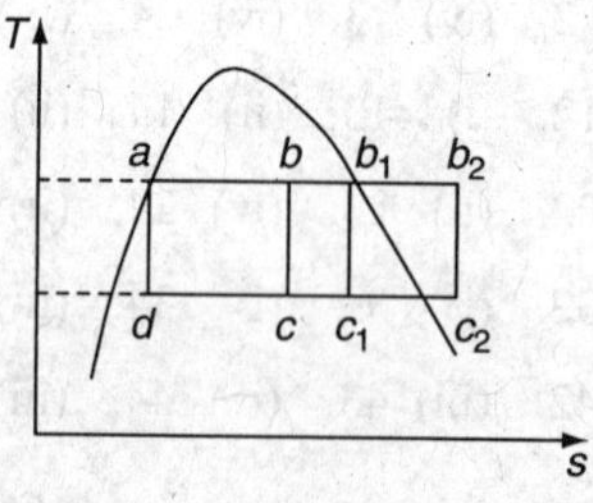

Figure 6.2 *T-s* Diagram of Carnot Vapour Cycle

The heat supplied at temperature T_1 is $T_1\Delta s$ or $T_1\Delta s_1$ or $T_1\Delta s_2$ where $\Delta s = s_b - s_a$ or

$$\Delta s_1 = (s)_{b_1} - (s)_a \quad \text{or} \quad \Delta s_2 = (s)_{b_2} - (s)_a.$$

The heat rejected at temperature T_2 is $T_2\Delta s$ or $T_2\Delta s_1$ or $T_2\Delta s_2$ where $\Delta s = s_c - s_d$ or

$$\Delta s_1 = (s)_{c_1} - (s)_d \quad \text{or} \quad \Delta s_2 = (s)_{c_2} - (s)_d.$$

The work done is represented by the area ***a-b-c-d*** or ***a-b$_1$-c$_1$-d*** or ***a-b$_2$-c$_2$-d*** and is equal to $(T_1 - T_2)\,\Delta s$ or $(T_1 - T_2)\,\Delta s_1$ or $(T_1 - T_2)\,\Delta s_2$.

$$\text{Thermal efficiency} = \frac{(T_1 - T_2)\Delta s}{T_1\Delta s} \text{ or } \frac{(T_1 - T_2)\Delta s_1}{T_1\Delta s_1} \text{ or } \frac{(T_1 - T_2)\Delta s_2}{T_1\Delta s_2} \text{ or } \frac{T_1 - T_2}{T_1}.$$

Although the cycle is thermodynamically simple, it is extremely difficult to realize in practice. The isothermal condensation must stop at d so that subsequent adiabatic compression restores the fluid to its initial state a. Hence, no practical prime-mover or engine operates on the Carnot cycle.

2. Sketch the Rankine cycle for steam on *p-v* and *T-s* co-ordinates and derive an expression for its thermal efficiency.

Ans: In steam power plant, the supply and rejection of heat is more easily performed at constant pressures than at constant temperatures. An attempt was made by Rankine to calculate the maximum possible work that could be developed by an engine using dry saturated steam between the pressure limits of the boiler and condenser. In the Rankine cycle, we commence with liquid at the lower temperature, T_1 and pressure, p_1 (see Figure 6.3). The pressure of the liquid is then raised to p_2 by adiabatic compression 1–2 in the feed pump. The increase in temperature consequent on this compression may be of the order of a few degrees centigrade. From 2 to 2' the liquid receives sensible heat at constant pressure p_2 to be followed by evaporation, which may be partial at 3, or complete at 3', or superheated to 3" (frequently a super heat is imparted to raise the temperature at constant pressure to 3") (see Figure 6.4 and Figure 6.5). The vapour expands adiabatically from state 3 (or 3' or 3") to 4 (or 4' or 4", the expansion curve being discontinuous at *f* because of the change from superheated to wet state).

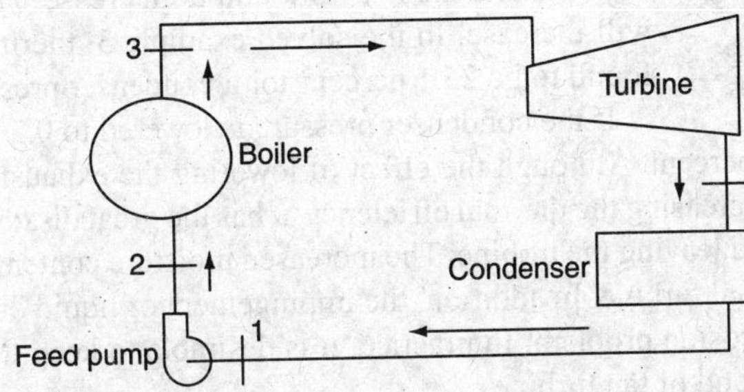

Figure 6.3 Elements of Rankine Cycle

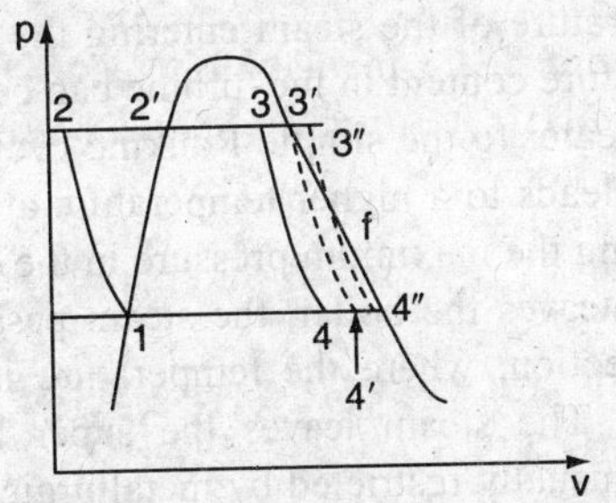

Figure 6.4 *P-v* Diagram of Rankine Cycle

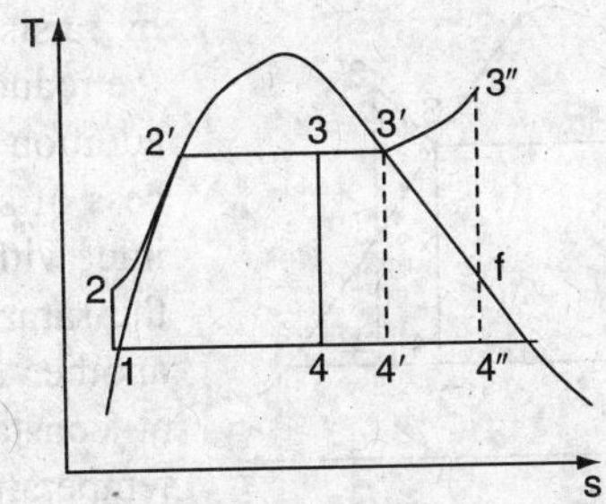

Figure 6.5 *T-s* Diagram of Rankine Cycle

Heat supplied in the boiler, $Q_s = (h_3 - h_2)$ kJ/kg of steam flowing through the boiler

Turbine work, $W_t = (h_3 - h_4)$ kJ/kg

Pump work, $W_p = v_1 (p_2 - p_1)$ kN.m/kg

Thermal efficiency, $\eta = \dfrac{W_t - W_p}{Q_s}$

While the enthalpies at points 3, 4 and 1 can be found from Mollier chart/ steam tables, the enthalpy at point 2 can be calculated from the relation $h_2 = h_1 + W_p$.

3. Illustrate with the aid of *T-s* diagram how the efficiency of Rankine cycle can be improved by decreasing the temperature (pressure) at which heat is rejected or by superheating the steam.

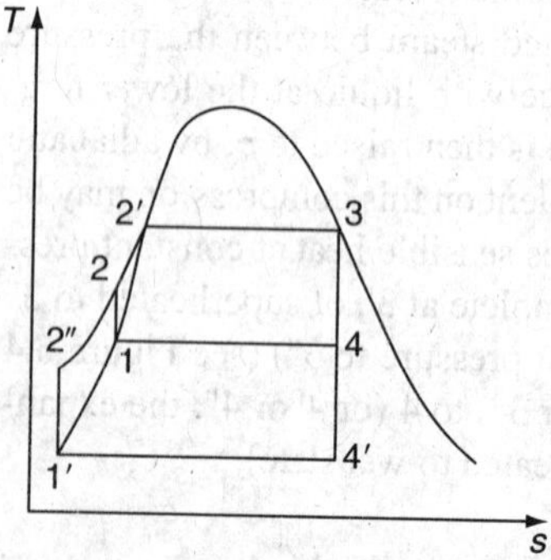

Figure 6.6 *T*-*s* Diagram of Rankine Cycle with Two Condenser Pressures

Ans: The efficiency of the simple Rankine cycle may be increased on the basis of theoretical Carnot cycle either by decreasing the temperature (pressure) at which heat is rejected or by increasing the average temperature at which heat is added. The first of these effects is accomplished by lowering the exhaust pressure, which in turn lowers the temperature T_4 (consequently T_1 also). This point is illustrated in Figure 6.6.

The area enclosed by the states 1'–2"–2–1–4–4'–1' is a measure of the decrease in heat rejected when the exhaust pressure is lowered from p_4 to $p_{4'}$. As a result, both the quantities of heat input required and the net work output in crease. The thermal efficiency will increase. In the solved example 3, thermal efficiency will be found to be 25.1 percent for a condenser pressure of 0.3 bar.

If the condenser pressure is lowered to 0.2 bar, the thermal efficiency will increase to 26.35 percent. Although the effect of lowering the exhaust pressure is advantageous from the standpoint of increasing the thermal efficiency, it has the great disadvantage of increasing the moisture content of the fluid leaving the turbine. The increased moisture content in the turbine stages decreases the efficiency of actual turbine. In addition, the impingement of liquid droplets on the turbine blades may lead to a serious erosion problem. In practice, it is desirable to keep the moisture less than 10 percent at the low pressure end of the turbine.

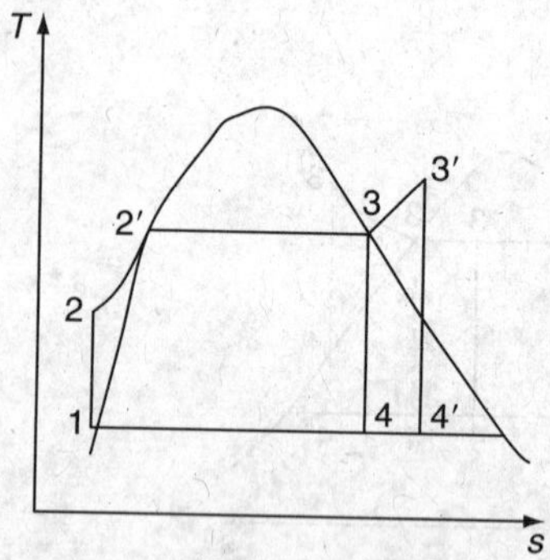

Figure 6.7 *T*-*s* Diagram of Rankine Cycle with Superheat

It was mentioned previously that increasing the average temperature at which heat is supplied would increase the efficiency of the Rankine cycle. Both the increase of the efficiency of the cycle by raising the temperature of the steam entering the turbine and the reduction of moisture content in the turbine can be met by the addition of a super heater to the simple Rankine cycle. The process of superheating leads to a higher temperature at the turbine inlet without increasing the maximum pressure in the cycle. After the saturated vapour leaves the boiler, the steam passes through another heat input section, where the temperature is increased at constant pressure. The steam leaves the super heater at a temperature which is usually restricted by metallurgical strength. Temperatures in the range of 550°C–650°C are generally permitted. The temperature-entropy diagram for a Rankine cycle with superheating is shown in Figure 6.7.

The area 4–3–3'–4'–4 represents additional work output available and the area beneath the curve 3–3' represents the additional heat supplied in the super heater section. The average temperature at which heat is supplied during the process 3–3' is greater than the heat addition process in the boiler section and the temperature at which heat rejection takes place is the same. The efficiency of the cycle would be expected to increase. The other point to be noted is that the quality of steam at state 4' is considerably higher than that at state 4. Hence, the moisture content in the turbine will be reduced.

4. What characterises reheat cycle? What are its advantages?

Ans: In the reheat cycle, the steam is not allowed to expand to condenser pressure in one stage. After partial expansion, the steam is withdrawn from the turbine and reheated at constant pressure. Then, it is returned to the turbine for further expansion to the exhaust pressure. The turbine may be considered to consist of two stages, a high pressure one and a low pressure one. Figure 6.8 illustrates the parts of the reheat cycle and the cycle is also shown on the *T-s* diagram in Figure 6.9.

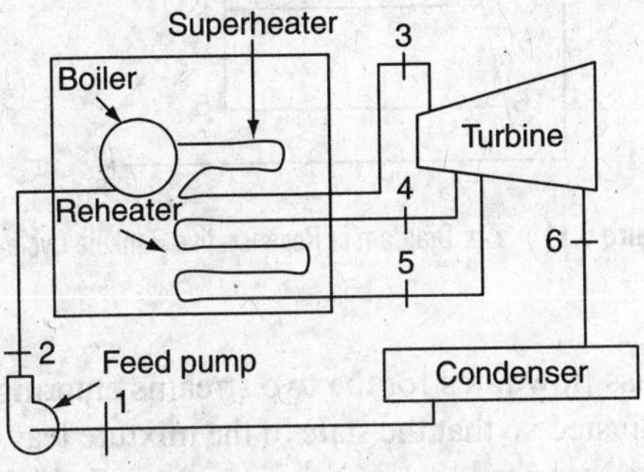

Figure 6.8 Reheat Rankine Cycle

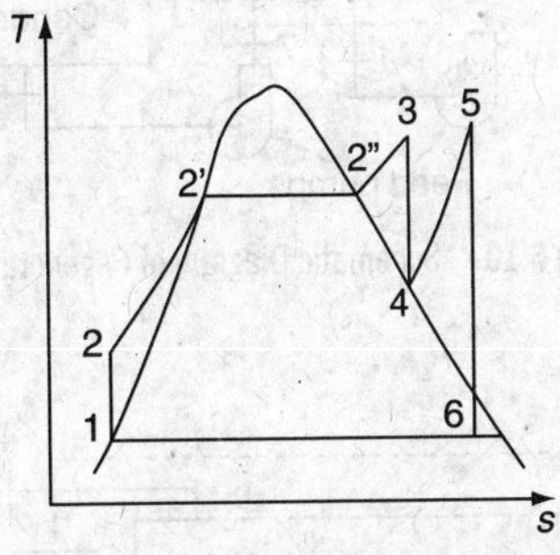

Figure 6.9 *T*-s Diagram of Reheat Rankine Cycle

Since the average temperature of heat addition during the reheating process 4–5 is not much different from that of heat addition during superheating 2"–3, the thermal efficiency is not greatly affected. Nevertheless, the reheat cycle does remove the objectionable feature of high moisture content at the turbine exhaust. In computing the thermal efficiency of reheat cycle, one must remember to account for the work output of the high pressure and low pressure turbines as well as the heat input in the boiler and super heater section and also in the reheater section.

5. Explain with the help of *T-s* diagram how the efficiency of Rankine cycle can be improved by regenerative feed water heating.

Ans: The simple Rankine cycle with superheating reveals a serious disadvantage of the basic cycle. For the portion of heat addition process 2–2' (see Figure 6.7), the average temperature at which heat is added is much below the temperature at which heat is added during the vapourization and the superheating process 3–3'. Hence, the cycle efficiency is reduced. If the average temperature for this portion of the heat addition process could be raised, the cycle efficiency would more nearly approach that of the Carnot cycle. One practical method of accomplishing this is by the use of a regeneration process internal to the overall cycle.

The ideal regeneration vapour power cycle (see Figure 6.10 and Figure 6.11) is accomplished in the following way. Part of the superheated steam which enters the turbine at state 3 is bled or extracted from the turbine at state 4 which is an intermediate state in the turbine expansion process. The extracted steam is directed into a feedwater heater or regenerator. The portion of the steam which is not extracted expands completely to the condenser pressure (state 5) and it is then condensed to saturated liquid state 6. The feed pump then increases the pressure of the liquid leaving the condenser isentropically to the same pressure as that of extracted steam. The compressed liquid at state 7 then enters the feedwater heater where it mixes directly with the flowing steam extracted from the turbine. Because of this direct mixing process, the feedwater heater is called an open- or direct-contact type. In the ideal

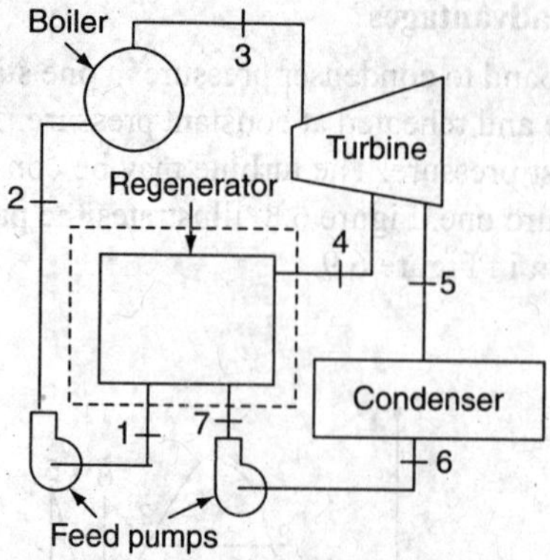

Figure 6.10 Schematic Diagram of Regenerative Rankine Cycle

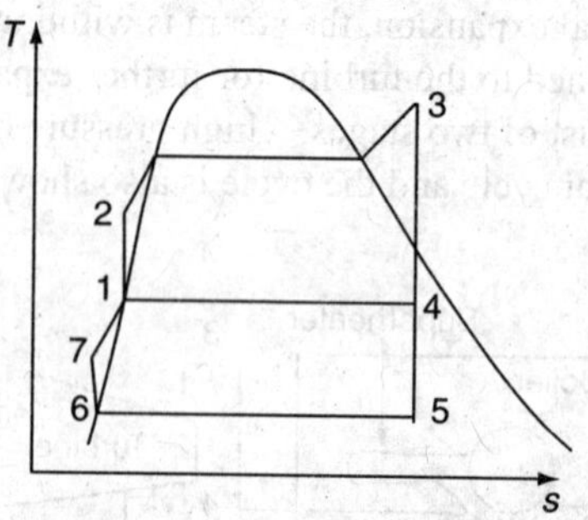

Figure 6.11 *T-s* Diagram of Regenerative Rankine Cycle

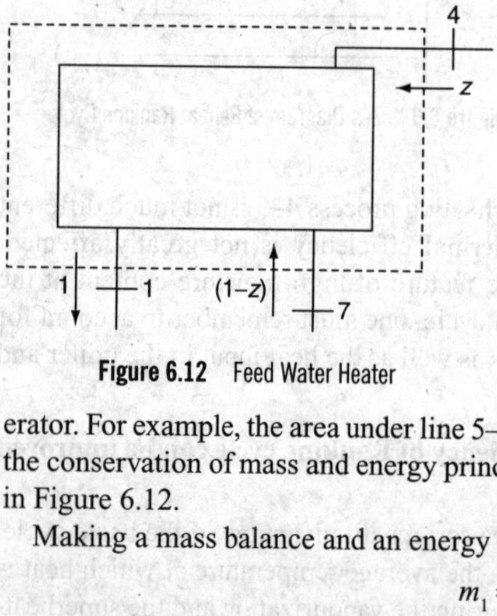

Figure 6.12 Feed Water Heater

situation, the mass flow rates for the two streams entering the heater are adjusted so that the state of the mixture leaving the heater is a saturated liquid at the pressure at which the steam is extracted from the turbine (state 4). A second pump then isentropically raises the pressure of the liquid to state 2, which corresponds to the steam generator pressure. For an ideal internally reversible cycle, the areas beneath the curves on the *T-s* diagram represent heat addition or heat rejection quantities. One must keep in mind, however, that the mass flow rates through certain parts of the cycle are not equal to the mass flow rate through the steam generator. For example, the area under line 5–6 is a measure of the heat rejected in the condenser. We employ the conservation of mass and energy principles to the control volume of open feedwater heater as shown in Figure 6.12.

Making a mass balance and an energy balance, we obtain

$$m_1 = m_4 + m_7,$$

$$m_1\, h_1 = m_4\, h_4 + m_7\, h_7.$$

Combining these equations, we get

$$h_1 = \frac{m_4\, h_4}{m_1} + \frac{(m_1 - m_4)\, h_7}{m_1}.$$

Assuming the fraction of steam extracted from the turbine as z, the earlier relation simplifies to

$$h_1 = z\, h_4 + (1 - z)\, h_7.$$

Since the enthalpies of the flowing streams entering and leaving the open feedwater heater are known for the ideal cycle, the earlier relation permits the evaluation of the quantity of steam which must be extracted from the turbine at state 4 in order to get the stream leaving the heater at saturated liquid state.

Solved Problems

1. A Carnot steam cycle operates between a source temperature of 320°C and a sink temperature of 15°C. Determine the thermal efficiency and work ratio (i) when all processes are reversible and (ii) when the internal efficiency of the turbine is 80 percent and that of compression is 60 percent.

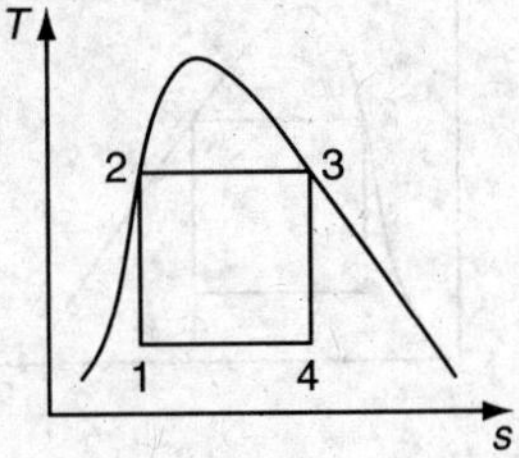

$p_2 = p_3 = 113$ bar (320°C) and $p_1 = p_4 = 0.017$ bar (15°C).

The following values are taken from steam tables:

$h_3 = 2702$ kJ/kg $s_3 = 5.540$ kJ/kg K,

$h_2 = 1461$ kJ/kg $s_2 = 3.45$ kJ/kg K,

$s_4 = s_3$ and $s_1 = s_2$.

From the aforesaid values, we can determine the dryness fraction of steam at point 1 and 4 as follows: Specific entropy of saturated liquid water and saturated steam at 0.017 bar are 0.224 and 8.781 kJ/kg K, respectively.

$$x_4 = \frac{5.54 - 0.224}{8.781 - 0.224} = 0.6212 \text{ and } x_1 = \frac{3.45 - 0.224}{8.781 - 0.224} = 0.377$$

The specific enthalpy of saturated liquid water and enthalpy of evaporation at 0.017 bar are 63 and 2465.9 kJ/kg, respectively.

Hence, $h_1 = 63 + 0.377 \times 2465.9 = 992.6$ kJ/kg and $h_4 = 63 + 0.6212 \times 2465.9 = 1594.8$ kJ/kg.

(i) When all processes are reversible:

Turbine work $= h_3 - h_4 = 2702 - 1594.8 = 1107.2$ kJ/kg

Pump work $= h_2 - h_1 = 1461 - 992.6 = 468.4$ kJ/kg

Heat supplied $= h_3 - h_2 = 2702 - 1461 = 1241$ kJ/kg

$$\text{Work ratio} = \frac{\text{Net work}}{\text{Turbine work}} = \frac{1{,}107.2 - 468.4}{1{,}107.2} = 0.576$$

$$\text{Thermal efficiency} = \frac{\text{Net work}}{\text{Heat added}} = \frac{1{,}107.2 - 468.4}{1{,}241} = 0.5147 \text{ or } 51.47\%.$$

(ii) Actual turbine work $= 0.8 \times 1107.2 = 885.76$ kJ/kg

$$\text{Actual pump work} = \frac{468.4}{0.6} = 780.66 \text{ kJ/kg}$$

$$\text{Work ratio} = \frac{(885.76 - 780.66)}{885.76} = 0.1186$$

$$\text{Thermal efficiency} = \frac{(885.76 - 780.66)}{1{,}241} = 0.0846 \text{ or } 8.46\%.$$

2. Compare the Carnot and Rankine efficiencies for a steam cycle working between steam pressures of 100 bar and 0.2 bar with inlet steam in dry state condition. Also calculate the work ratios for both the cycles.

(i) **Carnot cycle:**

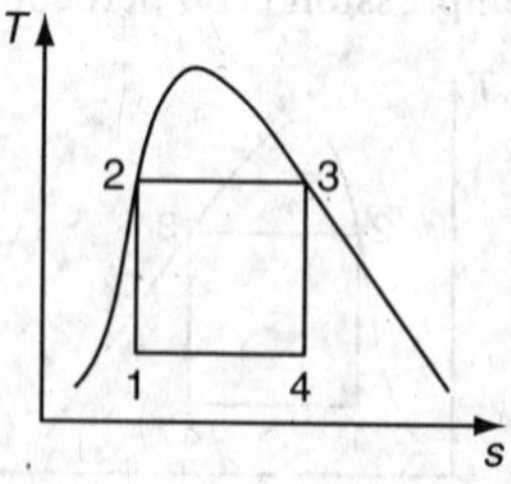

The following values are taken from steam tables:

$h_3 = 2727.7$; $s_3 = 5.6198$ (100 bar)

$h_2 = 1408.0$; $s_2 = 3.3605$.

Since specific entropy of saturated liquid water and saturated steam at 0.2 bar are 0.8321 and 7.9094 kJ/kg K, respectively, the condition of steam at points 4 and 1 are wet. The dryness fraction x_4 and x_1 can be evaluated as follows:

$$x_4 = \frac{5.6198 - 0.8321}{7.9094 - 0.8321} = 0.676 \text{ and } x_1 = \frac{3.3605 - 0.8321}{7.9094 - 0.8321} = 0.357.$$

The specific enthalpy of saturated liquid water and enthalpy of evaporation at 0.2 bar are 251.5 and 2358.4 kJ/kg, respectively.

Hence, $h_4 = 251.5 + 0.676 \times 2358.4 = 1845.7$ and $h_1 = 251.5 + 0.357 \times 2358.4 = 1093.4$.

Turbine work $= h_3 - h_4 = 2727.7 - 1845.7 = 882.0$

Pump work $= h_2 - h_1 = 1408.0 - 1093.4 = 314.6$

Heat supplied $= h_3 - h_2 = 2727.7 - 1408.0 = 1319.7$

$$\text{Work ratio} = \frac{882.0 - 314.6}{882.0} = 0.643 \text{ and}$$

$$\text{Thermal efficiency} = \frac{882.0 - 314.6}{1319.7} = 0.4299 \text{ or } 42.99\%$$

(ii) **Rankine cycle:**

The turbine work will be the same as earlier.

The specific volume of saturated liquid at 0.2 bar, $v_f = 0.001017$

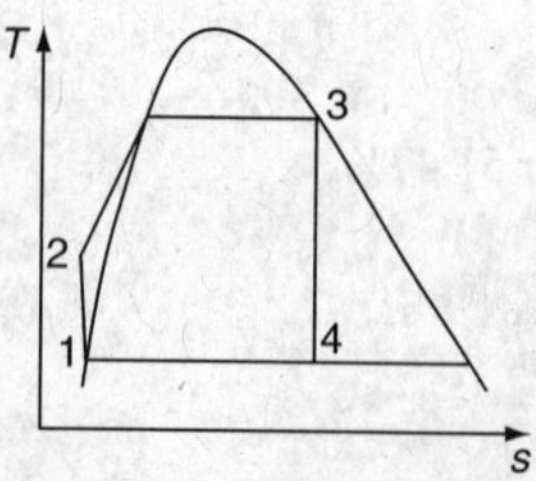

$$\text{Pump work} = 0.001017 \times (100 - 2) \times 100$$
$$= 10.149 \text{ kJ/kg}$$

Hence, $h_2 = 251.5 + 10.149 = 261.649$

$$\text{Heat supplied} = 2727.7 - 261.649$$
$$= 2466.051$$

$$\text{Work ratio} = \frac{882.0 - 10.149}{882.0} = 0.988$$

$$\text{Thermal efficiency} = \frac{882.0 - 10.149}{2466.051} = 0.3535 \text{ or } 35.35\%.$$

3. Obtain the thermal efficiency of an engine working on the Rankine cycle when receiving dry saturated steam at 15 bar and expanding it to 0.3 bar.

Using Mollier chart and steam tables, the following values are obtained. $h_3 = 2790$ kJ/kg; $h_4 = 2160$ kJ/kg; dryness fraction, $x_4 = 0.8$; $h_1 = 289.3$ kJ/kg; and specific volume, $v_1 = 0.001022$ m^3/kg.

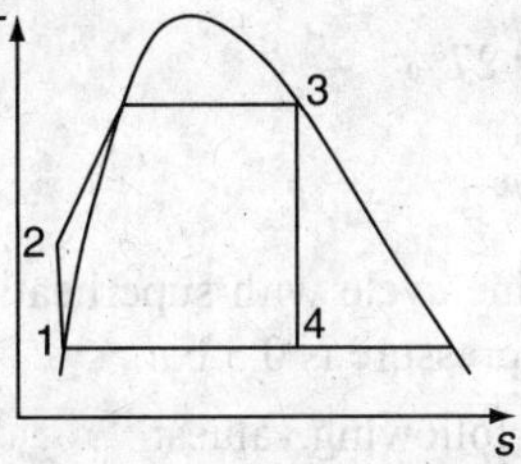

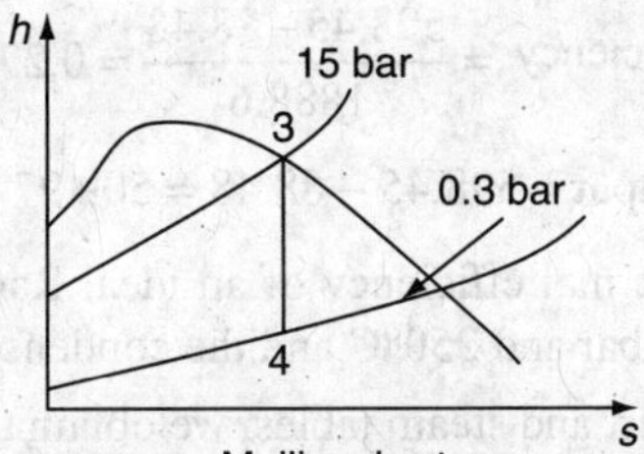

Mollier chart

Pump work, $W_p = v_1 (p_2 - p_1) = 0.001022 \times 14.7 \times 100 = 1.5$ kN.m/kg or 1.5 kJ/kg

Hence, $h_2 = h_1 + W_p = 289.3 + 1.5 = 290.8$ kJ/kg

Turbine work, $W_t = h_3 - h_4 = 2790 - 2160 = 630$ kJ/kg

Heat supplied, $Q_s = h_3 - h_2 = 2790 - 290.8 = 2499.2$ kJ/kg

$$\text{Thermal efficiency, } \eta = \frac{W_t - W_p}{Q_s} = \frac{630 - 1.5}{2{,}499.2} \times 100 = 25.1\%.$$

4. A Carnot engine using steam as a working fluid works between pressures 20 bar and 0.5 bar. The vapour is dry and saturated at the end of isothermal expansion. The liquid is saturated at the end of isentropic compression. Determine (i) the thermal efficiency and (ii) work done per kg of steam.

$p_2 = p_3 = 20$ bar (212.4°C)

$p_1 = p_4 = 0.5$ bar (81.3°C)

the following values are taken from steam tables

$h_3 = 2797.2$ kJ/kg $s_3 = 6.3366$ kJ/kg K

$h_2 = 908.6$ kJ/kg $s_2 = 2.4469$ kJ/kg K

$s_4 = s_3$ and $s_1 = s_2$

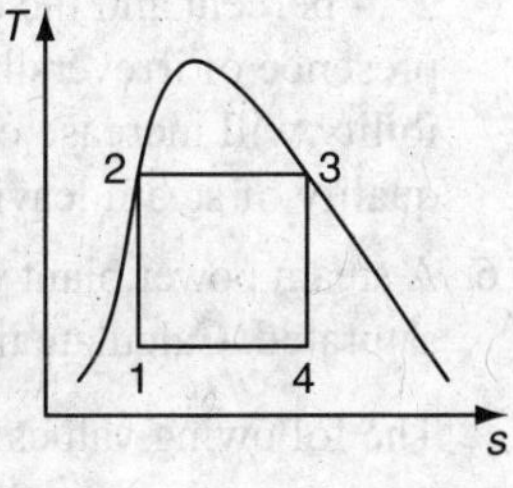

From the aforesaid values, we can determine the dryness fraction of steam at point 1 and 4 as follows:

Specific entropy of saturated liquid water and saturated steam at 0.5 bar are 1.0912 and 7.5947 kJ/kg K, respectively.

$$x_1 = \frac{2.4469 - 1.0912}{7.5947 - 1.0912} = 0.208 \text{ and } x_4 = \frac{6.3366 - 1.0912}{7.5947 - 1.0912} = 0.806.$$

The specific enthalpy of saturated liquid water and enthalpy of evaporation at 0.5 bar are 340.6 and 2305.4 kJ/kg, respectively.

Hence, $h_1 = 340.6 + 0.208 \times 2305.4 = 820.12$

$h_4 = 340.6 + 0.806 \times 2305.4 = 2198.75$

Work done during 3–4, $h_3 - h_4 = 2797.2 - 2198.75 = 598.45$

Work done during 1–2, $h_2 - h_1 = 908.6 - 820.12 = 88.48$

Heat supplied, $h_3 - h_2 = 2797.2 - 908.6 = 1888.6$

$$\text{Thermal efficiency} = \frac{598.45 - 88.48}{1888.6} = 0.27 \text{ or } 27\%.$$

Net work output = 598.45 – 88.48 = 509.97 kJ/kg.

5. Determine the thermal efficiency of an ideal Rankine cycle with superheating for which the inlet conditions are 15 bar and 250°C and the condenser pressure is 0.3 bar.

Using Mollier chart and steam tables, we obtain the following values:

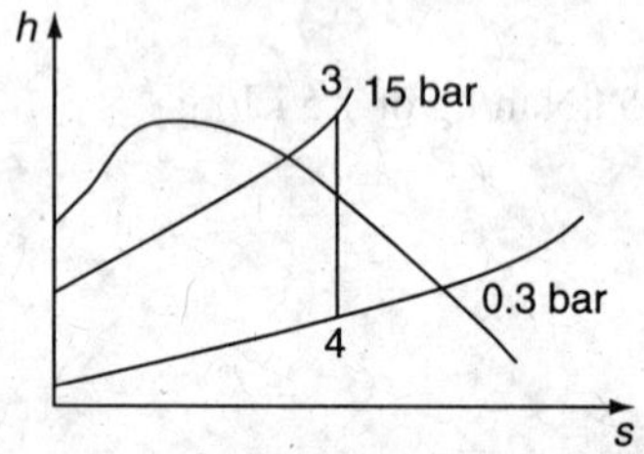

Mollier chart

$h_3 = 2920$ kJ/kg, $h_4 = 2250$ kJ/kg,

$h_1 = 289.3$ kJ/kg, $x_4 = 0.84$, and

$v_1 = 0.001022$ m³/kg.

Pump work, $W_p = v_1\,(p_2 - p_1) = 1.5$ kJ/kg

Hence, $h_2 = 289.3 + 1.5 = 290.8$ kJ/kg

Turbine work, $W_t = h_3 - h_4 = 670$ kJ/kg

Heat supplied, $Q_s = h_3 - h_2 = 2629.2$ kJ/kg

$$\text{Thermal efficiency, } \eta = \frac{W_t - W_p}{Q_s} = \frac{670 - 1.5}{2629.2} \times 100 = 25.4\%.$$

The result of superheating is to increase the thermal efficiency from 25.1 (refer to problem 3) to 25.4 percent and the quality of steam leaving the turbine has increased from 80 to 84 percent. The presence of irreversibility in the flow through turbine will decrease the work output. The irreversibility will increase outlet enthalpy as a result of frictional heating which in turn will improve the quality of steam leaving the turbine.

6. A steam power plant works between 30 bar and 0.4 bar. The steam supplied to the turbine is dry and saturated. Calculate the Rankine cycle efficiency using steam tables.

The following values are taken from steam tables:

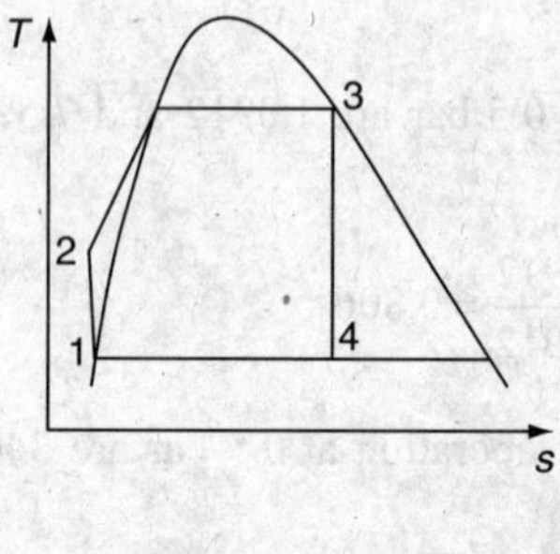

$h_3 = 2802.3$ (30 bar)

$s_3 = 6.1837$ (30 bar)

$h_1 = 317.7$ (0.4 bar)

$s_1 = 1.0261$ (0.4 bar)

$s_{fg} = 6.6448$ (0.4 bar)

$h_{fg} = 2319.2$ (0.4 bar)

$v_f = 0.001026$ (0.4 bar)

$$s_4 = s_3$$

$$6.1837 = 1.0261 + x_4(6.6448)$$

$$x_4 = \frac{6.1837 - 1.0261}{6.6448} = 0.776$$

$$h_4 = 317.7 + 0.776(2319.2) = 2117.4$$

Pump work = 0.001026 × (30 – 0.4) × 100 = 3.037 kJ/kg

$h_2 = 317.7 + 3.037 = 320.737$

Turbine work = $h_3 - h_4 = 2802.3 - 2117.4 = 684.9$

Heat supplied = $h_3 - h_2 = 2802.3 - 320.737 = 2481.563$

Net work output = 684.9 – 3.037 = 681.863

$$\text{Thermal efficiency} = \frac{681.863}{2481.563} = 0.2747 \text{ or } 24.47\%.$$

7. A Rankine cycle operates between 40 bar and 400°C and 0.1 bar. Determine the efficiency of the cycle.

The following values are obtained from Mollier diagram and steam tables:

$h_3 = 3213.6$; $s_3 = 6.769$

$s_f = 0.649$ (0.1 bar); $s_{fg} = 7.501$ (0.1 bar)

$h_1 = 191.8$ (0.1 bar); $h_{fg} = 2392.8$ (0.1 bar).

$$x_4 = \frac{6.769 - 0.649}{7.501} = 0.82$$

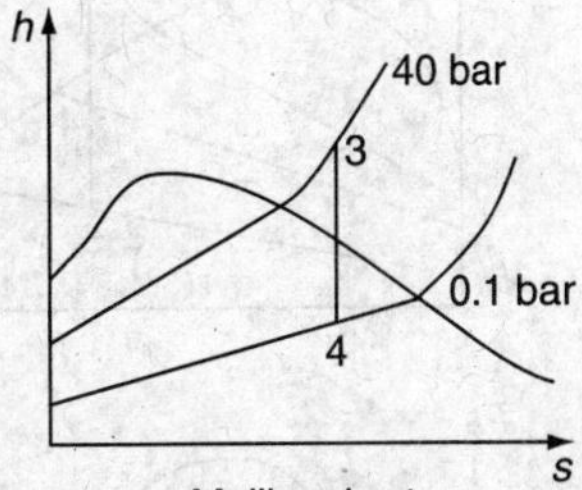

Mollier chart

$h_4 = 191.8 + 0.82 \times 2392.8 = 2153.8$

$v_1 = 0.00101$;

Pump work = 0.00101 (40 – 0.1) × 100

= 4.03

$h_2 = 191.8 + 4.03 = 195.83$

Turbine work = $h_3 - h_4 = 3213.6 - 2153.8 = 1059.8$

Heat supplied = $h_3 - h_2 = 3213.6 - 195.83 = 3017.77$

$$\text{Thermal efficiency} = \frac{1059.8 - 4.03}{3017.77} = 0.3498 \text{ or } 34.98\%.$$

8. In a steam power plant utilizing the reheat cycle, the turbine inlet conditions are 15 bar and 250°C. After expansion to 5 bar, the steam is reheated to 250°C and then expanded to the condenser pressure of 0.3 bar. Compute the cycle efficiency.

Using Mollier chart and steam tables, we obtain the following values:

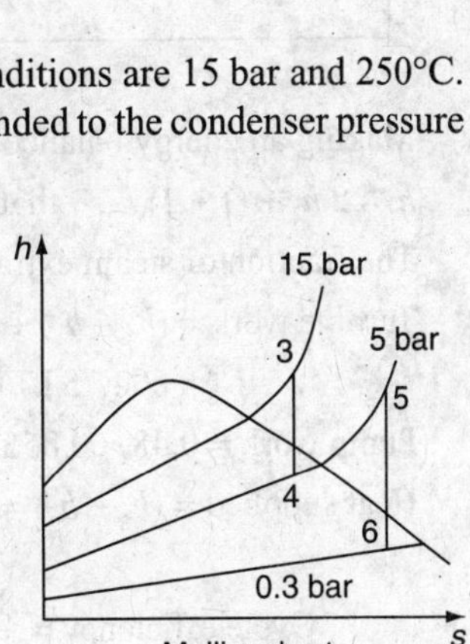

Mollier chart

$h_3 = 2920$ kJ/kg, $h_4 = 2700$ kJ/kg,

$h_5 = 2950$ kJ/kg, $h_6 = 2450$ kJ/kg,

$x_4 = 0.97$, $x_6 = 0.92$,

$h_1 = 289.3$ kJ/kg and $v_1 = 0.001022$ m³/kg.

Pump work, $W_p = v_1 (p_2 - p_1) = 1.5$ kJ/kg

$h_2 = 289.3 + 1.5 = 290.8$ kJ/kg

Turbine work $= (h_3 - h_4) + (h_5 - h_6)$
$= 220 + 500 = 720$ kJ/kg

Heat input $= (h_3 - h_2) + (h_5 - h_4)$
$= 2629.2 + 250 = 2879.2$ kJ/kg

$$\text{Cycle efficiency} = \frac{\text{Turbine work} - \text{Pump work}}{\text{Heatinput}} = \frac{(720 - 1.5)}{2879.2} \times 100 = 24.95\%.$$

[*Note*: Though there is no improvement in the cycle efficiency (see Example 5), the moisture content is reduced (quality of steam is improved from 0.84 to 0.92)].

9. An ideal regenerative steam power plant operates so that steam enters the turbine at 15 bar and 250°C and exhausts at 0.3 bar. A simple open feedwater heater is employed which operates at 5 bar. Compute the cycle efficiency.

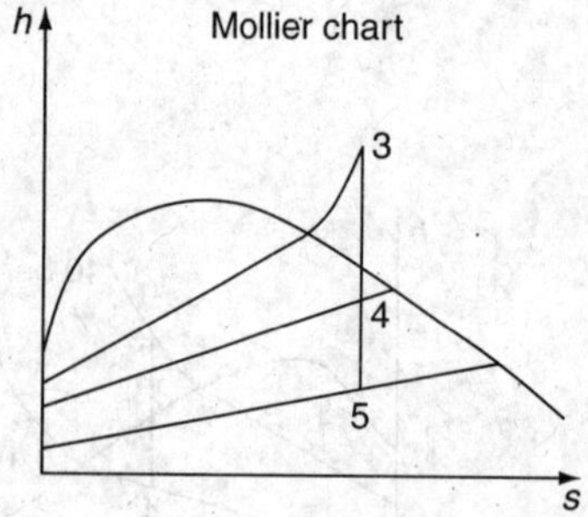

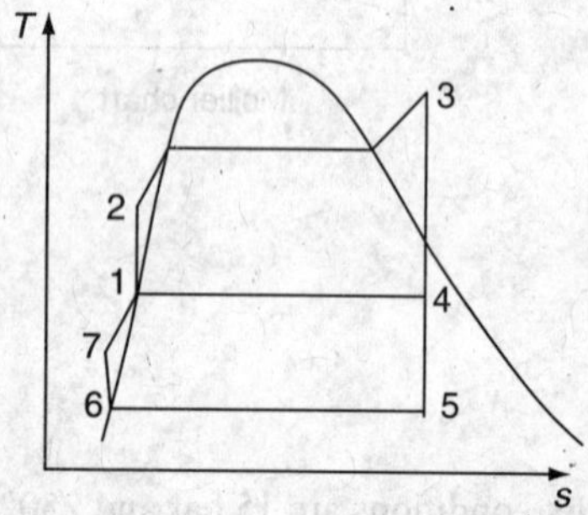

Refer to the following *T-s* diagram:

Using Mollier chart and steam tables, we obtain the following values:

$h_3 = 2920$ kJ/kg, $h_4 = 2700$ kJ/kg,
$h_5 = 2250$ kJ/kg, $h_6 = 289.3$ kJ/kg,
$h_1 = 640.1$ kJ/kg, $v_6 = 0.001022$ m³/kg and
$v_1 = 0.001093$ m³/kg.

$h_7 - h_6 = v_6\,(p_7 - p_6) = 0.001022 \times 4.7 \times 100$
$h_7 = 289.3 + 0.48 = 289.78$ kJ/kg
$h_2 - h_1 = v_1\,(p_2 - p_1) = 0.001093 \times 10 \times 100$
$h_2 = 640.1 + 1.09 = 641.19$ kJ/kg

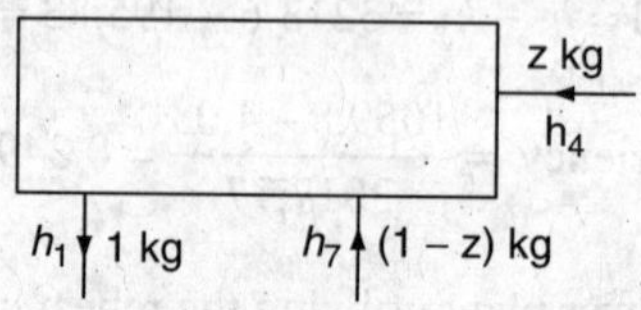

Making an energy balance for steam and water in the open feedwater heater

$h_1 = z\,h_4 + (1 - z)\,h_7$, that is, $640.1 = 2700\,z + 289.78\,(1 - z)$. Hence, $z = 0.145$.

The fraction of steam extracted at 5 bar = 0.145 kg/ kg of steam

Turbine work $= (h_3 - h_4) + (1 - z)\,(h_4 - h_5) = (2920 - 2700) + 0.855\,(2700 - 2250)$
$= 604.75$ kJ/kg

Pump work $= 0.48 \times 0.855 + 1.09 = 1.5004$ kJ/kg

Heat supplied $= (h_3 - h_2) = 2920 - 641.19 = 2278.81$ kJ/kg

$$\text{Cycle efficiency} = \frac{\text{Turbine work} - \text{Pump work}}{\text{Heat supplied}} = \frac{(604.75 - 1.5)}{2278.81} \times 100 = 26.47\%.$$

10. Prove that the thermal efficiency of a Rankine cycle using superheated steam is greater than the thermal efficiency of a corresponding Rankine cycle using steam without superheat. Both the cycles operate between the same boiler and condenser pressure limits.

Refer to the *T-s* diagram:

3–4–5–6–7 is the basic Rankine cycle and 1–2–3–4–5–6 represents the Rankine cycle with superheat.

Let 3–4–5–8–9 be another theoretical cycle which has the same heat supplied as the superheated cycle. Area 6–7–3–4–5 is common to the superheated cycle and the theoretical cycle. Thus, only the efficiency of part of cycles 6–1–2–7 and 6–8–9–7 needs to be investigated.

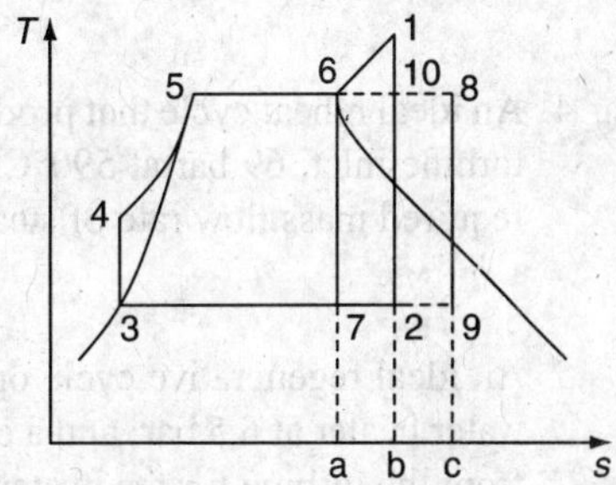

Considering cycle 6–1–2–7:

Work done = area 6–1–2–7 and heat input = area 6–1–b–a

$$\text{Efficiency, } \eta_1 = \frac{\text{area } 6-1-2-7}{\text{area } 6-1-b-a} \tag{1}$$

Considering cycle 6–8–9–7:

Work done = area 6–8–9–7 and heat input = area 6–8–c–a = area 6–1–b–a.

This is because both the cycles are supplied with the same amount of heat.

$$\text{Efficiency, } \eta_2 = \frac{\text{area } 6-8-9-7}{\text{area } 6-1-b-a} \tag{2}$$

Area 6–8–c–a = area 6–1–b–a.

Since, area 6–10–b–a is common to both, we can write:

Area 6–1–10 = area 10–8–c–b

or area 6–1–10 > area 10–8–9–2.

Adding area 6–10–2–7 to both sides, we get

Area 6–10–2–7 is greater than area 6–8–9–7.

Hence, from equations (1) and (2) we conclude that $\eta_1 > \eta_2$.

Practice Problems

1. Dry saturated steam is supplied to a turbine at 14 bar and the condenser pressure is 0.04 bar. Find Rankine efficiency (i) neglecting pump work and (ii) taking pump work into account.

   [(i) 31.56%; (ii) 31.52%]

2. A power plant operating on the Rankine cycle has steam entering the turbine at 38 bar and 425°C. If the turbine output is equivalent to 100 kW, determine the efficiency of the cycle and the mass flow rate of steam for condenser pressures of (i) 0.07 bar and (ii) 0.035 bar.

   [(i) 0.3617, 319.6 kg/hr; (ii) 0.3843, 296 kg/hr]

3. A Rankine cycle has an exhaust pressure of 0.07 bar. Determine the moisture content leaving the turbine and the thermal efficiency of a cycle which has a turbine inlet temperature of 540°C and a turbine inlet pressure of (i) 110 bar and (ii) 55 bar.

[(i) 0.825, 0.4275; (ii) 0.848, 0.4]

4. An ideal reheat cycle that produces 125000 kW of turbine power has the following given conditions: turbine inlet, 69 bar at 590°C; reheat, 14 bar to 540°C; condenser pressure 0.07 bar. Compute the required mass flow rate of steam in kg/hr and the cycle efficiency.

(262687 kg/hr, 0.4337)

5. An ideal regenerative cycle operates with throttle conditions at 27 bar, 370°C; a single open feed-water heater at 5.5 bar; and a condenser pressure of 0.07 bar. Calculate (i) the fraction of steam bled from the turbine for feedwater heating and (ii) the thermal efficiency of the cycle.

[(i) 0.186, (ii) 0.363]

Objective Type Questions

1. The lifeblood of any prime mover is

(i) air (ii) fuel
(iii) lubricant (iv) water

2. The efficiency of a Carnot cycle is the same between 750 K and 300 K and between X and 750 K (sink temperature), then the source temperature X, in Kelvin (K), is

(i) 2250 (ii) 1875
(iii) 1500 (iv) 1050

3. Which one of the following changes/sets of the changes in the source and sink temperatures (T_1 and T_2) of a reversible engine will result in the maximum improvement in efficiency?

(i) $T_1 + \Delta T$ (ii) $T_2 - \Delta T$
(iii) $(T_1 + \Delta T)$ and $(T_2 - \Delta T)$ (iv) $(T_1 - \Delta T)$ and $(T_2 + \Delta T)$

4. The efficiency of a reversible heat engine cyclic process undergone by a substance shown in the following *T-s* diagram is

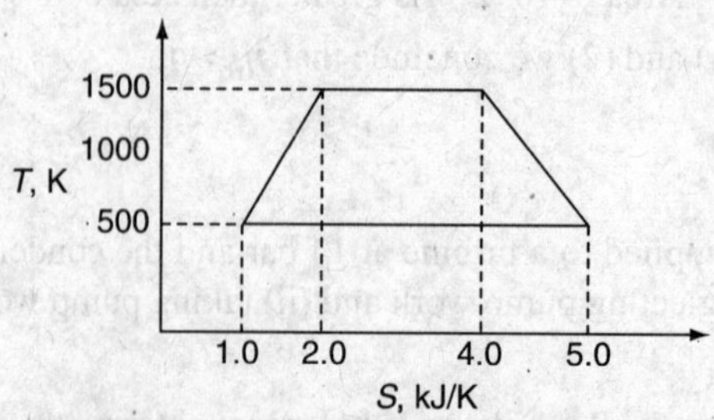

(i) 0.8 (ii) 0.6
(iii) 0.5 (iv) 0.4

5. Who demonstrated that it is more economical to use the high temperature heat produced by combustion of fuel in a heat engine and then to use the work so developed to pump heat from outdoors to a house than the direct heating of the house by the fuel?
   (i) Carnot
   (ii) Clausius
   (iii) Joule
   (iv) Kelvin
6. The heat interactions during the reversible processes '12' and '23' shown in the following *T-s* diagram, in kJ/kg, is

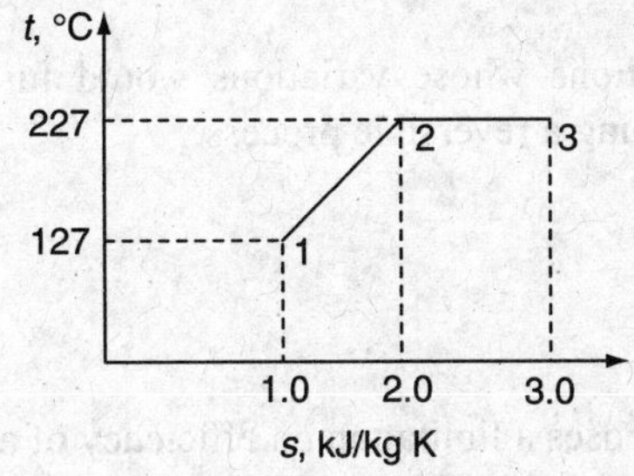

   (i) 850
   (ii) 950
   (iii) 1050
   (iv) 1150
7. A thermoelectric engine, which consists of two dissimilar electric conductors connected at two junctions and maintained at different temperatures, converts
   (i) electric energy into heat energy
   (ii) heat energy into electric energy
   (iii) mechanical work into electric energy
   (iv) electric energy into mechanical work
8. Which engine/machine violates the second law of thermodynamics?
   (i) Perpetual motion machine of the first kind
   (ii) Perpetual motion machine of the second kind
   (iii) Perpetual motion machine of the third kind
   (iv) Wankel engine
9. In which machine/engine, friction is inevitable?
   (i) Perpetual motion machine of the first kind
   (ii) Perpetual motion machine of the second kind
   (iii) Perpetual motion machine of the third kind
   (iv) Wankel engine
10. A reversible heat engine rejects 50 percent of the heat supplied during a cycle of operation. If this engine is reversed and operates as a heat pump, then its coefficient of performance shall be
    (i) 1.0
    (ii) 1.5
    (iii) 2.0
    (iv) 2.5
11. Which change is more effective in increasing the COP of a heat pump operating between the maximum temperature T_1 and the minimum temperature T_2?
    (i) $T_1 + \Delta T$ when T_2 is held constant
    (ii) $T_1 - \Delta T$ when T_2 is held constant
    (iii) $T_2 + \Delta T$ when T_1 is held constant
    (iv) $T_2 - \Delta T$ when T_1 is held constant

12. Assertion (A): A perpetual motion machine of third kind(PMM III) as compared to PMM I and PMM II is not a valuable contrivance for the thermodynamic cycle analysis.

    Reason (R): A PMM III produces no work output.

    (i) Both A and R are true and R is the correct explanation of A
    (ii) Both A and R are true and R is not the correct explanation of A
    (iii) A is true but R is false
    (iv) A is false but R is true

13. Identify the property combinations whose variations would furnish the heat interaction for a thermodynamic system undergoing a reversible process.

    (i) Pressure and volume
    (ii) Pressure and enthalpy
    (iii) Temperature and volume
    (iv) Temperature and entropy

14. Which thermodynamics law imposes a limitation on efficiency of all gadgets where thermal energy is encountered?

    (i) Zeroth law
    (ii) First law
    (iii) Second law
    (iv) Third law

15. Who originated the use of cyclic processes in thermodynamic analysis?

    (i) Sadi Carnot
    (ii) Max Planck
    (iii) James Joule
    (iv) Rudolf Clausius

16. Four hypothetical heat engine cycles working between temperatures T_1 and T_2 are as follows:

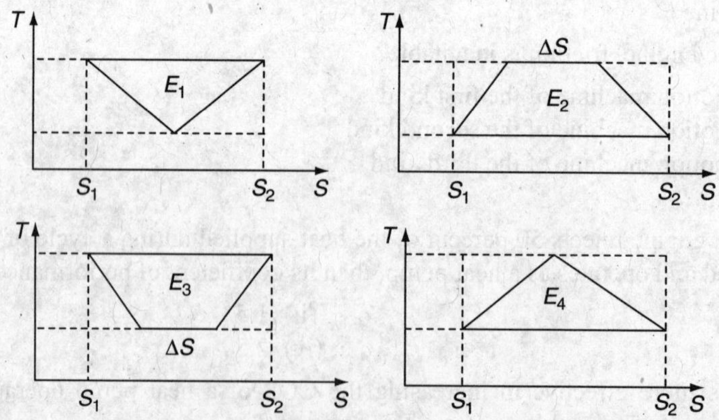

The engine that has the highest efficiency is

(i) E_1 (ii) E_2
(iii) E_3 (iv) E_4

17. Which one of the following is not true for the following Carnot cycle?

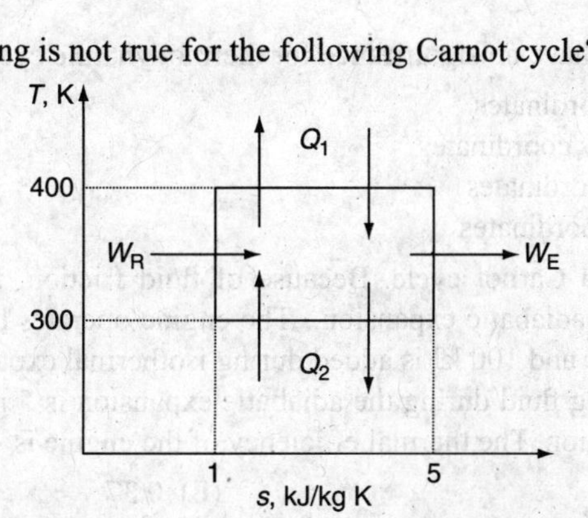

(i) $Q_1 = 1600$ kJ/kg
(ii) $Q_2 = 1200$ kJ/kg
(iii) Efficiency = 0.25
(iv) $(COP)_R = 5$

18. The thermal efficiency of the hypothetical cycle shown in the following is

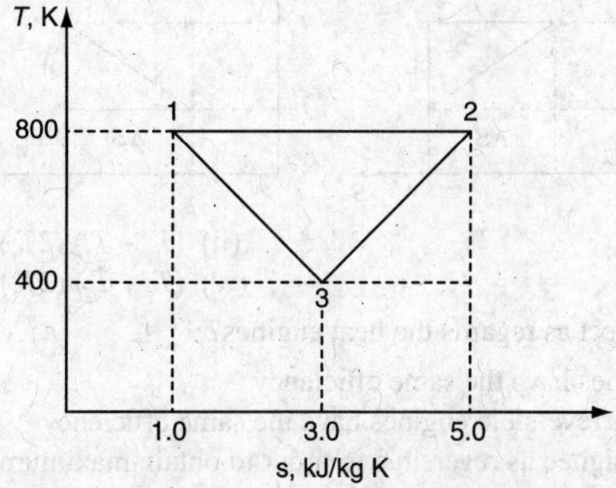

(i) 0.25
(ii) 0.35
(iii) 0.45
(iv) 0.55

19. Which hypothetical heat engine cycle has got the maximum efficiency?

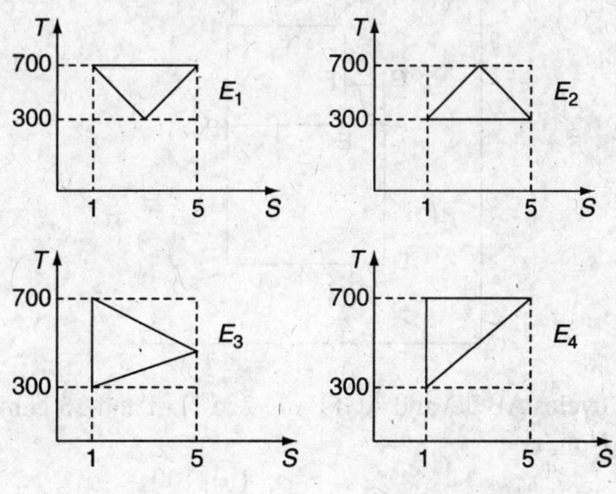

(i) E_1
(ii) E_2
(iii) E_3
(iv) E_4

20. Areas on which diagram have no significance for thermodynamic cycle analysis?
    (i) Pressure-volume coordinates
    (ii) Temperature-entropy coordinates
    (iii) Enthalpy-entropy coordinates
    (iv) Pressure-enthalpy coordinates
21. An engine operates on a Carnot cycle. Because of fluid friction, the entropy of working fluid increases slightly during adiabatic expansion. The engine operates between constant temperature limits of 227°C and 27°C and 100 kJ is added during isothermal expansion per cycle. The increase in entropy for the working fluid during the adiabatic expansion is 5 percent of the entropy change for the isothermal expansion. The thermal efficiency of the engine is
    (i) 0.38 (ii) 0.37
    (iii) 0.36 (iv) 0.35
22. The ratio of the efficiencies (η_I/η_{II}) for the two cycles shown in the following *T-s* diagram is

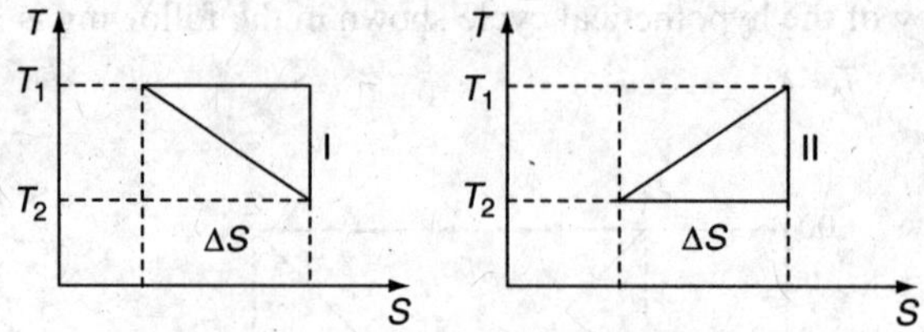

    (i) $(T_1 - T_2)/(2T_2)$ (ii) $(T_1 - T_2)/(2T_1)$
    (iii) $(T_1 + T_2)/(2T_2)$ (iv) $(T_1 + T_2)/(2T_1)$
23. Which statement is correct as regards the heat engines?
    (i) All reversible engines have the same efficiency
    (ii) All reversible and irreversible engines have the same efficiency
    (iii) All engines are designed as reversible in order to obtain maximum efficiency
    (iv) Irreversible engines have maximum efficiency
24. Three Carnot cycles, which are squares, are shown in the following diagram.

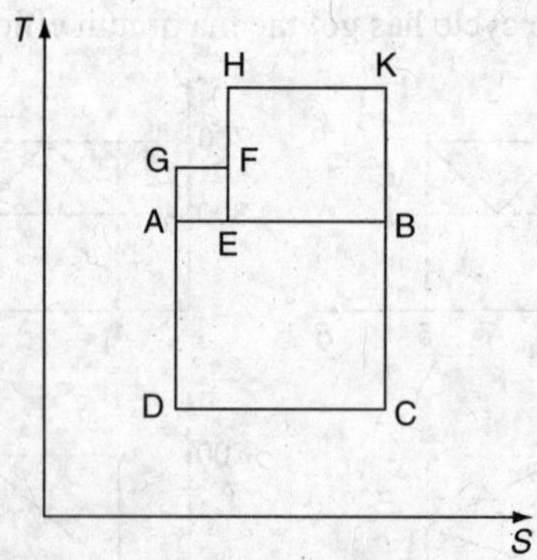

If the work outputs of cycles ABCD and AGFE are 225 N.m and 25 N.m, then the work output of the cycle HKBE, in N.m, is
    (i) 50 (ii) 100
    (iii) 125 (iv) 150

25. Which one of the modifications in a Rankine cycle would enhance its efficiency to approach that of Carnot cycle?
   (i) Incomplete expansion of steam
   (ii) Reheating of steam
   (iii) Partial condensation of steam
   (iv) Regenerative feed heating by steam
26. In a thermal power plant working on the Rankine cycle, the object of reheating the steam is
   (i) to reduce the irreversibility
   (ii) to reduce the heat loss
   (iii) to increase the output
   (iv) to prevent the steam leakage
27. Two ideal cycles ABCD and DEC are represented in the following figure together with the maximum and minimum temperatures.

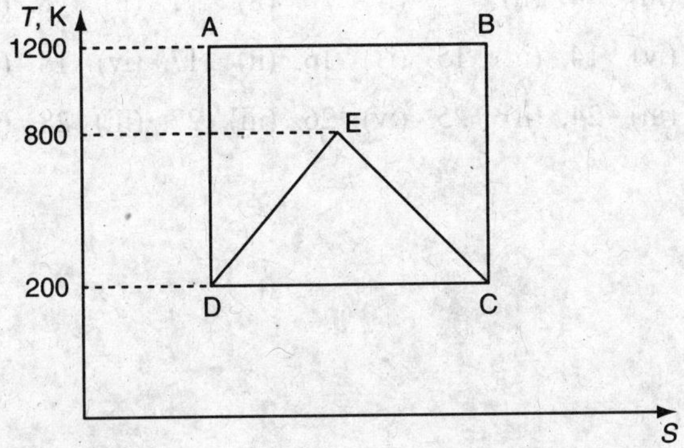

If the work output of the cycle ABCD is 4000 N.m, then the work output of the cycle DEC, in N.m, is
   (i) 1600 (ii) 1400
   (iii) 1200 (iv) 1000
28. A Carnot engine operating between reservoirs at T_1 and T_2 drives a Carnot refrigerator removing heat Q_R from a reservoir at T_R and rejecting heat at temperature T_2.
Assuming that $T_1 > T_2 > T_R$, the minimum heat that is taken away from the reservoir at T_1 to remove heat Q_R from reservoir at T_R is

   (i) $Q_R\left(\frac{T_2}{T_R}\right)\left(\frac{T_2 - T_R}{T_1 - T_2}\right)$ (ii) $Q_R\left(\frac{T_1}{T_R}\right)\left(\frac{T_2 - T_R}{T_1 - T_2}\right)$

   (iii) $Q_R\left(\frac{T_1}{T_R}\right)\left(\frac{T_1 - T_2}{T_2 - T_R}\right)$ (iv) $Q_R\left(\frac{T_2}{T_R}\right)\left(\frac{T_1 - T_2}{T_2 - T_R}\right)$

29. Two bodies of thermal capacities C_1 and C_2 at temperatures T_1 and T_2 are brought to the same temperature T by means of a reversible heat engine. Which one of the following relationships is true for the final temperature?

(i) $T = \dfrac{(C_1T_1 + C_2T_2)}{(C_1 + C_2)}$

(ii) $T = \dfrac{\sqrt{C_1T_1C_2T_2}}{(C_1 + C_2)}$

(iii) $\ell nT = \dfrac{(C_1\,\ell nT_1 + C_2\,\ell nT_2)}{(C_1 + C_2)}$

(iv) $\dfrac{(C_1 + C_2)}{T} = \dfrac{C_1}{T_1} + \dfrac{C_2}{T_2}$

30. Which one of the following gadgets does not require the applications of the Carnot's theorem/principle when used to operate a car?

(i) MHD generator
(ii) Fuel cell
(iii) Gas turbine
(iv) Petrol engine

Answers

1. (iii) 2. (ii) 3. (iii) 4. (ii) 5. (iv) 6. (ii) 7. (ii) 8. (ii) 9. (iii) 10. (iii)
11. (iii) 12. (i) 13. (iv) 14. (iii) 15. (i) 16. (ii) 17. (iv) 18. (i) 19. (ii) 20. (iii)
21. (ii) 22. (iv) 23. (iii) 24. (ii) 25. (iv) 26. (iii) 27. (iii) 28. (ii) 29. (iii) 30. (ii)

7

Gas Power Cycles

1. Derive an expression for the air-standard efficiency of Otto cycle in terms of compression ratio. Sketch the cycle on *p-v* and *T-s* coordinates.

Ans: The air-standard Otto cycle is an ideal cycle that approximates a spark-ignition internal combustion engine. This cycle is shown on the *p-v* and *T-s* diagrams (Figure 7.1). Process 1–2 is an isentropic compression of the air as the piston moves from crank-end dead centre to head-end dead centre. Heat is then added at constant volume during the process 2–3 while the piston is momentarily at rest at head-end dead centre. This process corresponds to the ignition of the fuel–air mixture by the spark and subsequent burning in the actual engines. Process 3–4 is an isentropic expansion and process 4–1 is the rejection of heat from gas while the piston is at crank-end dead centre.

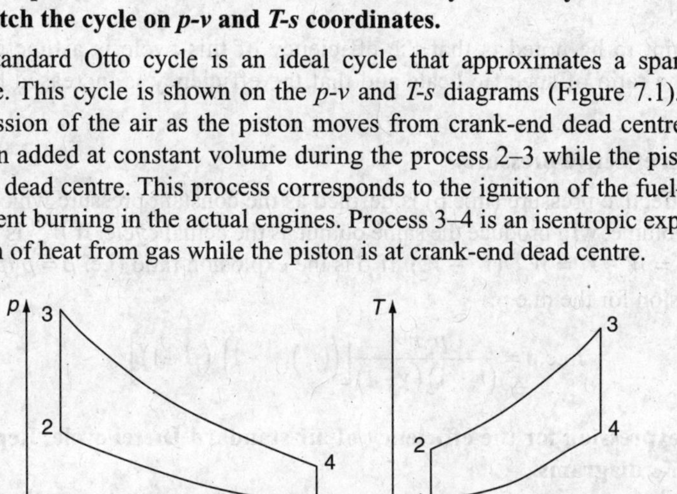

Figure 7.1 Otto Cycle

For unit mass of air and constant specific heats, the following relations are obtained.

(i) For the process 1–2 (isentropic compression)

$$\frac{T_2}{T_1} = \left(\frac{v_1}{v_2}\right)^{\gamma-1}, \text{ i.e. } T_2 = T_1 (r_c)^{\gamma-1}, \text{ where } r_c \text{ is the compression ratio} = \frac{V_s + V_c}{V_c},$$

V_s is the swept/stroke volume ($= V_1 - V_2$) and V_c is the clearance volume ($= V_2$).

(ii) For the process 2–3 (heat addition at constant volume)

$$Q_s = C_v (T_3 - T_2)$$

(iii) For the process 3–4 (isentropic expansion)

$$\frac{T_4}{T_3} = \left(\frac{v_3}{v_4}\right)^{\gamma-1}, \text{ i.e. } T_4 = \frac{T_3}{(r_c)^{\gamma-1}}, \text{ where } r_c = \frac{v_4}{v_3} = \frac{v_1}{v_2}. \text{ Hence, we can show that}$$

$$\frac{T_2}{T_1} = \frac{T_3}{T_4} \text{ or } \frac{T_4}{T_1} = \frac{T_3}{T_2}.$$

(iv) For process 4–1 (heat rejection at constant volume)

$$Q_r = C_v\,(T_4 - T_1)$$

$$\text{Thermal efficiency} = \frac{Q_s - Q_r}{Q_s} = 1 - \frac{Q_r}{Q_s} = 1 - \frac{(T_4 - T_1)}{(T_3 - T_2)} = 1 - \frac{T_1}{T_2}\frac{\left(\frac{T_4}{T_1} - 1\right)}{\left(\frac{T_3}{T_2} - 1\right)} = 1 - \frac{T_1}{T_2}$$

Hence, $\eta = 1 - \frac{1}{(r_c)^{\gamma-1}}$, where r_c is compression ratio (also expansion ratio for this cycle).

The important point to be noted is that the efficiency of this cycle is a function only of compression ratio and the ratio of specific heats and that the efficiency is increased by increasing the compression ratio.

2. Define mean effective pressure.

Ans: The mean effective pressure (m.e.p) is defined as the constant pressure which, acting over the whole swept/stroke volume, will produce the same output as the actual cycle. If W_{net} is the net output per cycle, then the m.e.p = $W_{net}/V_s = W_{net}/(V_1 - V_2)$. If β is the explosion ratio (i.e. $\beta = p_3/p_2$), we can obtain the following expression for the m.e.p.

$$\text{m.e.p} = \frac{p_1\, r_c}{(r_c - 1)(\gamma - 1)}\left[\left((r_c)^{\gamma-1} - 1\right)(\beta - 1)\right]$$

3. Derive the expression for the efficiency of air-standard Diesel cycle. Represent the cycle on *p-v* and *T-s* diagrams.

Ans: The air-standard Diesel cycle is the ideal cycle for diesel engines, which is also called the compression-ignition engine. This cycle is shown on the *p-v* and *T-s* diagrams (Figure 7.2). In this cycle the heat is transferred to the working fluid at constant pressure. This process corresponds to the injection and burning of the fuel in the actual engine. Since the gas is expanding during the heat addition in the air-standard cycle, the heat transfer must be just sufficient to maintain constant pressure. When state 3 is reached the heat addition stops and the gas undergoes an isentropic expansion 3–4, until the piston reaches crank-end dead centre. As in the Otto cycle, a constant volume rejection of heat at crank-end dead centre replaces the exhaust and intake processes of the actual engines.

The following relations are needed in the derivation of thermal efficiency expression:

compression ratio, $r_c = \frac{v_1}{v_2}$; expansion ratio, $r_e = \frac{v_4}{v_3}$ and cut-off ratio, $\rho = \frac{v_3}{v_2}$

From the above relations we can show $r_c = r_e\,\rho$.

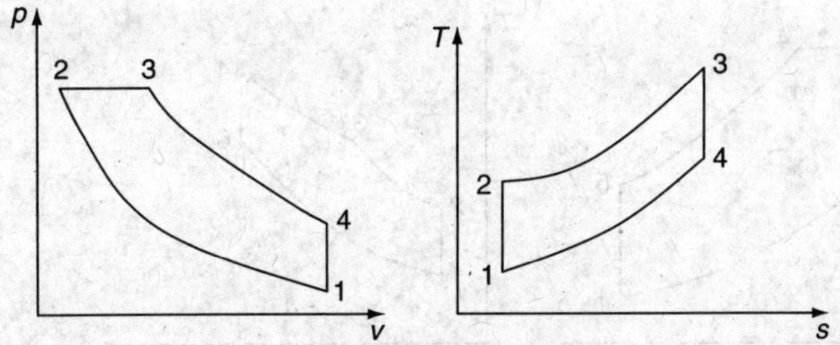

Figure 7.2 Diesel Cycle

(i) For the isentropic compression 1–2

$$\frac{T_2}{T_1} = (r_c)^{\gamma-1} \text{ i.e. } T_2 = T_1 (r_c)^{\gamma-1}$$

(ii) For heat addition at constant pressure 2–3

$$Q_s = C_p (T_3 - T_2) \text{ and } \frac{T_3}{T_2} = \frac{v_3}{v_2} = \rho \text{ i.e. } T_3 = T_2\rho = T_1 (r_c)^{\gamma-1} \rho$$

(iii) For the isentropic expansion 3–4

$$\frac{T_4}{T_3} = \left(\frac{v_3}{v_4}\right)^{\gamma-1} = \left(\frac{1}{r_e}\right)^{\gamma-1} \text{ i.e. } T_4 = \frac{T_3}{(r_e)^{\gamma-1}} = \frac{T_1 (r_c)^{\gamma-1} \rho}{r_e^{\gamma-1}} = T_1 \rho^{\gamma}$$

Heat addition, $Q_s = C_p (T_3 - T_2) = C_p (T_1(r_c)^{\gamma-1}\rho - T_1(r_c)^{\gamma-1}) = C_p T_1 (r_c)^{\gamma-1}(\rho - 1)$

Heat rejection, $Q_r = C_v (T_4 - T_1) = C_v (T_1\rho^{\gamma} - T_1) = C_v T_1 (\rho^{\gamma} - 1)$

$$\text{Thermal efficiency, } \eta = \frac{Q_s - Q_r}{Q_s} = 1 - \frac{Q_r}{Q_s} = 1 - \frac{C_v T_1 (\rho^{\gamma} - 1)}{C_p T_1 (r_c)^{\gamma-1} (\rho - 1)}$$

$$\text{or } \eta = 1 - \frac{1}{(r_c)^{\gamma-1}} \left[\frac{1}{\gamma} \frac{(\rho^{\gamma} - 1)}{(\rho - 1)}\right]$$

If W_{net} is the net output per cycle, then the m.e.p $= W_{net}/V_s = W_{net}/(v_2 - v_1)$. We can obtain the following expression for the m.e.p.

$$\text{m.e.p} = \frac{p_1 r_c}{(r_c - 1)(\gamma - 1)} \left\{(r_c)^{\gamma-1} \gamma (\rho - 1) - ((\rho)^{\gamma} - 1)\right\}$$

4. Derive an expression for the air-standard efficiency of dual-combustion cycle.

Ans: The air-standard Diesel cycle does not simulate exactly the pressure–volume variation in an actual compression-ignition engine, where the fuel injection is started before the end of compression stroke. A closer approximation is the limited pressure cycle or the dual cycle in which some part of heat is added to air at constant volume and the remainder at constant pressure. The cycle is shown on the *p-v* and *T-s* diagrams (Figure 7.3).

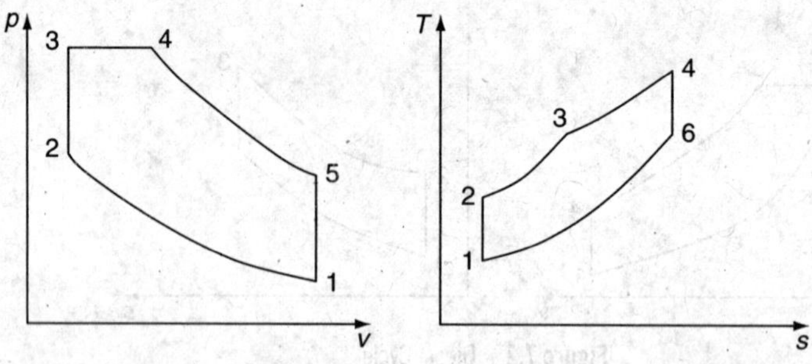

Figure 7.3 Dual Cycle

The following relations are required in the derivation of thermal efficiency expression:

Compression ratio, $r_c = \frac{v_1}{v_2}$; expansion ratio, $r_e = \frac{v_5}{v_4}$;

Cut-off ratio, $\rho = \frac{v_4}{v_3}$; explosion ration, $\beta = \frac{p_3}{p_2}$

We can derive the following relations for temperatures at cardinal points

$$\frac{T_2}{T_1} = \left(\frac{v_1}{v_2}\right)^{\gamma-1} \quad \text{i.e. } T_2 = T_1 (r_c)^{\gamma-1}$$

$$\frac{T_3}{T_2} = \frac{p_3}{p_2} = \beta \quad \text{i.e. } T_3 = T_2\beta = T_1 (r_c)^{\gamma-1}\ \beta$$

$$\frac{T_4}{T_3} = \frac{v_4}{v_3} = \rho \quad \text{i.e. } T_4 = T_3\ \rho = T_1 (r_c)^{\gamma-1}\ \beta\ \rho$$

$$\frac{T_5}{T_4} = \left(\frac{v_4}{v_5}\right)^{\gamma-1} = \left(\frac{v_4}{v_3}\right)^{\gamma-1}\left(\frac{v_3}{v_5}\right)^{\gamma-1} = \frac{\rho^{\gamma-1}}{(r_c)^{\gamma-1}} \quad \text{i.e. } T_5 = T_4\frac{\rho^{\gamma-1}}{(r_c)^{\gamma-1}} = T_1\ \beta\ \rho^{\gamma}$$

Heat addition, $Q_s = C_v (T_3 - T_2) + C_p (T_4 - T_3)$

Heat rejection, $Q_r = C_v (T_5 - T_1)$

Hence, $\eta = 1 - \frac{Q_r}{Q_s} = 1 - \frac{C_v (T_5 - T_1)}{C_p (T_4 - T_3) + C_v (T_3 - T_2)}$

Substituting for temperatures T_5, T_4, T_3 and T_2 in terms of T_1, r_c, ρ and β, we get

$$\eta = 1 - \frac{(\beta\rho^{\gamma} - 1)}{(r_c)^{\gamma-1}\{\gamma\beta(\rho-1) + (\beta-1)\}}$$

By putting $\rho = 1$ in the above expression gives $\eta = 1 - \frac{1}{(r_c)^{\gamma-1}}$ which is the Otto cycle efficiency.

By putting $\beta = 1$ in the above expression gives $\eta = 1 - \frac{(\rho^{\gamma} - 1)}{(r_c)^{\gamma-1}\gamma(\rho - 1)}$ which is the Diesel cycle efficiency.

5. Draw the flow, *p-v* and *T-s* diagrams of the simple gas turbine unit working on Brayton cycle. Derive the expression for the efficiency.

Ans: A simple air-standard Brayton cycle is shown in Figure 7.4 together with the state diagrams on *p-v* and *T-s* coordinates. Air is first drawn in and compressed adiabatically during the process 1–2. The air then enters the combustion chamber where fuel is injected and burned essentially at constant pressure during the process 2–3. The products of combustion expand in the turbine to the ambient pressure during the process 3–4 and exhausted to the surroundings.

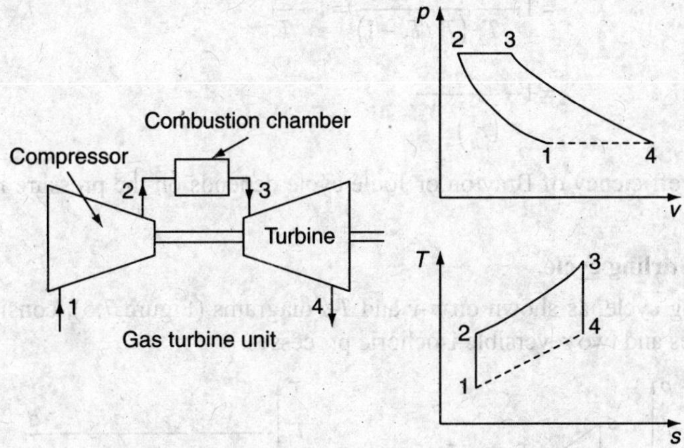

Figure 7.4 Brayton Cycle that Simulate Gas-Turbine Unit

The steady flow energy equation is

$$\dot{Q} + \dot{m}\left[h_1 + \frac{V_1^2}{2} + g Z_1\right] = \dot{W}_S + \dot{m}\left[h_2 + \frac{V_2^2}{2} + g Z_2\right].$$

When applied to the compressor with negligible changes in kinetic and potential energy, we get, $0 + 1[h_1 + 0 + 0] = \dot{W}_c + 1[h_2 + 0 + 0]$, that is, $W_c = -C_p(T_2 - T_1)$.

Similarly, we can get the work done by the turbine as $W_t = C_p(T_3 - T_4)$. Heat supplied, $Q_s = C_p(T_3 - T_2)$

For the isentropic compression process 1–2,

$$\frac{T_2}{T_1} = \left(\frac{p_2}{p_1}\right)^{\gamma-1/\gamma} \text{ i.e. } T_2 = T_1\left(r_p\right)^{\gamma-1/\gamma}$$

Similarly for the isentropic expansion process 3–4, $\frac{T_3}{T_4} = \left(\frac{p_3}{p_4}\right)^{\gamma-1/\gamma}$ i.e. $T_3 = T_4\left(r_p\right)^{\gamma-1/\gamma}$

From the above relations we can show

$$\frac{T_2}{T_1}=\frac{T_3}{T_4} \quad \text{or} \quad \frac{T_4}{T_1}=\frac{T_3}{T_2}$$

and using this expression, the efficiency of the gas turbine unit can be simplified as

$$\eta=\frac{W_t-W_c}{Q_s}=\frac{C_p(T_3-T_4)-C_p(T_2-T_1)}{C_p(T_3-T_2)}$$

$$=\frac{C_p(T_3-T_2)-C_p(T_4-T_1)}{C_p(T_3-T_2)}=1-\frac{(T_4-T_1)}{(T_3-T_2)}$$

$$=1-\frac{T_1}{T_2}\frac{(T_4/T_1-1)}{(T_3/T_2-1)}=1-\frac{T_1}{T_2}$$

$$=1-\frac{1}{(r_p)^{\gamma-1/\gamma}}$$

The air-standard efficiency of Brayton or Joule cycle depends on the pressure ratio, r_p and the ratio of specific heats, γ.

6. Describe Stirling cycle.

Ans: The Stirling cycle, as shown on *p-v* and *T-s* diagrams (Figure 7.5), consists of two reversible isothermal processes and two reversible isochoric processes.

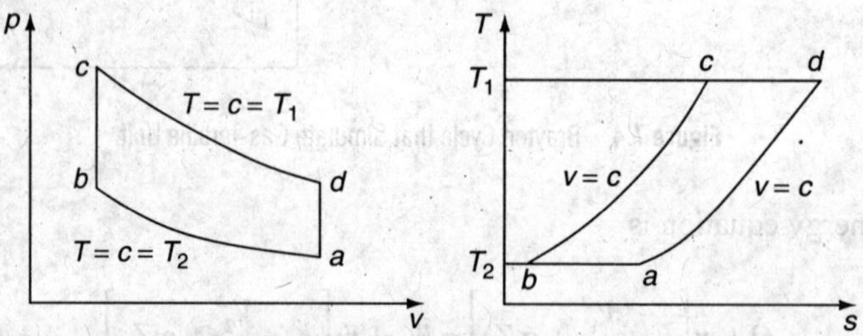

Figure 7.5 Stirling Cycle

For 1 kg of ideal gas, the heat and work interactions during the processes *c-d*, *d-a*, *a-b* and *b-c* are as given below:

$${}_cQ_d={}_cW_d=p_cv_c\,\ell n\left(\frac{v_d}{v_c}\right)=RT_1\,\ell n\left(\frac{v_d}{v_c}\right)$$

$${}_dQ_a=C_v(T_2-T_1),\quad {}_dW_a=0$$

$${}_aQ_b={}_aW_b=p_av_a\,\ell n\left(\frac{v_b}{v_a}\right)=-RT_2\,\ell n\left(\frac{v_a}{v_b}\right)$$

$$=-RT_2\,\ell n\left(\frac{v_d}{v_c}\right)$$

$${}_bQ_c=C_v(T_1-T_2),\quad {}_bW_c=0$$

If a regenerative arrangement is used such that the heat rejection during the process *d-a* is equal to the heat absorbed during the process *b-c*, then ${}_dQ_a = {}_bQ_c$. Then the net work output will be equal to

$$\Sigma Q = RT_1\, \ell n\left(\frac{v_d}{v_c}\right) - RT_2\, \ell n\left(\frac{v_d}{v_c}\right) \text{ and the heat supplied, } {}_cQ_d = RT_1\, \ell n\left(\frac{v_d}{v_c}\right).$$

Hence, the thermal efficiency, $\eta = \Sigma Q/{}_cQ_d$ and on simplification we will get $\eta = \dfrac{T_1 - T_2}{T_1}$, which is the same as that of Carnot cycle operating between the same temperatures. So the regenerative Stirling cycle has the same efficiency as that of the Carnot cycle. The cycle is identical with the Carnot cycle except that the adiabatic operations are replaced by constant volume operations.

7. Describe the Ericsson cycle.

Ans: The Ericsson cycle, as shown on *p-v* and *T-s* coordinates (Figure 7.6), is made up of two reversible isothermal process and two reversible isobaric processes.

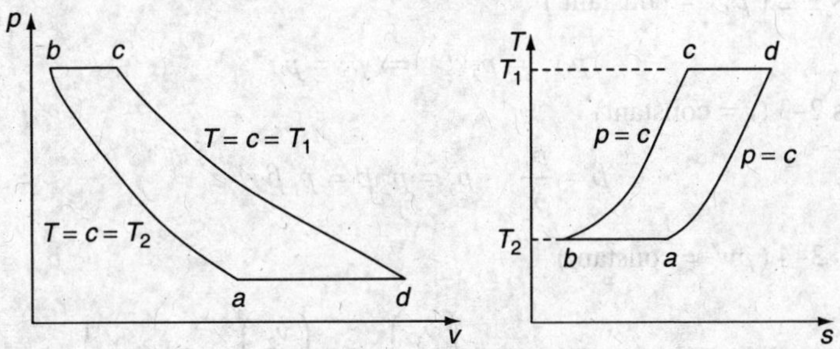

Figure 7.6 Ericsson Cycle

For 1 kg of ideal gas, the heat and work interactions during the processes *a-b*, *b-c*, *c-d* and *d-a* are as given below:

$${}_aQ_b = {}_aW_b = RT_2\, \ell n\left(\frac{p_a}{p_b}\right) = -RT_2\, \ell n\left(\frac{p_b}{p_a}\right)$$

$${}_bQ_c = C_p(T_1 - T_2),\ {}_bW_c = p_c(v_c - v_b) = R(T_1 - T_2)$$

$${}_cQ_d = {}_cW_d = RT_1\, \ell n\left(\frac{p_c}{p_d}\right) = RT_1\, \ell n\left(\frac{p_b}{p_a}\right)$$

$${}_dQ_a = C_p(T_2 - T_1),\quad {}_dW_a = p_d(v_a - v_d) = R(T_2 - T_1)$$

For ideal regeneration, ${}_bQ_c = {}_dQ_a$ and the net work output will be equal to ΣQ.

That is, $\Sigma Q = RT_1\, \ell n\left(\dfrac{p_b}{p_a}\right) - RT_2\, \ell n\left(\dfrac{p_b}{p_a}\right)$ and $Q_s = RT_1\, \ell n\left(\dfrac{p_b}{p_a}\right)$

Hence, the thermal efficiency, $\eta = \Sigma Q/Q_s$ and on simplification we will get $\eta = \dfrac{T_1 - T_2}{T_1}$, which is the same as that of Carnot cycle operating between the same temperatures. So the regenerative Ericsson cycle has the same efficiency as that of the Carnot cycle. The cycle is identical with the Carnot cycle except that the adiabatic operations are replaced by constant pressure operations.

Solved Problems

1. Derive an expression for the mean effective pressure (m.e.p) of Otto cycle in terms of the initial pressure at the start of compression, compression ratio, explosion ratio and the ratio of specific heats.

For the Otto cycle shown in the *p-v* diagram,

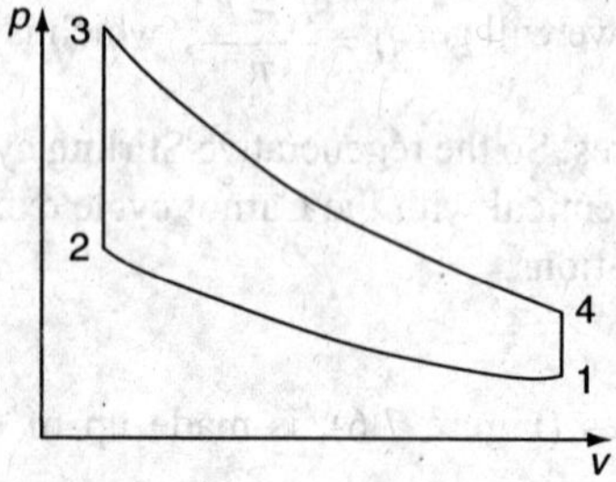

Compression ratio, $r_c = \dfrac{v_1}{v_2}$

Expansion ratio, $r_e = \dfrac{v_4}{v_3}$

For Otto cycle, $r_c = r_e$

Explosion ratio, $\beta = \dfrac{p_3}{p_2}$

For the process 1–2 (pv^γ = constant)

$$p_1 v_1^\gamma = p_2 v_2^\gamma \quad \Rightarrow p_2 = p_1 r_c^\gamma$$

For the process 2–3 (v = constant)

$$\beta = \frac{p_3}{p_2}; \quad p_3 = p_2\,\beta = p_1\,\beta\, r_c^\gamma$$

For the process 3–4 (pv^γ = constant)

$$p_3 v_3^\gamma = p_4 v_4^\gamma \quad \Rightarrow p_4 = p_3\left(\frac{v_3}{v_4}\right)^\gamma = p_3\left(\frac{v_2}{v_1}\right)^\gamma = p_1\,\beta\, r_c^\gamma \frac{1}{r_c^\gamma} = p_1\,\beta.$$

$$\text{Work done/cycle} = \frac{(p_3 v_3 - p_4 v_4)}{(\gamma-1)} - \frac{(p_2 v_2 - p_1 v_1)}{(\gamma-1)}$$

$$= \frac{1}{(\gamma-1)}\left[p_1\,\beta r_c^\gamma \frac{v_1}{r_c} - p_1\,\beta\, v_1\right] - \frac{1}{(\gamma-1)}\left[p_1\, r_c^\gamma \frac{v_1}{r_c} - p_1\, v_1\right]$$

$$= \frac{p_1\, v_1}{(\gamma-1)}\left[\beta r_c^{\gamma-1} - \beta - r_c^{\gamma-1} + 1\right]$$

$$= \frac{p_1\, v_1}{(\gamma-1)}\left[\left(r_c^{\gamma-1} - 1\right)(\beta - 1)\right]$$

$$\text{Stroke/swept volume, } v_s = v_1 - v_2 = v_1 - \frac{v_1}{r_c} = \frac{v_1\,(r_c - 1)}{r_c}$$

$$\text{m.e.p} = \frac{\text{work done /cycle}}{\text{stroke volume}} = \frac{p_1\, v_1}{(\gamma-1)}\left[\left(r_c^{\gamma-1} - 1\right)(\beta - 1)\right]\frac{r_c}{v_1\,(r_c - 1)}$$

$$= \frac{p_1\, r_c}{(\gamma-1)(r_c-1)}\left[\left(r_c^{\gamma-1} - 1\right)(\beta - 1)\right]$$

2. An engine works on Otto cycle. The pressure, temperature and volume of air at the beginning of compression are 1 bar, 40°C and 2172 cm^3, respectively. The compression ratio is 6. The peak pressure is limited to 50 bar and compression and expansion follow the law $pv^{1.4}$ = constant. Find (i) the pressure and temperature at all points and (ii) the thermal efficiency.

Data given:

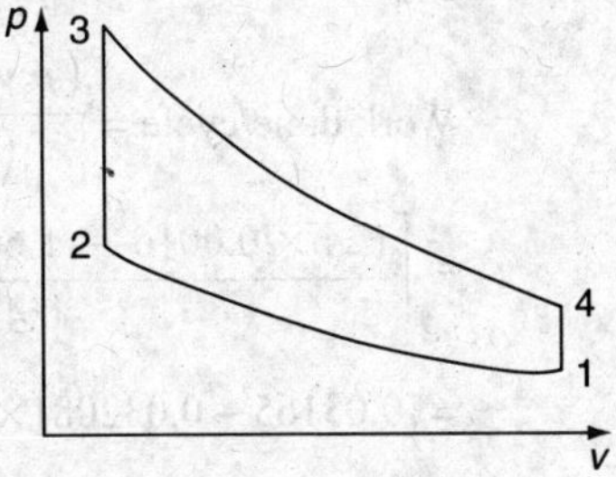

$p_1 = 1$ bar, $v_1 = 0.002172$ m^3, $T_1 = 313$ K

$r_c = 6$, $p_3 = 50$ bar

$p_2 = p_1\,(r_c)^{\gamma} = 1\,(6)^{1.4} = 12.286$ bar

$T_2 = T_1\,(r_c)^{\gamma-1} = 313\,(6)^{0.4} = 640.9$ K

$$\frac{p_3}{p_2} = \frac{T_3}{T_2};\quad T_3 = 640.9\frac{50}{12.286} = 2608.3\text{ K}$$

$$\frac{T_4}{T_3} = \left(\frac{v_3}{v_4}\right)^{\gamma-1};\quad T_4 = \frac{2608.3}{(6)^{0.4}} = 1273.8\text{ K}$$

$$\frac{p_4}{p_3} = \left(\frac{v_3}{v_4}\right)^{\gamma};\quad p_4 = \frac{50}{(6)^{1.4}} = 4.069\text{ bar}$$

$$\text{Thermal efficiency, } \eta = 1 - \frac{1}{(r_c)^{\gamma-1}} = 1 - \frac{1}{(6)^{0.4}} = 0.5116 \text{ or } 51.16\%.$$

3. An engine 200 mm bore and 300 mm stroke works on Otto cycle. The clearance volume is 0.0016 m^3. The initial pressure and temperature are 1 bar and 60°C. If the maximum pressure is limited to 24 bar, find the air-standard efficiency and the m.e.p.

Data given:

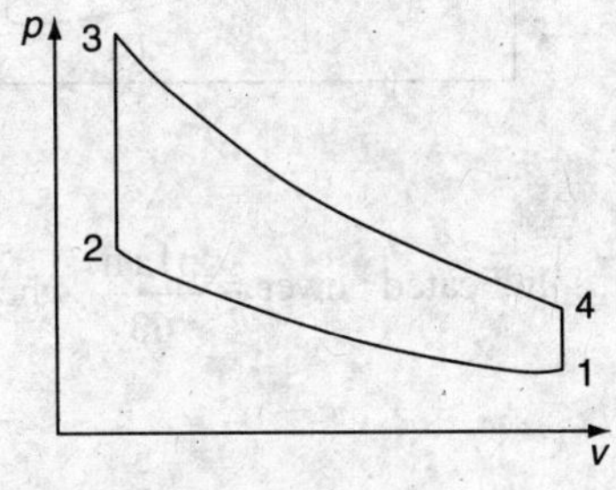

Stroke, $L = 0.3$ m,

Diameter, $D = 0.2$ m,

Clearance volume, v_c or $v_2 = 0.0016$ m^3,

$p_1 = 1$ bar,

$T_1 = 333$ K,

$p_3 = 24$ bar.

$$\text{Swept volume} = \frac{\pi}{4} \times (0.2)^2 \times (0.3) = 0.0094247\text{ m}^3$$

$v_1 = 0.0094247 + 0.0016 = 0.0110247$ m^3, $v_4 = v_1 = 0.0110247$ m^3, $v_3 = v_2 = 0.0016$ m^3

$$\text{Compression ratio, } r_c = \frac{v_1}{v_2} = \frac{0.0110247}{0.0016} = 6.89$$

$p_2 = p_1 (r_c)^{\gamma} = 1 (6.89)^{1.4} = 14.91$ bar

$T_2 = T_1 (r_c)^{\gamma-1} = 333 (6.89)^{0.4} = 720.6$ K

$$\frac{p_3}{p_2} = \frac{T_3}{T_2};\; T_3 = 720.6\frac{24}{14.91} = 1159.9\text{ K};\; \frac{p_4}{p_3} = \left(\frac{v_3}{v_4}\right)^{\gamma};\; p_4 = \frac{24}{(6.89)^{1.4}} = 1.609\text{ bar}$$

$$\text{Thermal efficiency, } \eta = 1 - \frac{1}{(r_c)^{\gamma-1}} = 1 - \frac{1}{(6.89)^{0.4}} = 0.5379 \text{ or } 53.79\%$$

$$\text{Work done/cycle} = \frac{(p_3v_3 - p_4v_4)}{(\gamma-1)} - \frac{(p_2v_2 - p_1v_1)}{(\gamma-1)}$$

$$= \left[\frac{(24 \times 0.0016 - 1.609 \times 0.0110247)}{0.4} - \frac{(14.91 \times 0.0016 - 1 \times 0.0110247)}{0.4}\right] \times 10^5 \text{ Nm}$$

$$= [0.05165 - 0.03208] \times 10^5 \text{ Nm} = 0.01957 \times 10^5 \text{ Nm}$$

$$\text{Mean effective pressure, m.e.p} = \frac{\text{Work done}}{v_s} = \frac{0.01957 \times 10^5}{0.0094247} = 2.0765 \times 10^5 \text{ N/m}^2$$

$$= 2.0765 \text{ bar.}$$

4. The diameter and stroke of a single cylinder two-stroke engine working on the constant volume cycle are 200 mm and 300 mm, respectively, with a clearance volume of 0.0028 m³. While running at 135 rpm, the fuel consumption is 8.8 m³/h and the indicated m.e.p is 5.2 bar. Find the air-standard efficiency, the indicated power and the indicated thermal efficiency. Assume the calorific value of the fuel as 16,350 kJ/m³.

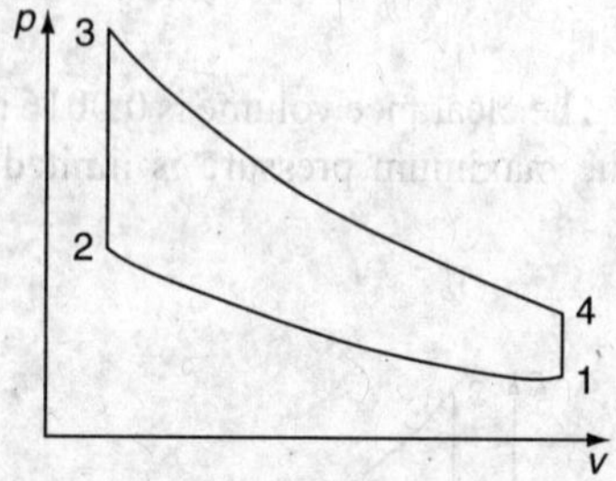

$$v_s = \frac{\pi}{4} \times 0.2^2 \times 0.3 = 0.0094247 \text{ m}^3$$

$$v_c = 0.0028 \text{ m}^3$$

$$r_c = \frac{0.0094247 + 0.0028}{0.0028} = 4.37$$

Air-standard efficiency,

$$\eta = 1 - \frac{1}{(r_c)^{\gamma-1}} = 1 - \frac{1}{(4.37)^{0.4}} = 0.4456 \text{ or } 44.56\%.$$

$$\text{Indicated power} = \frac{plan}{60}, \text{ where } n = N \text{ for two stroke engine}$$

$$= \frac{5.2 \times 10^5 \times 0.3 \times (\pi \times 0.01) \times 135}{60 \times 1000} = 11.027 \text{ kW}$$

$$\text{Indicated thermal efficiency} = \frac{\text{indicated power}}{\text{heat input}} = \frac{11.027}{(8.8 \times 16350/3600)} = 0.276 \text{ or } 27.6\%.$$

5. An air-standard Diesel cycle has a compression ratio of 16 and the maximum cycle temperature of 1600 K. The compression begins at 1 bar and 15°C. Calculate (i) the values of p, v, and T at the end of each process, (ii) heat supplied, (iii) heat rejected, (iv) net work output and (v) thermal efficiency.

Given data:

$p_1 = 1$ bar; $T_1 = 288$ K; $v_1/v_2 = 16$; $T_3 = 1600$ K

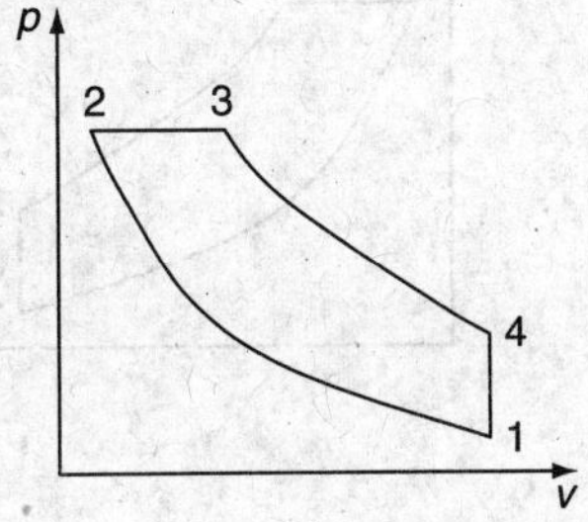

Assume 1 kg of air is present in the cycle.

$$v_1 = \frac{287 \times 288}{10^5} = 0.826 \text{ m}^3$$

$$v_2 = v_1/16 = 0.0516 \text{ m}^3$$

$$p_2 = p_1 (v_1/v_2)^\gamma = 1 \times 16^{1.4} = 48.5 \text{ bar}$$

$$T_2 = \frac{p_2 v_2}{R} = \frac{48.5 \times 10^5 \times 0.0516}{287} = 872 \text{ K}$$

$$v_3 = v_2 \times \frac{T_3}{T_2} = 0.0516 \times \frac{1600}{872} = 0.0947 \text{ m}^3; \; p_3 = p_2 = 48.5 \text{ bar}; \; v_4 = v_1 = 0.826 \text{ m}^3$$

$$p_4 = p_3 \left(\frac{v_3}{v_4}\right)^\gamma = 48.5 \times \left(\frac{0.0947}{0.826}\right)^{1.4} = 2.338 \text{ bar};$$

$$T_4 = \frac{p_4 v_4}{R} = \frac{2.338 \times 10^5 \times 0.826}{287} = 672.8 \text{ K}$$

For 1 kg of air,

Heat supplied $= C_p (T_3 - T_2) = 1.005 (1600 - 872) = 731.64$ kJ;

Heat rejected $= C_v (T_4 - T_1) = 0.718 (672.8 - 288) = 276.29$ kJ;

Net work output = Heat supplied − Heat rejected = 731.64 − 276.29 = 455.35 kJ and

$$\text{Thermal efficiency} = \frac{\text{Net work output}}{\text{Heat supplied}} = \frac{455.35}{731.64} = 0.6224 \text{ or } 62.24\%.$$

The above value can also be checked with the formula derived for the cycle.

$$\text{Cut-off ratio, } \rho = \frac{v_3}{v_2} = \frac{0.0947}{0.0516} = 1.835$$

$$\eta = 1 - \frac{1}{(r_c)^{\gamma-1}}\left[\frac{1}{\gamma}\frac{(\rho^\gamma - 1)}{(\rho - 1)}\right] = 1 - \frac{1}{(16)^{0.4}}\left[\frac{1}{1.4}\frac{(1.835^{1.4} - 1)}{(1.835 - 1)}\right] = 0.622 \text{ or } 62.2\%$$

6. A Diesel cycle operates with inlet conditions of 1 bar and 30°C. If the maximum pressure is limited to 44.31 bar and the maximum temperature to 980°C, determine the cycle efficiency and the mean effective pressure.

Given data:

$$p_1 = 1 \text{ bar}, T_1 = 303 \text{ K}, p_2 = p_3 = 44.31 \text{ bar}, T_3 = 1253 \text{ K},$$

$$\frac{p_2}{p_1} = \left(\frac{v_1}{v_2}\right)^\gamma, \text{ i.e. } \left(\frac{v_1}{v_2}\right) = \left(\frac{p_2}{p_1}\right)^{1/\gamma} = (44.31)^{1/1.4} = 14.999 \text{ or } 15$$

$$T_2 = T_1 (r_c)^{\gamma-1} = 303 (15)^{0.4} = 895 \text{ K}$$

$$\rho = \frac{v_3}{v_2} = \frac{T_3}{T_2} = \frac{1253}{895} = 1.4$$

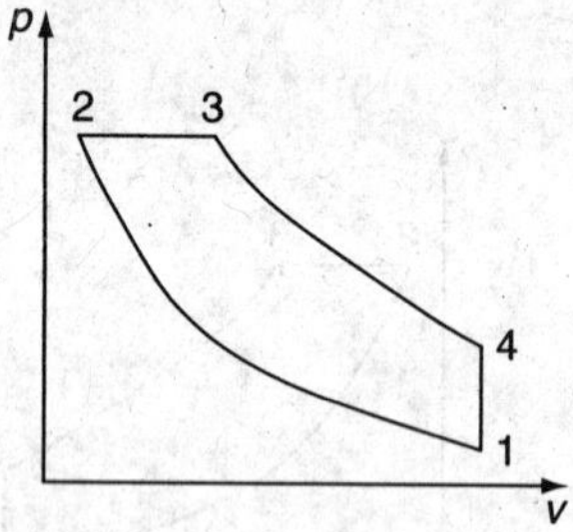

$$\eta = 1 - \frac{1}{(r_c)^{\gamma-1}}\left[\frac{1}{\gamma}\frac{(\rho^{\gamma}-1)}{(\rho-1)}\right]$$

$$= 1 - \frac{1}{(15)^{0.4}}\left[\frac{1}{1.4}\frac{(1.4^{1.4}-1)}{(1.4-1)}\right]$$

$$= 0.6363 \text{ or } 63.63\%$$

$$\text{m.e.p} = \frac{p_1 r_c}{(r_c-1)(\gamma-1)}\left\{(r_c)^{\gamma-1}\gamma(\rho-1)-((\rho)^{\gamma}-1)\right\}$$

$$= \frac{1\times 15}{(15-1)(1.4-1)}\left\{(15)^{0.4}\times 1.4\times(1.4-1)-((1.4)^{1.4}-1)\right\}.$$

$$= 2.8196 \text{ bar.}$$

7. An air-standard Diesel cycle has a compression ratio of 8 and the heat transferred to the working fluid is 1800 kJ/kg. The pressure and temperature at the beginning of the compression stroke are, respectively, 1 bar and 300 K. Calculate (i) the pressure and temperature at each point of the cycle, (ii) the air-standard efficiency and (iii) the m.e.p. Assume the values of C_p and R as 1.005 and 0.287.

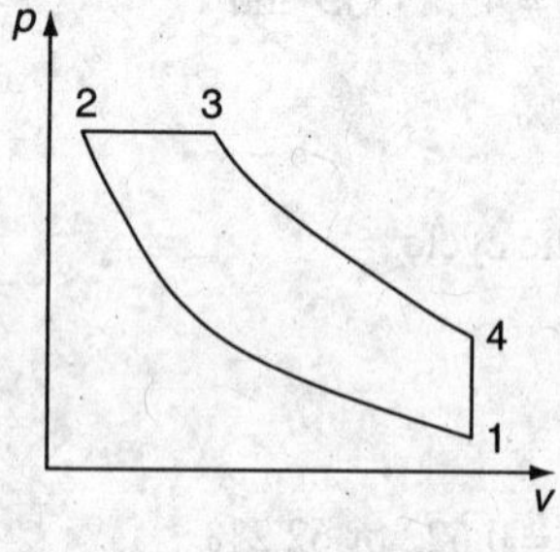

Given data:

$r_c = 8$, $Q = 1800$ kJ/kg, $T_1 = 300$ K, $p_1 = 1$ bar

$p_2 = p_1 (r_c)^{\gamma} = 1 (8)^{1.4} = 18.379$ bar,

$p_3 = p_2 = 18.379$ bar,

$T_2 = T_1 (r_c)^{\gamma-1} = 300 (8)^{0.4} = 689.2$ K

$1800 = C_p (T_3 - T_2)$;

Hence, $T_3 - T_2 = (1800/1.005) = 1791$,

$T_3 = 1791 + 689.2 = 2480.2$ K

$$\text{Cut-off ratio, } \rho = \frac{v_3}{v_2} = \frac{T_3}{T_2} = \frac{2480.2}{689.2} = 3.598$$

$$T_4 = T_3\left(\frac{v_3}{v_4}\right)^{\gamma-1} = T_3\left(\frac{v_3}{v_2}\right)^{\gamma-1}\left(\frac{v_2}{v_4}\right)^{\gamma-1} = T_3\frac{\rho^{\gamma-1}}{r_c^{\gamma-1}} = 2480.2\times\frac{3.598^{0.4}}{8^{0.4}} = 1801.66 \text{ K}$$

$$\frac{p_4}{T_4} = \frac{p_1}{T_1} \quad \text{or} \quad p_4 = \frac{T_4}{T_1}p_1 = \frac{1801.66}{300}\times 1 = 6.0056 \text{ bar}$$

The pressure and temperature at the cardinal points are as follows:

$p_1 = 1$ bar, $p_2 = 18.379$ bar, $p_3 = 18.379$ bar, $p_4 = 6.0056$ bar

$T_1 = 300$ K, $T_2 = 689.2$ K, $T_3 = 2480.2$ K, $T_4 = 1801.66$ K

$$\text{Air-standard efficiency} = 1-\frac{C_v(T_4-T_1)}{C_p(T_3-T_2)} = 1-\frac{(1801.66-300)}{1.4(2480.2-689.2)} = 0.4011 \text{ or } 40.11\%$$

$$\text{m.e.p} = \frac{p_1 r_c}{(r_c-1)(\gamma-1)}\left\{(r_c)^{\gamma-1}\gamma(\rho-1)-\left((\rho)^{\gamma}-1\right)\right\}$$

$$= \frac{1\times 8}{(8-1)(1.4-1)}\left\{(8)^{0.4}\times 1.4\times(3.598-1)-\left((3.598)^{1.4}-1\right)\right\} = 9.575 \text{ bar.}$$

8. An indicator diagram taken on a diesel engine shows that the compression curve follows $p\,v^{1.4}$ = constant. At two points on the compression curve at $^1\!/_8$ th and $^7\!/_8$ th of the stroke, the pressures are 1.6 bar and 16 bar. Find the compression ratio of the engine. If the cut off occurs at 6 per cent of the stroke, calculate the air-standard efficiency.

Given data:

$v_1 - v_a = 1/8\ (v_1 - v_2),\ p_a = 1.6$ bar (1)

$v_1 - v_b = 7/8\ (v_1 - v_2),\ p_b = 16$ bar (2)

$\rho = \dfrac{v_3}{v_2},\ (v_3 - v_2) = 0.06\ (v_1 - v_2)$ (3)

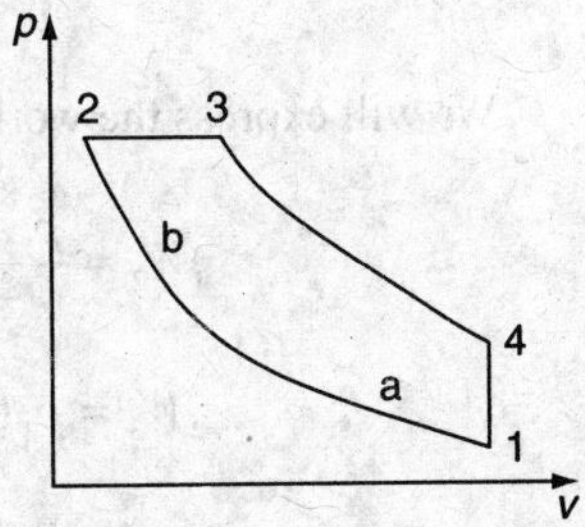

From the above relations, we get

$v_a = v_1 - 1/8\ (v_1 - v_2)$ and $v_b = v_1 - 7/8\ (v_1 - v_2)$ and in terms of the compression ratio, r_c

$$v_a = v_2\,[r_c - 1/8\ (r_c - 1)] \text{ and } v_b = v_2\,[r_c - 7/8\ (r_c - 1)].$$

$$\text{We know } p_a v_a^{\gamma} = p_b v_b^{\gamma} \text{ or } \frac{\left[r_c - \frac{1}{8}(r_c-1)\right]}{\left[r_c - \frac{7}{8}(r_c-1)\right]} = \left(\frac{16}{1.6}\right)^{1/1.4} = 5.1795$$

$$r_c - \frac{r_c}{8} + \frac{1}{8} = 5.1795\left[r_c - \frac{7}{8}r_c + \frac{7}{8}\right] = 0.64743\,r_c + 4.532, \text{ or } r_c = 19.365,$$

the compression ratio is 19.365.

From the third relation, we get $(v_3 - v_2) = 0.06\ (v_1 - v_2)$ or in terms of r_c

$$\rho - 1 = 0.06\ (r_c - 1) \text{ or } \rho = 1 + 0.06\ (19.365 - 1) = 2.1019.$$

The cut-off ratio, $\rho = 2.1019$

The air-standard efficiency,

$$\eta = 1-\frac{1}{(r_c)^{\gamma-1}}\left[\frac{1}{\gamma}\frac{(\rho^{\gamma}-1)}{(\rho-1)}\right] = 1-\frac{1}{(19.365)^{0.4}}\left[\frac{1}{1.4}\frac{(2.1019^{1.4}-1)}{(2.1019-1)}\right] = 0.6376 \text{ or } 63.76\%.$$

9. In an air-standard Diesel cycle the compression ratio is 12, the expansion ratio is 4.8 and the pressure at the beginning of compression is 1 bar. Determine (i) the pressure at the end of compression and

expansion and (ii) the mean effective pressure. Assume the specific heats of air as $C_p = 1.005$ kJ/kg K and $C_v = 0.718$ kJ/kg K.

Gas constant, $R = C_p - C_v = 1.005 - 0.718 = 0.287$ kJ/kg K.

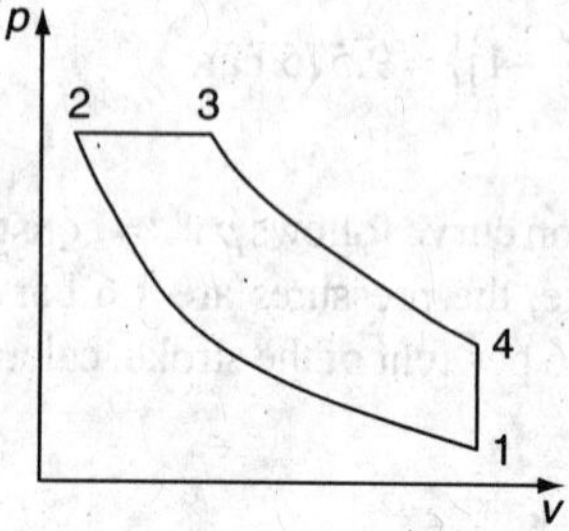

Given data:

$$\frac{v_1}{v_2} = 12 \text{ and } \frac{v_4}{v_3} = 4.8\,;\, p_1 = 1 \text{ bar}$$

$$\frac{p_2}{p_1} = \left(\frac{v_1}{v_2}\right)^{\gamma};\quad p_2 = 1(12)^{1.4} = 32.4 \text{ bar}$$

$$p_3 = p_2 = 32.4 \text{ bar}$$

$$\frac{p_3}{p_4} = \left(\frac{v_4}{v_3}\right)^{\gamma};\, \frac{p_3}{p_4} = (4.8)^{1.4};\, p_4 = \frac{32.4}{(4.8)^{1.4}} = 3.6 \text{ bar}$$

$$\frac{v_3}{v_2} = \frac{v_3}{v_4} \times \frac{v_4}{v_2} = \frac{1}{4.8} \times 12 = 2.5$$

We will express the work done expression in terms of v_2 as detailed below:

$$_1W_2 = \frac{p_1 v_1 - p_2 v_2}{\gamma - 1} = \frac{(1 \times 12 v_2 - 32.4 \times v_2) \times 10^5}{0.4} = -51 \times 10^5 v_2$$

$$_2W_3 = p_3(v_3 - v_2) = 32.4(2.5 - 1)v_2 \times 10^5 = 48.6 \times 10^5 v_2$$

$$_3W_4 = \frac{p_3 v_3 - p_4 v_4}{\gamma - 1} = \frac{(32.4 \times 2.5 v_2 - 3.6 \times 12 v_2) \times 10^5}{0.4} = 94.5 \times 10^5 v_2$$

Total work done (T W D) = $-51 \times 10^5\, v_2 + 48.6 \times 10^5\, v_2 + 94.5 \times 10^5\, v_2 = 92.1 \times 10^5\, v_2$

$$\text{m.e.p} = \frac{\text{T W D}}{v_1 - v_2} = \frac{92.1 \times 10^5 v_2}{(12 - 1) v_2} = 8.373 \times 10^5 \text{ N/m}^2 \text{ or } 8.373 \text{ bar.}$$

10. In an air-standard Diesel cycle, the compression ratio is 15 and at the beginning of isentropic compression the temperature is 20°C and the pressure is 1 bar. Heat is added at constant pressure until the temperature is 1360°C. Calculate (i) the heat supplied, (ii) the cut-off ratio and (iii) the air-standard efficiency.

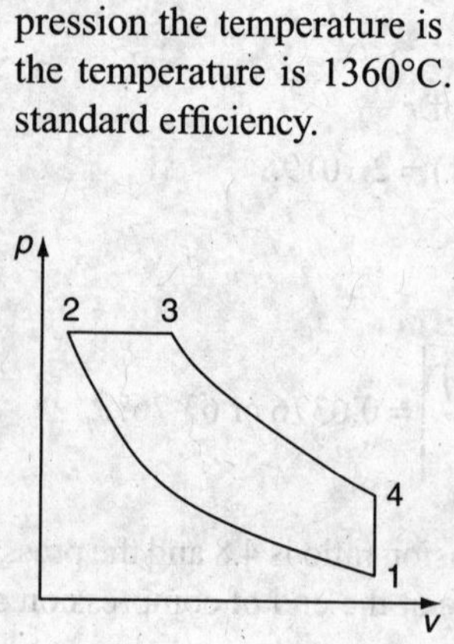

Given data:

$$\frac{v_1}{v_2} = 15,\; p_1 = 1 \text{ bar},\; T_1 = 293\,K,\; T_3 = 1633 \text{ K}$$

$$\frac{T_2}{T_1} = \left(\frac{v_1}{v_2}\right)^{\gamma - 1} = (15)^{0.4},\; T_2 = 293 \times (15)^{0.4} = 865.6 \text{ K}$$

Heat supplied = 1.005(1633 – 865.6) = 771.237 kJ/kg.

We know, $\dfrac{T_3}{T_2} = \dfrac{v_3}{v_2};\quad \dfrac{v_3}{v_2} = \dfrac{1633}{865.6} = 1.8865$

The cut-off ratio, $\rho = 1.8865$

$$\text{A.S.E, } \eta = 1-\frac{1}{(r_c)^{\gamma-1}}\left[\frac{1}{\gamma}\frac{(\rho^{\gamma}-1)}{(\rho-1)}\right] = 1-\frac{1}{(15)^{0.4}}\left[\frac{1}{1.4}\frac{(1.8865^{1.4}-1)}{(1.8865-1)}\right] = 0.6095 \text{ or } 60.95\,\%$$

11. In a dual combustion cycle, the pressure and temperature at the beginning of the cycle are 1 bar and 15°C. The specific volume at the end of isentropic compression is 0.068 m³/kg of the working substance. The highest temperature reached in the cycle is 1440 K. The heat supplied at constant volume is 230 kJ/kg. Calculate (i) the states of the substance at the end of each process in the cycle, (ii) the compression ratio, (iii) the cut-off ratio, (iv) the explosion ratio and (v) the thermal efficiency of the cycle.

Given data:

$p_1 = 1$ bar; $T_1 = 288$ K; $v_2 = 0.068$ m³/kg; $T_4 = 1440$ K; ${}_2Q_3 = 230$ kJ/kg

Assuming 1 kg of air is present in the cycle,

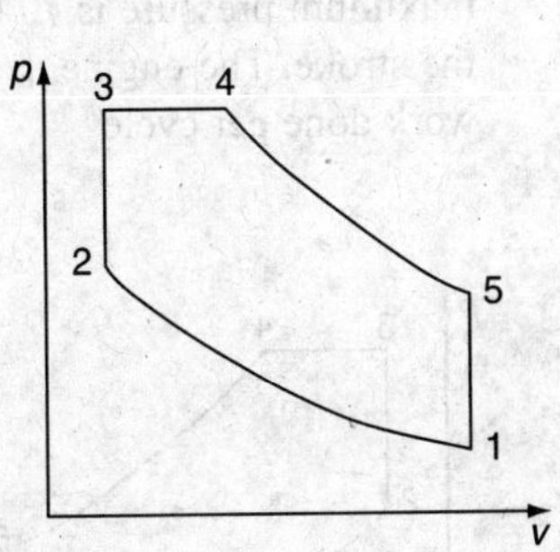

$$v_1 = \frac{287\times 288}{100000} = 0.826 \text{ m}^3\text{/kg}$$

$$r_c = \frac{v_1}{v_2} = \frac{0.826}{0.068} = 12.15$$

$$p_2 = p_1\left(\frac{v_1}{v_2}\right)^{\gamma} = 1\times(12.15)^{1.4} = 32.99 \text{ bar}$$

$$T_2 = T_1(r_c)^{\gamma-1} = 288\times(12.15)^{0.4} = 782 \text{ K}$$

$${}_2Q_3 = C_v(T_3 - T_2),\quad T_3 = T_2 + \frac{{}_2Q_3}{C_v} = 782 + \frac{230}{0.718} = 1102\ K$$

$${}_3Q_4 = C_p(T_4 - T_3) = 1.005(1440-1102) = 339.7 \text{ kJ}$$

$$p_3 = \frac{p_2}{T_2}\times T_3 = \frac{32.99}{782}\times 1102 = 46.49 \text{ bar}$$

$$v_4 = \frac{v_3}{T_3}\times T_4 = \frac{0.068}{1102}\times 1440 = 0.08886 \text{ m}^3$$

$$\frac{T_5}{T_4} = \left(\frac{v_4}{v_5}\right)^{\gamma-1} = \left(\frac{0.08886}{0.826}\right)^{0.4} = 0.4098;\quad T_5 = 1440\times 0.4098 = 590.1 \text{ K}$$

$$p_5 = \frac{p_1}{T_1}\times T_5 = \frac{1}{288}\times 590.1 = 2.0489 \text{ bar}$$

$$\text{Compression ratio, } r_c = \frac{v_1}{v_2} = \frac{0.826}{0.068} = 12.15$$

$$\text{Explosion ratio, } \beta = \frac{p_3}{p_2} = \frac{46.49}{32.99} = 1.41$$

$$\text{Cut-off ratio, } \rho = \frac{v_4}{v_3} = \frac{0.08886}{0.068} = 1.307$$

$$\eta = 1-\frac{(\beta\rho^{\gamma}-1)}{(r_c)^{\gamma-1}\{\gamma\beta(\rho-1)+(\beta-1)\}} = 1-\frac{(1.41\times1.307^{1.4}-1)}{(12.15)^{0.4}\{1.4\times1.41\times(1.307-1)+(1.41-1)\}}$$

$$= 1-\frac{1.051185}{2.71537\times1.016018} = 0.6189 \text{ or } 61.89\%$$

The thermal efficiency can also be calculated from the heat supplied and rejection rate as

$$\eta = 1-\frac{\text{heat rejection}}{\text{heat supplied}} = 1-\frac{0.718(590.1-288)}{0.718(1102-782)+1.005(1440-1102)} = 0.619 \text{ or } 61.9\%.$$

12. An engine works on Dual cycle. It takes air at a pressure of 1 bar and a temperature of 27°C, the maximum pressure is 70 bar. The heat addition during constant pressure is limited to 5 per cent of the stroke. The engine uses 10 litres of air per cycle and the compression ratio is 16. Find out the work done per cycle.

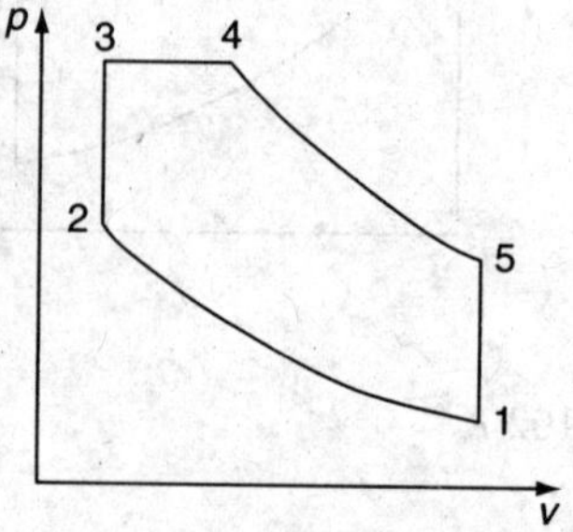

Given data:

$p_1 = 1$ bar, $T_1 = 300$ K, $v_1 = 0.01$ m³, $r_c = 16$,

$p_3 = p_4 = 70$ bar, $(v_4 - v_3) = 0.05(v_1 - v_2)$,

$$r_c = \frac{v_1}{v_2}, \quad v_2 = \frac{v_1}{r_c} = \frac{0.01}{16} = 0.000625 \text{ m}^3$$

$$p_2 = p_1(r_c)^{\gamma} = 1\times(16)^{1.4} = 48.5 \text{ bar}$$

$$T_2 = T_1(r_c)^{\gamma-1} = 300\times(16)^{0.4} = 909 \text{ K}$$

$(v_4 - v_3) = 0.05\,(v_2 - v_1)$ or

$$\left(\frac{v_4}{v_3}-1\right) = 0.05\left(\frac{v_1}{v_2}-1\right) = 0.05(16-1) = 0.75$$

$$\frac{v_4}{v_3} = 1.75; \; v_4 = 1.75\times0.000625 = 0.001093 \text{ m}^3$$

$$\frac{p_3}{T_3} = \frac{p_2}{T_2}, \quad T_3 = \frac{p_3}{p_2}\times T_2 = \frac{70}{48.5}\times909 = 1311.9 \text{ K}$$

$$\frac{T_4}{T_3} = \frac{v_4}{v_3}, \quad T_4 = \frac{0.001093}{0.000625}\times1311.9 = 2294 \text{ K}$$

$$\frac{T_5}{T_4} = \left(\frac{v_4}{v_5}\right)^{\gamma-1} = \left(\frac{0.001093}{0.01}\right)^{0.4} = 0.4125, \quad T_5 = 0.4125\times2294 = 946 \text{ K}$$

$$\frac{p_5}{T_5} = \frac{p_1}{T_1}, \quad p_5 = \frac{1}{300}\times946 = 3.15 \text{ bar}$$

$$\text{Work done/cycle} = \frac{(p_4v_4 - p_5v_5)}{(\gamma-1)} + p_4(v_4 - v_3) - \frac{(p_2v_2 - p_1v_1)}{(\gamma-1)}$$

$$\frac{(p_4v_4 - p_5v_5)}{(\gamma-1)} = \frac{(70 \times 0.001093 - 3.15 \times 0.01)}{0.4} \times 10^5 = 0.11252 \times 10^5\,\text{Nm}$$

$$p_4(v_4 - v_3) = 70 \times (0.001093 - 0.000625) \times 10^5 = 0.03276 \times 10^5\,\text{Nm}$$

$$\frac{(p_2v_2 - p_1v_1)}{(\gamma-1)} = \frac{(48.5 \times 0.000625 - 1 \times 0.01)}{(0.4)} \times 10^5 = 0.05078 \times 10^5\ \text{Nm}$$

Hence, the net work done = $0.11252 \times 10^5 + 0.03276 \times 10^5 - 0.05078 \times 10^5$

= 0.0945×10^5 Nm.

13. A gas turbine power plant working on the Brayton cycle operates under the following conditions: Air enters the compressor at 1 bar and 15°C and leaves the compressor at 4 bar. Air leaves the combustion chamber at 600°C. Calculate (i) the compressor work, (ii) the turbine work, (iii) the net work output and (iv) the thermal efficiency of the cycle.

Given data:

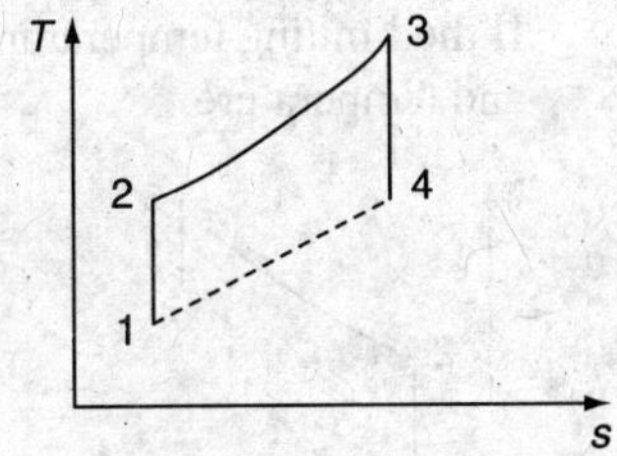

$p_1 = 1$ bar, $T_1 = 288$ K, $p_2 = 4$ bar, $T_3 = 873$ K,

$$T_2 = T_1(r_p)^{(\gamma-1)/\gamma} = 288(4)^{0.4/1.4} = 427.9\,\text{K}$$

$$T_4 = T_3/(r_p)^{(\gamma-1)/\gamma} = 873/(4)^{0.4/1.4} = 587.5\,\text{K}$$

The compressor work, $W_c = 1.005(427.9 - 288)$

= 140.6 kJ/kg.

The turbine work, $W_t = 1.005(873 - 587.5)$

= 286.9 kJ/kg.

The net work output = $W_t - W_c = 286.9 - 140.6 = 146.3$ kJ/kg.

The heat supplied = $1.005(873 - 427.9) = 447.3$ kJ/kg.

$$\text{The thermal efficiency} = \frac{\text{net Work output}}{\text{heat supplied}} = \frac{146.3}{447.3} = 0.327 \text{ or } 32.7\%$$

$$\text{As a check, the thermal efficiency} = 1 - \frac{1}{(r_p)^{(\gamma-1)/\gamma}} = 1 - \frac{1}{(4)^{0.4/1.4}} = 0.327 \text{ or } 32.7\%$$

14. A Brayton cycle operates with air entering the compressor at 1 bar, 20°C, at the rate of 32.0 m³/s and air entering the turbine at 8 bar, 1000°C. Determine the power output and the cycle thermal efficiency.

Given data:

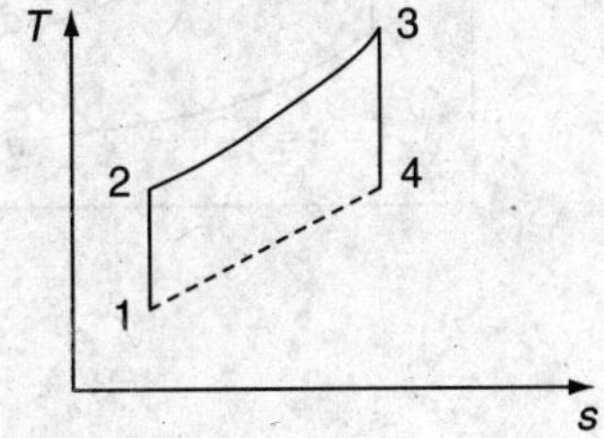

$p_1 = 1$ bar, $T_1 = 293$ K, $p_2 = 8$ bar, $T_3 = 1273$ K

$$T_2 = T_1(r_p)^{(\gamma-1)/\gamma} = 293(8)^{0.4/1.4} = 530.7\,\text{K}$$

$$T_4 = T_3/(r_p)^{(\gamma-1)/\gamma} = 1273/(8)^{0.4/1.4} = 702.7\,\text{K}$$

Mass of air flowing in the cycle

$$m = \frac{p_1 V_1}{RT_1} = \frac{1 \times 10^5 \times 32}{287 \times 293} = 38.05 \text{ kg/s}$$

Power output $= m\, C_p\, (T_3 - T_4) - m\, C_p\, (T_2 - T_1)$

$= 38.05 \times 1.005 \times (1273 - 702.7) - 38.05 \times 1.005 \times (530.7 - 293)$

$= 21808.4 - 9089.7 = 12718.7$ kJ/kg.

Heat supplied $= m\, C_p\, (T_3 - T_2) = 38.05 \times 1.005 \times (1273 - 530.7) = 28385.7$ kJ/kg.

$$\text{Thermal efficiency} = \frac{\text{power output}}{\text{heat supplied}} = \frac{12718.7}{28385.7} \times 100 = 44.8\%.$$

$$\text{As a check, the thermal efficiency} = 1 - \frac{1}{(r_p)^{(\gamma-1)/\gamma}} = 1 - \frac{1}{(8)^{0.4/1.4}} = 0.448 \text{ or } 44.8\%.$$

15. A simple gas turbine cycle operates so that there is no net work output. The inlet pressure and temperature to the isentropic compressor are 1 bar and 15°C and the compressor pressure ratio is 6. If the limiting temperature of the isentropic turbine is 840°C, determine the turbine outlet pressure and temperature.

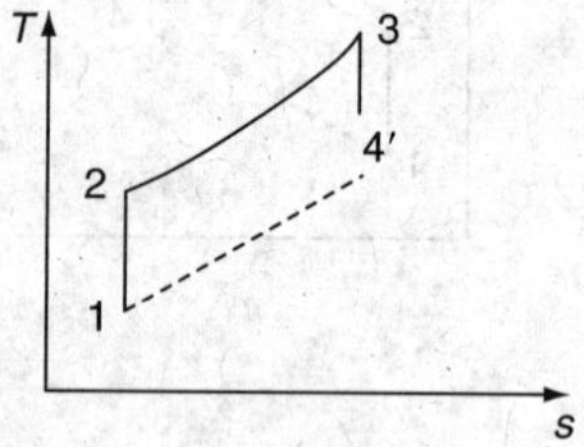

Given data:

$p_1 = 1$ bar, $T_1 = 288$ K, $p_2 = 6$ bar, $T_3 = 1113$ K

$$T_2 = T_1 (p_2/p_1)^{(\gamma-1)/\gamma} = 288(6)^{0.4/1.4} = 480.5 \text{ K}$$

Equating the compressor work to the turbine work, we get

$C_p\,(T_2 - T_1) = C_p\,(T_3 - T_{4'})$

or $T_{4'} = T_3 - (T_2 - T_1) = 1113 - (480.5 - 288) = 920.5$ K

$$\frac{p_{4'}}{p_3} = \left(\frac{T_{4'}}{T_3}\right)^{\gamma/(\gamma-1)}; \quad p_{4'} = 6 \times \left(\frac{920.5}{1113}\right)^{1.4/0.4} = 3.0867 \text{ bar}$$

The pressure and temperature at the turbine outlet are 3.0867 bar and 920.5 K.

16. In a Stirling cycle the volume varies between 0.03 and 0.06 m³, the maximum pressure is 2 bar and the temperature varies between 540°C and 270°C. The working fluid is air (an ideal gas). Find the efficiency and the work done per cycle for the simple cycle.

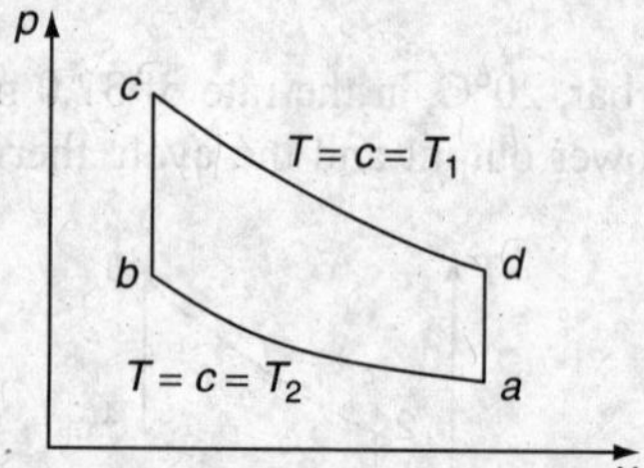

Given data:

$T_a = T_b = T_2 = 543$ K

$T_c = T_d = T_1 = 813$ K

$p_c = 2$ bar

$v_a = v_d = 0.06$ m³

$v_b = v_c = 0.03$ m³

Work done during the process *a-b*:

$${}_aW_b = R\, T_2\, \ell n\left(\frac{v_b}{v_a}\right) = 0.287 \times 543 \times \ell n\left(\frac{0.03}{0.06}\right) = -108 \text{ kJ/kg}.$$

Work done during the process *c-d*:

$$ {}_cW_d = RT_1\,\ell n\left(\frac{v_d}{v_c}\right) = 0.287\times 813\times \ell n\left(\frac{0.06}{0.03}\right) = 161.7\text{ kJ/kg.} $$

Net work done = 161.7 – 108 = 53.7 kJ/kg.

Heat supplied during the process b-c = 0.718(813 – 543) = 193.86 kJ/kg.

Total heat supplied = 161.7 + 193.86 = 355.56 kJ/kg.

$$ \text{Thermal efficincy} = \frac{\text{Net work done}}{\text{Total heat supplied}} = \frac{53.7}{355.56} = 0.151 \text{ or } 15.1\%. $$

Practice Problems

1. In an air-standard Otto cycle, the compression ratio is 8 and the pressure and temperature of air at the beginning of compression stroke are 1 bar and 15°C. Heat transferred to air is 1500 kJ/kg. Determine (i) maximum pressure and temperature in the cycle and (ii) thermal efficiency.

   (76.41 bar; 2750.6 K; 56.47%)

2. An internal combustion engine working on the ideal Otto cycle has temperature and pressure of 25°C and 1 bar, respectively, at the beginning of compression. If its air-standard efficiency is 48 per cent, determine pressure and temperature at the end of compression.

   (9.86 bar; 573 K)

3. An air-standard Otto cycle has a clearance volume of 17 per cent of stroke volume. The initial pressure is 0.75 bar and temperature is 30°C. If the peak pressure of the cycle is 28 bar, find (i) the air-standard efficiency, (ii) peak cycle temperature and (iii) work done per kg of air.

   (53.6%; 1644 K; 381.49 kN.m)

4. A four stroke spark-ignition engine having a swept volume of 0.13 m^3 operates on constant volume air cycle. The compression ratio is 6 and the conditions at the beginning of compression are 1 bar and 20°C. The heat supplied is 150 kJ/cycle. Calculate the pressure and temperature at the salient points. Determine the cycle efficiency and the m.e.p of the cycle.

   (12.29 bar; 35.36 bar; 2.877 bar; 599.96 K; 1726.2 K; 843 K; 51.16%; 5.903 bar)

5. A diesel engine working on air-standard Diesel cycle takes in air at 1 bar and 25°C. The specific volume of air at the inlet is 0.8 m^3/kg. The compression ratio is 14 and 840 kJ/kg of heat is added at constant pressure. Determine (i) pressure, temperature and specific volume at each point of the cycle; (ii) thermal efficiency and (iii) mean effective pressure.

   (1 bar, 40.24 bar, 40.24 bar, 2.595 bar; 0.8 m^3, 0.057 m^3, 0.1126 m^3, 0.8 m^3: 298 K, 856.4 K, 1692.2 K, 772.4 K; 59.38%; 6.713 bar)

6. In an air-standard Diesel cycle the compression ratio is 12, the expansion ratio is 4.8 and the pressure at the beginning of compression is 1 bar. Determine (i) the pressure at the end of compression and expansion and (ii) the mean effective pressure.

   (32.42 bar, 3.6 bar; 8.365 bar]

7. A Diesel cycle operates with inlet conditions of 1 bar and 30°C. If the maximum temperature is limited to 1980°C and the maximum pressure to 45 bar, determine the cycle efficiency and the mean effective pressure.

   (57.9%; 9.73 bar)

8. A four-stroke diesel engine has a cylinder of 15 cm bore and 25 cm of stroke. The compression ratio is 14. The cut off takes place at 43 per cent of the stroke. The m.e.p is 460 kPa. The engine runs at 200 rpm. The fuel consumption is 1.13 kg/h and its calorific value is 41,800 kJ/kg. Calculate the relative efficiency of the engine.

(61.35%)

9. The pressures on the compression curve of a diesel engine are found to be at ¼ stroke, 1.61 bar and at ⅞ stroke, 11.1 bar. Estimate the compression ratio. Calculate the ideal efficiency of the engine if cut off occurs at one-fifteenth of the stroke.

(12.71; 58.85%)

10. The fuel cut off in a diesel engine of compression ratio 16 takes place at 5 per cent of the stroke. What will be the loss in ideal efficiency if the engine cut off is delayed by 3 per cent.

(2.245)

11. An oil engine working on the dual combustion cycle has a cylinder 20 cm bore and a stroke of 40 cm. The compression ratio is 13.5 and the explosion ratio obtained from the indicator card is 1.41. From the indicator it was also found that the cut off occurred at 4.9 per cent of the stroke. Find the ideal efficiency.

(61.78%)

12. In an air-standard dual cycle, the conditions at the start of the compression stroke are 1 bar and 300 K. The compression ratio is 15. The maximum temperature of the cycle is 3000 K and the maximum pressure is 70 bar. Determine the work done per kg of air, the energy added per kg of air and the mean effective pressure.

(1203 kJ; 1976 kJ; 14.97 bar)

13. The pressure ratio of an air-standard Brayton cycle is 5. The conditions at the start of the compression are 1 bar and 300 K. The maximum allowed temperature in the cycle is 1200 K. Determine the compressor work per kg of air, the turbine work per kg of air and the thermal efficiency.

(175.97 kJ; 444.42 kJ; 36.86%)

14. The pressure ratio of an air-standard Brayton cycle is 4 and the inlet conditions to the compressor are 1 bar and 15°C. The turbine is limited to a temperature of 840°C and the flow rate is 4.8 kg/s. Determine (i) the compressor work, (ii) the turbine work and (iii) the thermal efficiency.

(674.4 kJ/s; 1756.4 kJ/s; 32.7%)

Objective Type Questions

1. Identify the gas power cycle consisting of four processes in which work alone is transferred during two processes and heat alone is transferred during the other two processes.

   (i) Atkinson cycle (ii) Carnot cycle
   (iii) Diesel cycle (iv) Otto cycle

2. The number of cardinal points for the Lenoir gas power cycle when represented on the temperature–entropy diagram is

   (i) 3 (ii) 4
   (iii) 5 (iv) 6

3. Which working fluid gives maximum efficiency to an ideal Otto cycle?

   (i) Air (ii) Argon
   (iii) Hydrogen (iv) Nitrogen

4. The four reversible ideal gas processes 1-2, 2-3, 3-4 and 4-1 shown in the following volume–temperature diagram constitute

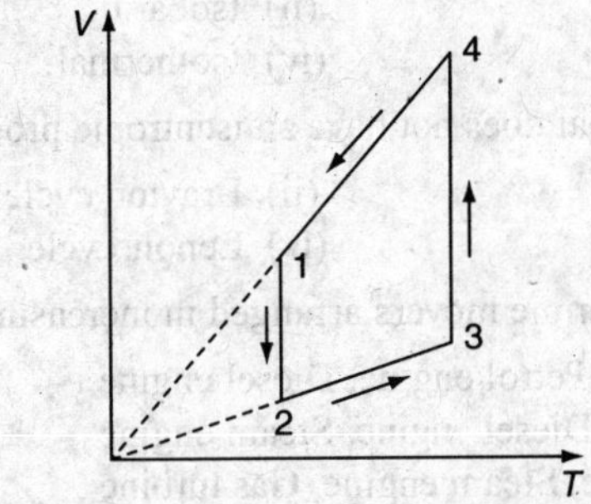

   (i) the Brayton cycle (ii) the Diesel cycle
   (iii) the Ericsson cycle (iv) the Otto cycle

5. The four reversible ideal gas processes 1-2, 2-3, 3-4 and 4-1 shown in the following pressure–temperature diagram constitute

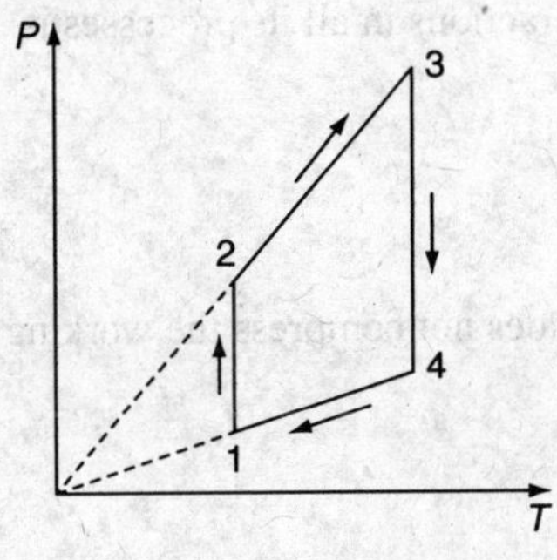

   (i) the Brayton cycle (ii) the Diesel cycle
   (iii) the Otto cycle (iv) the Stirling cycle

6. If the shape of an ideal cycle on the pressure–volume coordinates is a rectangle, then the same cycle when plotted on temperature–volume plane shall appear as

   (i) parallelogram (ii) rectangle
   (iii) rhombus (iv) trapezium

7. The number of cardinal points in a dual air-standard cycle is

   (i) 3 (ii) 4
   (iii) 5 (iv) 6

8. Which ideal gas cycle takes the shape of trapezium when represented on pressure–enthalpy coordinates?

   (i) Atkinson cycle (ii) Carnot cycle
   (iii) Diesel cycle (iv) Stirling cycle

9. Which prime mover basic cycle has the lowest maximum operating pressure?
   (i) Diesel engine (ii) Gas turbine
   (iii) Petrol engine (iv) Steam turbine
10. Which process is to be incorporated in any air-standard cycle to approach the efficiency of the Carnot cycle?
    (i) Isenthalpic (ii) Isobaric
    (iii) Isometric (iv) Isothermal
11. Identify the air-standard cycle that does not have an isentropic process in it.
    (i) Atkinson cycle (ii) Brayton cycle
    (iii) Ericsson cycle (iv) Lenoir cycle
12. The thermal efficiencies of the prime movers arranged in increasing order are
    (i) Steam engine, Gas turbine, Petrol engine, Diesel engine
    (ii) Gas turbine, Petrol engine, Diesel engine, Steam engine
    (iii) Petrol engine, Diesel engine, Steam engine, Gas turbine
    (iv) Diesel engine, Steam engine, Gas turbine, Petrol engine
13. The addition of a regenerator in a gas turbine cycle increases its
    (i) specific output (ii) work ratio
    (iii) thermal efficiency (iv) mechanical efficiency
14. The ideal cycle that has heat interactions in all its processes is
    (i) Bell Coleman cycle
    (ii) Carnot cycle
    (iii) Joule cycle
    (iv) Stirling cycle
15. Identify the ideal gas cycle that does not compress the working substance.
    (i) Atkinson cycle
    (ii) Ericsson cycle
    (iii) Lenoir cycle
    (iv) Stirling cycle
16. Which machine does not require air for combustion?
    (i) Ramjet engine
    (ii) Rocket engine
    (iii) Turbojet engine
    (iv) Turboprop engine
17. The pressure ratio in an ideal gas turbine cycle for the given maximum and minimum temperatures T_3 and T_1 can be expressed as, pressure ratio = $(T_3/T_1)^{\alpha}$. The output will be maximum when the exponent α, expressed in terms of γ, is
    (i) $\dfrac{\gamma}{(\gamma-1)}$ (ii) $\dfrac{(\gamma-1)}{\gamma}$
    (iii) $\dfrac{\gamma}{2(\gamma-1)}$ (iv) $\dfrac{2(\gamma-1)}{\gamma}$

18. Which one of the following changes would improve the thermal efficiency of the Diesel cycle?
   (i) Increasing the compression ratio and decreasing the cut-off ratio
   (ii) Decreasing the compression ratio and increasing the cut-off ratio
   (iii) Decreasing both the compression ratio and the cut-off ratio
   (iv) Increasing both the compression ratio and the cut-off ratio

19. Identify the ideal heat engine cycle which has four distinct processes on the pressure–volume coordinates, but has only three process representations on the temperature–entropy diagram.
   (i) Atkinson cycle (ii) Ericsson cycle
   (iii) Lenoir cycle (iv) Stirling cycle

20. Identify the reversible cycles with ideal regeneration which are as efficient as that of the Carnot when operating under the same temperature range.
   (i) The Stirling and the Ericsson cycles
   (ii) The Otto and the Brayton cycles
   (iii) The Diesel and the Stirling cycles
   (iv) The Brayton and the Atkinson cycles

21. The air-standard efficiency of Diesel cycle is always less than that of Otto cycle for the same compression ratio and heat input since the former cycle as compared to latter
   (i) has less maximum temperature (ii) has less maximum pressure
   (iii) has more heat rejection rate (iv) has more maximum temperature

22. In a gas turbine complex cycle, the transfer of heat from the turbine exhaust to the compressor exit is called
   (i) reheating (ii) intercooling
   (iii) aftercooling (iv) regeneration

23. Match List I (ideal gas cycle) with List II (thermodynamic coordinates) such that the cycle when represented in the thermodynamic coordinates assumes the shape of a rectangle using the codes given below the lists:

List I	*List II*
A. Carnot cycle	1. Pressure–entropy
B. Brayton cycle	2. Pressure–temperature
C. Ericsson cycle	3. Temperature–volume
D. Stirling cycle	4. Temperature–entropy

Codes:

	A	B	C	D		A	B	C	D
(i)	2	3	1	4	(ii)	4	3	1	2
(iii)	4	1	2	3	(iv)	3	2	4	1

24. Assertion (A): For the same heat input and peak pressure, the Diesel cycle thermal efficiency is more than that of the Otto cycle.

    Reason (R): The heat rejection in Diesel cycle is less than that of Otto cycle.

    (i) Both A and R are true and R is the correct explanation of A.
    (ii) Both A and R are true and R is not the correct explanation of A.
    (iii) A is true but R is false.
    (iv) A is false but R is true.

25. The ideal Brayton gas turbine cycle efficiency, which depends on the pressure ratio, decreases when operated at pressure ratios greater than the optimum one for the given temperature limits. The reduction in efficiency when compared to the cycle working on the optimum pressure ratio for the same temperature limits is due to

    (i) low heat intake by the cycle
    (ii) high heat rejection from the cycle
    (iii) low power output of the turbine
    (iv) low average/mean temperature of heat addition

26. The parameters whose numerical values are always more than unity in the operation of a gas turbine unit that incorporates reheating and regeneration are

    (i) power input factor and reheat factor
    (ii) reheat factor and slip factor
    (iii) slip factor and heat exchanger effectiveness
    (iv) heat exchanger effectiveness and power input factor

27. The theoretical air-standard cycle shown in the pressure–volume coordinates is

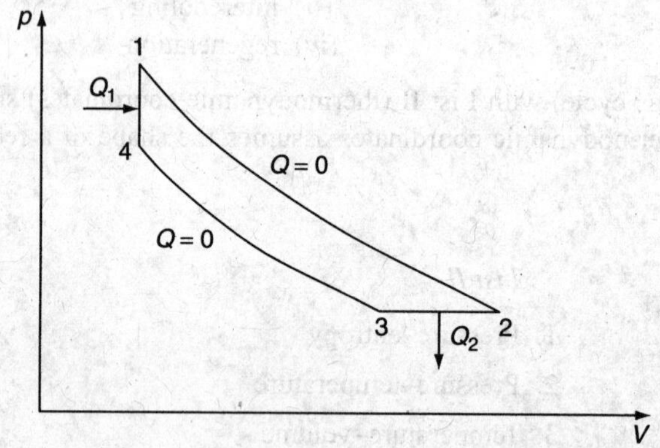

    (i) Atkinson cycle
    (ii) Ericsson cycle
    (iii) Lenoir cycle
    (iv) Striling cycle

28. Which air cycle has its thermal efficiency approaching that of Carnot efficiency with the addition of several stages of intercooling and several stages of reheating and ideal regeneration?

    (i) Brayton cycle
    (ii) Diesel cycle
    (iii) Dual cycle
    (iv) Otto cycle

29. The air-standard efficiency of the ideal cycle shown in the *T-s* diagram is

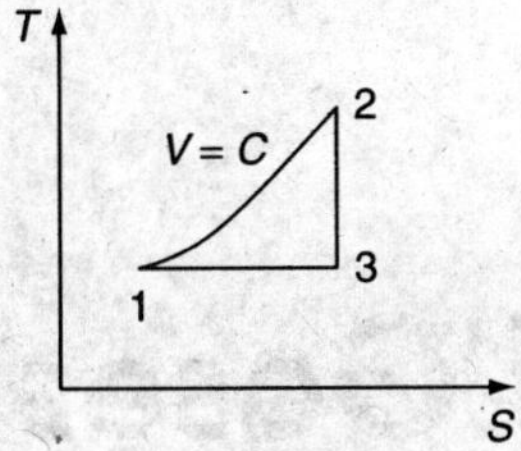

(i) $1-\frac{T_1}{T_2}$ (ii) $1-\frac{2\,T_1}{(T_2+T_1)}$

(iii) $1-\frac{T_1}{(T_2-T_1)}$ (iv) $1-\frac{T_1}{(T_2-T_1)}\ln\left(\frac{T_2}{T_1}\right)$

30. The thermal efficiency of the cycle shown on the *p-v* diagram given below is

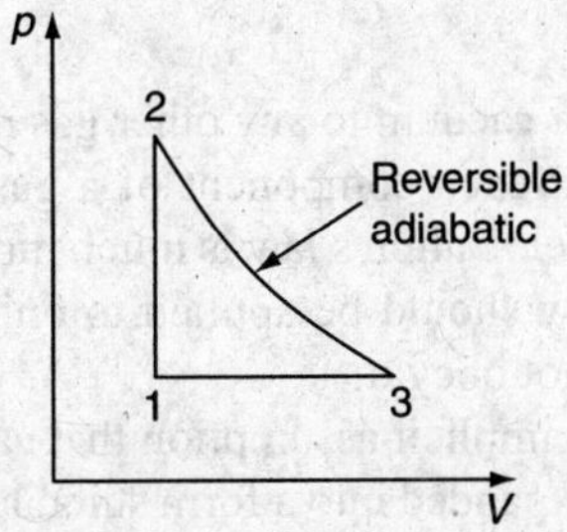

(i) $1-\frac{(V_3/V_1-1)^{\gamma}}{(p_2/p_1-1)^{\gamma}}$ (ii) $1-\gamma\frac{(V_3/V_1-1)}{(p_2/p_1-1)}$

(iii) $1-\frac{1}{\gamma}\frac{(V_3/V_1-1)}{(p_2/p_1-1)}$ (iv) $1-\frac{(V_3-V_1)}{(p_2-p_1)}\frac{p_1}{V_1}$

Answers

1. (iv) 2. (i) 3. (ii) 4. (iii) 5. (iv) 6. (iv) 7. (iii) 8. (iv) 9. (ii) 10. (iv)
11. (iii) 12. (i) 13. (iii) 14. (iv) 15. (iii) 16. (ii) 17. (iii) 18. (i) 19. (iii) 20. (i)
21. (iii) 22. (iv) 23. (iii) 24. (i) 25. (i) 26. (i) 27. (i) 28. (i) 29. (iv) 30. (ii)

8

Mixtures of Gases and Psychrometry

1. State the Dalton's law.

Ans: Dalton's law: 'Any gas is as a vacuum to any other gas mixed with it'.

This compact statement means that each component of a gaseous mixture acts and has the same properties, as if it alone filled the vessel. Dalton's law is much more closely obeyed by most substances than is the ideal gas rule. Dalton's law should be applied to only the thermodynamic properties of a mixture. The transport properties do not obey it.

Dalton's law, together with Dalton's implicit assumption that each constituent acts as if it were at the temperature of the mixture, has been expanded into a form suitable for engineering use by Gibbs (1875), whose reformulation of the principle is known as the Gibbs-Dalton law.

2. Explain how the Gibbs-Dalton law can be used to evaluate the properties of a mixture of non-reacting gases.

Ans: Gibbs-Dalton law:

(i) The pressure of a gaseous mixture is the sum of the pressures which each constituent would exert if it alone occupied the volume of the mixture at the temperature of the mixture.

(ii) The internal energy, the enthalpy and the entropy of a gaseous mixture are respectively equal to the sum of the internal energies, the enthalpies and the entropies of the constituents. The internal energy and the enthalpy of an ideal gas are functions of temperature only and the temperature used in evaluating properties is the mixture temperature. The entropy of an ideal gas depends on mixture temperature and volume of the mixture. Hence, the individual properties are evaluated as if each constituent occupies the volume of the mixture at the temperature of mixture.

The Gibbs-Dalton law is expressed symbolically by the following equations, in which the subscripts $a, b \ldots\ldots n$ each will refer to a particular component of the mixture. The symbol for the mixture property is left without suffix.

$$t = t_a = t_b = \ldots\ldots = t_n \qquad (1)$$

Since by Dalton's law each component is thought of as occupying the whole volume V, we write

$$V = V_a = V_b = \ldots\ldots = V_n \tag{2}$$

or in terms of the specific volumes, v

$$mv = m_a v_a = m_b v_b = \ldots\ldots = m_n v_n \tag{3}$$

where the m's are the masses.

Since the mass of the mixture is equal to the sum of the masses of the components, that is,

$$m = m_a + m_b + m_c + \ldots\ldots\ m_n \tag{4}$$

or $$\frac{1}{v} = \frac{1}{v_a} + \frac{1}{v_b} + \ldots\ldots + \frac{1}{v_n} \tag{5}$$

or $$\rho = \rho_a + \rho_b + \ldots\ldots + \rho_n \tag{6}$$

which means that the density of the mixture is equal to the sum of the densities of the components.

The first part of the Gibbs-Dalton law is concerned with the intensive property, pressure and is expressed symbolically as

$$p = p_a + p_b + \ldots\ldots + p_n \tag{7}$$

The second part of the Gibbs-Dalton is concerned with the extensive properties of internal energy, enthalpy and entropy and is written

$$U = U_a + U_b + \ldots\ldots + U_n$$

$$H = H_a + H_b + \ldots\ldots + H_n$$

$$S = S_a + S_b + \ldots\ldots + S_n$$

$$mu = m_a u_a + m_b u_b + \ldots\ldots + m_n u_n \tag{8}$$

$$mh = m_a h_a + m_b h_b + \ldots\ldots + m_n h_n \tag{9}$$

$$ms = m_a s_a + m_b s_b + \ldots\ldots + m_n s_n \tag{10}$$

Application of the Gibbs-Dalton law to mixtures of ideal gases:

$$p_a v_a = R_a T \tag{11}$$

$$p_b v_b = R_b T \tag{12}$$

$v_a = v_b = v$ each component occupies the mixture volume at the mixture temperature.

$$p = p_a + p_b \tag{13}$$

$$p = p_a + p_b = \frac{R_a T}{v_a} + \frac{R_b T}{v_b} = \frac{m_a}{mv} R_a T + \frac{m_b}{mv} R_b T = \left(\frac{m_a R_a + m_b R_b}{m}\right)\frac{T}{v} = R\frac{T}{v}$$

$$pv = RT \text{ where } R = \left(\frac{m_a R_a + m_b R_b}{m}\right) = \text{gas constant for the mixture.}$$

The aforesaid equation asserts that the mixture of ideal itself is an ideal gas, and that R is its gas constant

3. What is meant by psychrometry?

Ans: Psychrometry: The most important mechanical engineering example of mixtures of ideal gases with a condensable vapour is the air-water system. Dry air and water vapour form a binary mixture. If the vapour is superheated, the procedure can be simplified. When the pressure is sufficiently low for the Gibbs-Dalton law to be valid, a superheated vapour can be treated as a perfect gas. A mixture of superheated vapour and gas can then be regarded as a mixture of perfect gases and the method of previous section is applicable. Moist atmospheric air is an example of a mixture which can be treated in this way.

Normally the atmosphere is not saturated with water vapour but is a mixture of air and superheated vapour. Many industrial processes, for example, in the paper and textile industries, require close control of the vapour content of the atmosphere as well as its temperature. The air conditioning of buildings for comfort, the design of cooling towers and the science of meteorology are other instances that involve a study of the properties of moist air. This study has been given the name of psychrometry or hygrometry. A mixture of two substances required three properties to completely define its thermodynamic state, unlike a pure substance which requires only two. One of the three properties can be the composition. The properties of moist air are called psychrometric properties and the subject that deals with the behaviour of moist air is known as psychrometry.

Water vapour is present in the atmosphere at a very low pressure. At this low pressure and atmospheric temperature, the water vapour behaves as a perfect gas. The partial pressure of dry air is also below one atmosphere which may also be considered to behave very much as a perfect gas. The Gibbs-Dalton law can, therefore, be applied to the case of moist air. Psychrometry makes use of certain technical terms that arise either from the need to specify the composition of the mixture or from the method of measuring the composition. In addition, all calculations are based on the dry air part since the water vapour part is continuously variable.

4. Define the following terms as applied to psychrometry.

Ans: (i) *Specific humidity or humidity ratio*: Specific humidity or absolute humidity or humidity ratio or moisture content (w) is defined as the ratio of the mass of water vapour to the mass of dry air in a given volume of the mixture.

$$w = \frac{\text{mass of } H_2O}{\text{mass of } (dry) \text{ air}} = \frac{\dfrac{V}{v_v}}{\dfrac{V}{v_{\text{air}}}} = \frac{v_{\text{air}}}{v_v} = \frac{R_a T}{p_a}\frac{p_v}{R_v T} = \frac{8.3143}{28.9} \times \frac{18}{8.3143} \times \frac{p_v}{p_a}$$

$$= 0.622 \frac{p_v}{p - p_v}.$$

(ii) *Relative humidity*: It is the ratio of the actual partial pressure p_v of the vapour to the partial pressure p_s of the vapour when the air is saturated at the same temperature. Thus,

$$\phi = \frac{p_v}{p_s}.$$

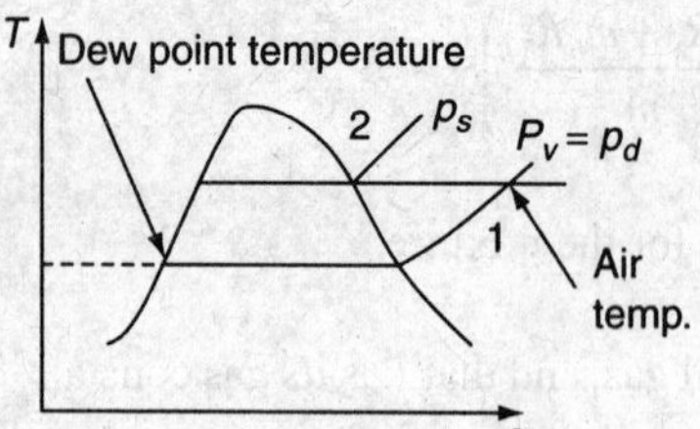

(iii) *Degree of saturation*: Consider now that the water vapour at temperature T is added to the moist air. The partial pressure p will go on increasing with the addition of water vapour until it reaches a value p_s corresponding to state 2 after which it cannot increase further as p_s is the saturation pressure or maximum possible pressure of water at temperature T. The thermodynamic state of water vapour is now saturated at point 2. The air containing moisture in such a state is called saturated air. Hence,

$$w_s = 0.622\frac{p_s}{p - p_s}.$$

The ratio of actual specific humidity w to the specific humidity w_s of saturated air at temperature T is termed as the degree of saturation denoted by μ.

$$\mu = \frac{w}{w_s} = \frac{p_v}{p_s}\frac{(p - p_s)}{(p - p_v)} = \frac{p_v}{p_s}\frac{(1 - p_s/p)}{(1 - p_v/p)}$$

(iv) *Dew point temperature*: It is the temperature to which unsaturated air must be cooled at constant pressure for it to become saturated (or for condensation to begin). For the vapour in the mixture at temperature T we have

$$p_v = \frac{m_v R_v T}{V}$$

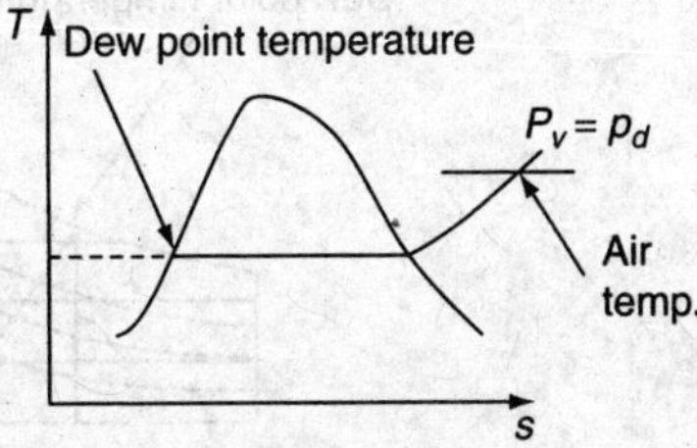

and for the mixture

$$p = \frac{m R T}{V}.$$

Therefore,

$$\frac{p_v}{p} = \frac{m_v R_v}{m R}.$$

Since the ratio m_v/m remains unchanged as the unsaturated mixture is cooled at constant pressure, p_v must remain constant. The cooling process undergone by the vapour constituent is shown in the earlier diagram. The partial pressure of the vapour at the dew point is the saturation pressure p_d corresponding to the dew point temperature.

(v) *Dry-bulb temperature*: The dry-bulb temperature of a steam-air mixture is simply the temperature of the mixture. The special term arises from the practice of measuring the moisture content of a mixture by means of a wet-and-dry-bulb hygrometer. This consists of two thermometers exposed to a stream of the mixture as shown in Figure 8.1. The bulb of one of the thermometers is kept wet by a wick dipping into water; the other is dry and gives the dry-bulb temperature.

(vi) *Wet-bulb temperature*: The temperature registered by the thermometer wetted by the wick is known as the wet-bulb temperature. This is normally lower than the dry-bulb temperature; the difference between the two is a measure of the humidity of the air.

The relation between the two temperatures is fixed by processes that lie outside thermodynamics: the so-called rate processes of heat transfer from the mixture stream to the wick surface and of mass transfer (diffusion of steam) from the surface into the stream.

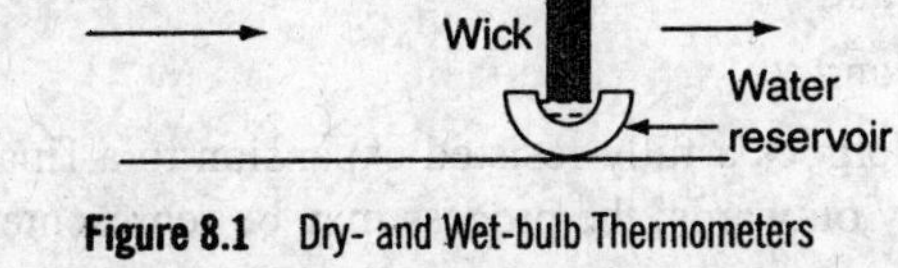

Figure 8.1 Dry- and Wet-bulb Thermometers

5. **Sketch a psychrometric chart and show the relative humidity lines, specific humidity lines, dry-bulb and wet-bulb temperature lines.**

Ans: Psychrometric chart: The ordinate and abscissa of the psychrometric chart are, as may be seen from Figure 8.2, respectively the specific humidity w and the dry-bulb temperature; lines of constant φ and constant wet-bulb temperature are plotted on the chart. The chart may therefore be used to obtain values of w and φ directly from the readings of the wet-and-dry-bulb thermometers.

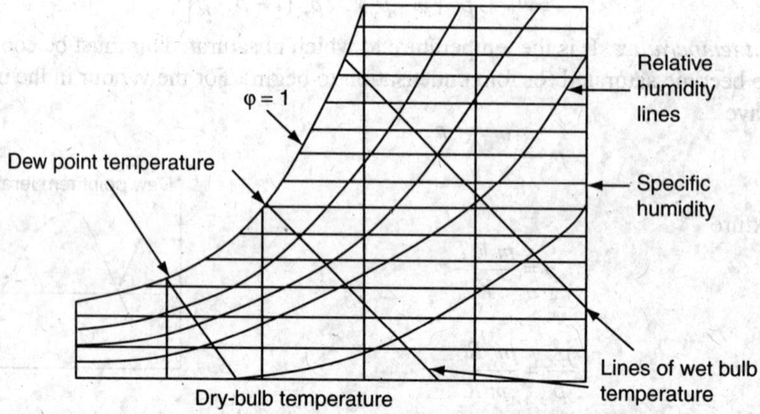

Figure 8.2 Psychometric Chart Thermomenters

Solved Problems

1. Assuming that air may be treated as a mixture of ideal gases which has 23 percent O_2 and 77 percent N_2 by mass, find the gas constant and the equivalent molecular mass.

Gas constant R for oxygen $= \dfrac{8314.3}{32} = 259.8$ and for nitrogen $= \dfrac{8314.3}{28} = 296.9$.

The gas constant for the mixture $R = \dfrac{0.23 \times 259.8 + 0.77 \times 296.9}{0.23 + 0.77} = 288.3 \dfrac{\text{N.m}}{\text{kg K}}$.

Equivalent molecular mass $= \dfrac{8314.3}{288.3} = 28.8 \dfrac{\text{kg}}{\text{kmole}}$.

2. (i) A cylinder, fitted with a piston, contains 0.01 kg of a perfect gas ($C_p = 800$ J/kg K, $M = 15$) at a pressure of 1 bar and a temperature of 15°C. Calculate volume of the gas.

(ii) With the piston held stationary, 0.015 kg of a second perfect gas ($C_p = 1040$ J/kg K, $M = 40$) is introduced into the cylinder. It is found that the temperature of the contents of the cylinder has increased to 40°C. For the mixture evaluate

(a) the final pressure and

(b) the specific heats and γ.

(iii) The mixture now undergoes a fully-resisted expansion to a final specific volume of 2 m³/kg as the piston moves slowly outwards; the process may be represented by $pV^{1.4} = c$, where p is the mixture pressure and V the corresponding mixture volume. Evaluate the final pressure, the work done by the mixture and the magnitude and sign of the heat transfer.

(i) Gas constant for the first gas $= \dfrac{8314.3}{15} = 554.286$ J/kg K.

Hence, $C_v = C_p - R = 800 - 554.286 = 245.714$ J/kg K.

$$pV = m\,R\,T;\; V = \frac{0.01 \times 554.286 \times 288}{10^5} = 0.01596 \text{ m}^3.$$

Volume of the first gas = 0.01596 m³.

(ii) Gas constant for the second gas $= \dfrac{8314.3}{40} = 207.86$ J/kg K.

Hence, $C_v = C_p - R = 1040 - 207.86 = 832.14$ J/kg K.

$$M \text{ for the mixture } = \frac{0.01 + 0.015}{\dfrac{0.01}{15} + \dfrac{0.015}{40}} = \frac{0.025}{0.0006667 + 0.000375} = 24 \text{ kg/Kmol.}$$

$$\text{Gas constant } R \text{ for the mixture } = \frac{8314.3}{24} = 346.43 \text{ J/kg K.}$$

$$\text{Final pressure, } p = \frac{0.025 \times 346.43 \times 313}{0.01596} = 1.698 \times 10^5 \text{ N/m}^2.$$

The specific heats for the mixture:

$$0.025 \times C_p = 0.01 \times 800 + 0.015 \times 1040.$$

Hence, $C_p = 944$ J/kg K.

$$0.025 \times C_v = 0.01 \times 245.714 + 0.015 \times 832.14.$$

Hence, $C_v = 597.5$ J/kg K.

$$\text{Ratio of specific heats, } \gamma = \frac{944}{597.5} = 1.579.$$

(iii) Final volume, $V_2 = 0.025 \times 2 = 0.05$ m³

Initial volume, $V_1 = 0.01596$ m³, initial pressure, $p_1 = 1.698 \times 10^5$ N/m²

$$p_1 V_1^{1.4} = p_2 V_2^{1.4} \qquad p_2 = 1.698 \times 10^5 \times \frac{(0.01596)^{1.4}}{(0.05)^{1.4}} = 0.343 \times 10^5 \text{ N/m}^2.$$

Final pressure, $p_2 = 0.343$ bar

$$\text{Work done} = \frac{p_1V_1 - p_2V_2}{1.4 - 1} = \frac{(1.698 \times 0.01596 - 0.343 \times 0.05) \times 10^5}{(0.4) \times 10^3}$$

$$= 2.49 \text{ kN.m.}$$

$$\text{Final temperature, } T_2 = T_1\left(\frac{V_1}{V_2}\right)^{1.4-1} = 313\left(\frac{0.01596}{0.05}\right)^{0.4} = 198.2 \text{ K.}$$

Change in internal energy

$$\Delta U = m\,C_v\,(T_2 - T_1) = 0.025 \times 0.5975 \times (198.2 - 313) = -1.715 \text{ kJ.}$$

Heat transfer = ΔU + Work done = −1.715 + 2.490 = + 0.775 kJ (to the gas).

3. One m³ of H_2 at 1 bar and 25°C is mixed with one m³ of N_2 at 1 bar and 25°C. For the mixture at the same condition, find

   (i) mole fractions of components,
   (ii) partial pressure of components,
   (iii) mass fraction of components,
   (iv) molecular mass of the mixture,
   (v) gas constant of the mixture and
   (vi) volume of the mixture.

Molecular mass, M_1 (hydrogen) = 2.016 and M_2 (nitrogen) = 28.02.

Gas constants, $R_1 = \dfrac{8314.3}{2.016} = 4124$ J/kg K; $R_2 = \dfrac{8314.3}{28.02} = 296.7$ J/kg K.

$$m_1 = \frac{1\times10^5\times1}{4124\times298} = 0.08137 \text{ kg}; \quad m_2 = \frac{1\times10^5\times1}{296.7\times298} = 1.131 \text{ kg.}$$

$$m = m_1 + m_2 = 0.08137 + 1.131 = 1.21237 \text{ kg.}$$

No. of moles, $n_1 = \dfrac{m_1}{M_1} = \dfrac{0.08137}{2.016} = 0.0403$; $n_2 = \dfrac{m_2}{M_2} = \dfrac{1.131}{28.02} = 0.0403.$

(i) *Mole fractions:*

$$x_1 = \frac{n_1}{n_1+n_2} = \frac{0.0403}{0.0806} = 0.5; \quad x_2 = \frac{n_2}{n_1+n_2} = \frac{0.0403}{0.0806} = 0.5.$$

(ii) *Partial pressures:*

$$p_1 = x_1\,p = 0.5 \times 1 = 0.5 \text{ bar}; \; p_2 = x_2\,p = 0.5 \times 1 = 0.5 \text{ bar.}$$

(iii) *Mass fractions:*

$$y_1 = \frac{m_1}{m} = \frac{0.08137}{1.21237} = 0.0671; \quad y_2 = \frac{m_2}{m} = \frac{1.131}{1.21237} = 0.9329.$$

(iv) *Molecular mass of the mixture:*

$$M = x_1 M_1 + x_2 M_2 = 0.5 \times 2.016 + 0.5 \times 28.02 = 15.018 \text{ kg/kmol.}$$

(v) *Gas constant for the mixture:*

$$R = \frac{8314.3}{15.018} = 553.6 \text{ J/kg K.}$$

(vi) *Volume of the mixture:*

$$V = \frac{mRT}{p} = \frac{1.21237 \times 553.6 \times 298}{10^5} = 2\ \text{m}^3$$

(*Note*: If two components existing at the same pressure and temperature having volumes V_1^* and V_2^* are mixed, the mixture volume V will be obtained by adding them)

4. A gaseous mixture consists of 1 kg of O_2 and 2 kg of N_2 at a pressure of 150 kPa and a temperature of 20°C. Determine the change in internal energy, enthalpy and entropy of the mixture when the mixture is heated to a temperature of 100°C (i) by constant volume process and (ii) by constant pressure process.

The gas constant for the mixture $R = \frac{(m)_{O_2}(R)_{O_2} + (m)_{N_2}(R)_{N_2}}{(m)_{O_2} + (m)_{N_2}}$

$$= \frac{1 \times \frac{8314.3}{32} + 2 \times \frac{8314.3}{28}}{1+2} = 284.56\frac{\text{N.m}}{\text{kgK}}.$$

$$(R)_{O_2} = \frac{8314.3}{32} = 259.82; \quad O_2: \ \gamma = 1.4; \quad C_v = \frac{259.82}{0.4} = 649.55\frac{\text{Nm}}{\text{kgK}}.$$

$$(R)_{N_2} = \frac{8314.3}{28} = 296.93; \quad N_2: \ \gamma = 1.4; \quad C_v = \frac{296.93}{0.4} = 742.325\frac{\text{Nm}}{\text{kgK}}.$$

$$(C_p)_{O_2} = \frac{259.82}{0.4} \times 1.4 = 909.37\frac{\text{Nm}}{\text{kgK}}.$$

$$(C_p)_{N_2} = \frac{296.93}{0.4} \times 1.4 = 1039.25\frac{\text{Nm}}{\text{kgK}}.$$

For the mixture:

$$C_v = \frac{(m)_{O_2}(C_v)_{O_2} + (m)_{N_2}(C_v)_{N_2}}{(m)_{O_2} + (m)_{N_2}} = \frac{1 \times 649.55 + 2 \times 742.325}{1+2} = 711.4\frac{\text{J}}{\text{kgK}}.$$

$$C_p = \frac{1 \times 909.37 + 2 \times 1039.25}{1+2} = 995.96\frac{\text{J}}{\text{kgK}}.$$

(i) *If the mixture is heated at constant volume*:

$$U_2 - U_1 = m\,C_v(T_2 - T_1) = 3 \times 0.7114 \times (80) = 170.736\ \text{kJ}.$$
$$H_2 - H_1 = m\,C_p(T_2 - T_1) = 3 \times 0.99596 \times (80) = 239.03\ \text{kJ}.$$

$$S_2 - S_1 = mC_v\,\ell n\frac{T_2}{T_1} + mR\,\ell n\frac{V_2}{V_1} = mC_v\,\ell n\frac{T_2}{T_1} = 3 \times 0.7114 \times \ell n\frac{373}{293} = 0.5152\frac{\text{kJ}}{\text{K}}.$$

(ii) *If the mixture is heated at constant pressure*:

$$U_2 - U_1 = m\,C_v(T_2 - T_1) = 3 \times 0.7114 \times (80) = 170.736\ \text{kJ}.$$
$$H_2 - H_1 = m\,C_p(T_2 - T_1) = 3 \times 0.99596 \times (80) = 239.03\ \text{kJ}.$$

$$S_2 - S_1 = mC_p \ell n\frac{T_2}{T_1} - mR\ell n\frac{p_2}{p_1} = mC_p \ell n\frac{T_2}{T_1} = 3 \times 0.99596 \times \ell n\frac{373}{293} = 0.72129\frac{\text{kJ}}{\text{K}}.$$

5. What is the specific humidity of atmospheric air at 760 mm of Hg (1.013 bar) and 21°C, having a relative humidity of 34 percent? Find also the dew point temperature.

From steam tables, at 21°C, $p_s = 0.0248$ bar.

Therefore, $p_v = 0.34 \times 0.0248 = 0.00843$ bar.

Specific humidity, $w = 0.622\dfrac{p_v}{p - p_v} = 0.622\dfrac{0.00843}{1.013 - 0.00843} = 0.0052$ kg/kg of d.a.

From steam tables, at 0.00843 bar, $t_{\text{dew}} = 4.5$°C.

6. The temperature in a room is 30°C and the relative humidity is 30 percent. Determine:

(i) the partial pressure of water vapour and the dew point, (ii) the specific humidity, (iii) the density of each constituent and (iv) degree of saturation. Barometer reads 760 mmHg.

At 30°C, $p_s = 0.042415$ bar, $v = 32.929$ m³/kg.

$$P_v = 0.3 \times 0.042415 = 0.012725 \text{ bar.}$$

Partial pressure of vapour = 0.012725 bar and at 0.012725 bar, $t_{\text{dew}} = 10.5$°C.

Specific humidity $= 0.622\dfrac{0.012725}{1.01325 - 0.012725} = 0.0079$ kg/kg of d.a.

Density of saturated steam at 30°C, $p_s = \dfrac{1}{v} = \dfrac{1}{32.929} = 0.03037$ kg/m³.

Since the relative humidity, $\varphi = 30\%$, density of vapour, $\rho_v = \varphi\,\rho_s = 0.3 \times 0.03037 = 0.00911$ kg/m³. This value can be checked using the ideal gas equation with $R_v = 461.5$ J/kg K.

$$\rho_v = \frac{p_s}{R_v T} = \frac{0.012725 \times 10^5}{461.5 \times 303} = 0.0091 \text{ kg/m}.$$

The result closely agrees with the answer.

Partial pressure of dry air = 1.01325 – 0.012725 = 1.000525 bar.

Density of dry air, $\rho_a = \dfrac{p_a}{R_a T} = \dfrac{1.000525 \times 10^5}{287 \times 303} = 1.1505$ kg/m³.

Check:

Specific humidity, $w = \dfrac{\rho_v}{\rho_a} = \dfrac{0.00911}{1.1505} = 0.0079$ kg/kg of d.a.

Degree of saturation $= \phi\dfrac{(p - p_s)}{(p - p_v)} = 0.3\dfrac{(1.01325 - 0.042415)}{(1.01325 - 0.012725)} = 0.2911$ or 29.11%.

7. An air water vapour mixture at 21°C and 1.01325 bar has a relative humidity of 80 percent. Determine (i) the partial pressures of vapour and dry air, (ii) the dew point, (iii) the specific humidity, (iv) the density of the mixture and (v) the amount of water vapour condensed, if the mixture is cooled at constant pressure to a temperature of 10°C.

The example is solved using a psychrometric chart drawn for a pressure of 1.01325 bar or 760 mm of Hg.

On the psychrometric chart, locate the point of intersection of 21°C dry bulb with 80 percent relative humidity line.

(i) Through the point of intersection, read the pressure of water vapour on the ordinate.

$$P_v = 0.02 \text{ bar; therefore, } p_a = p - p_v = 1.01325 - 0.02 = 0.99325 \text{ bar.}$$

(ii) The intersection of the horizontal line with the 100 percent relative humidity line gives the value of dew point temperature as 17.4°C.

(iii) Follow the same horizontal line and read the specific humidity on the relevant scale.

$$w = 0.0163 \text{ kg/kg of dry air.}$$

(iv) The point of intersection of 21°C dry bulb line with 80 percent relative humidity curve gives a specific volume of 0.86 m³/kg.

$$\text{Therefore, the density of the mixture, } p = \frac{1}{0.86} = 1.163 \text{ kg/m}^3.$$

(v) Since 10°C is below the dew point, some of the water vapour will condense and leave an air vapour mixture of 100 percent relative humidity.

The point of intersection of 10°C with 100 percent relative humidity curve gives a specific humidity of 0.0074 kg/kg of d.a. Hence, amount of water vapour condensed = 0.0163 – 0.0074 = 0.0089 kg/kg of d.a.

8. 250 kg/hr of air at 20°C DBT and 15°C WBT is mixed with 50 kg/hr of air at 35°C and 80 percent relative humidity. Determine the final state of air.

The following values are read from the chart.

For the first stream:

$w_1 = 0.0085$ kg/kg of dry air

$h_1 = 42$ kJ/kg of dry air

For the second stream:

$w_2 = 0.029$ kg/kg of dry air

$h_2 = 110$ kJ/kg of dry air

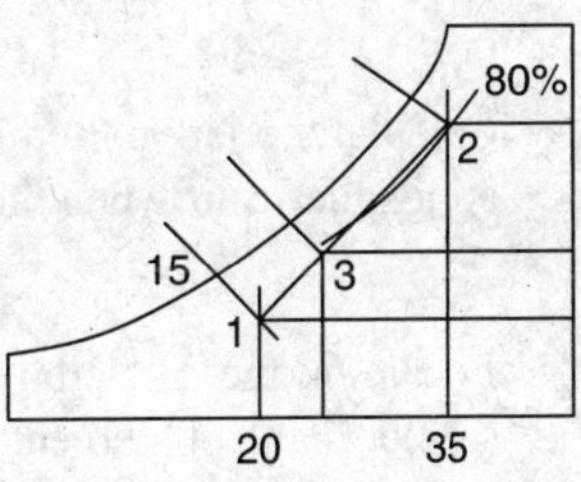

For the mixture:

$m_{a3} = 250 + 50 = 300$ kg/hr

$$\text{Specific humidity, } w_3 = \frac{w_1 m_{a1} + w_2 m_{a2}}{m_{a1} + m_{a2}} = \frac{0.0085 \times 250 + 0.029 \times 50}{300} = 0.0119 \text{ kg/kg of d.a.}$$

$$\text{Enthalpy, } h_3 = \frac{h_1 m_{a1} + h_2 m_{a2}}{m_{a1} + m_{a2}} = \frac{42 \times 250 + 110 \times 50}{300} = 53.3 \text{ kJ/kg of d.a.}$$

We can now locate the point 3 on the psychrometric chart and read the following values: DBT of the mixture, $t_{d3} = 24$°C and WBT of the mixture, $t_{w3} = 18.5$°C.

Practice Problems

1. A rigid vessel contains a mixture of 1 kg of carbon monoxide (CO) and 1 kg of hydrogen (H_2) at a pressure of 2 bar and at a temperature of 18°C. Assuming CO and H_2 to be ideal gases, evaluate

for the mixture: (i) the molar mass, (ii) the specific gas constant, (iii) the partial pressures of the constituents, (iv) the volume and (v) the specific volume.

(3.761 kg/kmol; 2210 J/kg K; 0.1344 bar; 1.8656 bar; 6.43 m³; 3.215 m³/kg)

2. Heat transfer to the mixture of the previous problem at constant volume raises its pressure to 4 bar. Assuming the mixture to be a perfect gas, evaluate for the mixture: (i) the final temperature, (ii) the increases in the specific internal energy, the specific enthalpy and the specific entropy and (iii) the magnitude of heat transfer. Atomic masses: C = 12.01, O = 16 and H = 1.008. Specific heat at constant pressure: for CO, C_p = 1042 J/kg K; for H_2, C_p = 14310 J/kg K. The molar gas constant R = 8314 J/kmol K.

(309°C; 1591 kJ/kg; 2234 kJ/kg; 3.79 kJ/kg K; 3181 kJ)

3. Atmospheric air at 1.0132 bar has a dry-bulb temperature of 32°C and a wet-bulb temperature of 26°C. Find (i) the partial pressure of water vapour, (ii) the specific humidity, (iii) the relative humidity, (iv) the dew point temperature, (v) the degree of saturation, (vi) the density of dry air in the mixture and (vii) the enthalpy of the mixture.

(0.03 bar; 0.0186 kg/kg of dry air; 62.5%; 24°C; 0.614; 1.12 kg/m³; 80.5 kJ/kg)

4. A sling psychrometer reads 30°C DBT and 20°C WBT. Find the humidity ratio, relative humidity, dew point temperature, specific volume and enthalpy of air.

(0.0105 kg/kg of dry air; 40%; 15°C; 0.875 m³/kg, 57 kJ/kg)

5. 30 m³/min of a stream of moist air at 15°C DBT and 12°C WBT are mixed with 12 m³/min of a second stream at 25°C DBT and 18°C WBT. Barometric pressure is one standard atmosphere. Determine the dry bulb and wet bulb temperatures of the resulting mixture.

(16°C and 14.5°C)

6. Calculate relative humidity, humidity ratio, dew point temperature, density and enthalpy of atmospheric air when the DBT is 35°C, WBT is 23°C and the barometer reads 760 mm Hg.

(36%; 0.0128 kg/kg of d.a.; 17.7°C; 1.1229 kg/m³; 68.15 kJ/kg of d.a)

7. Locate on the skeleton psychrometric chart five states: (i) 30°C and 80 percent relative humidity, (ii) 20°C and 60 percent relative humidity, (iii) 30°C and a wet-bulb temperature of 15°C, (iv) 30°C and a wet-bulb temperature of 20°C, (v) 30°C and a dew point of 15°C.

Objective Type Questions

1. Which of the following thermodynamic laws/relations are directly involved in determining the final properties during an adiabatic mixing process?
   (i) The first and second laws of thermodynamics
   (ii) The first law of thermodynamics and perfect gas relationship
   (iii) The second law of thermodynamics and the steady flow energy equation
   (iv) The steady flow energy equation and perfect gas relationship

2. Which property remains unchanged during an adiabatic saturation of atmospheric air?
   (i) Dew point temperature
   (ii) Dry-bulb temperature
   (iii) Wet-bulb temperature
   (iv) Relative humidity

3. Which property of a gaseous mixture can be evaluated using the Gibbs-Dalton law?
   (i) Diffusion coefficient
   (ii) Pressure
   (iii) Thermal conductivity
   (iv) Viscosity
4. What is the minimum temperature to which the hot water entering a natural draft cooling tower may be cooled?
   (i) Dry-bulb temperature of atmospheric air
   (ii) Wet-bulb temperature of atmospheric air
   (iii) Dew point temperature of atmospheric air
   (iv) Freezing temperature of water at atmospheric pressure
5. Which one of the following items is significant in the construction of psychrometric chart for the study of properties of moist air?
   (i) Lewis number
   (ii) Dalton's law of partial pressure
   (iii) The law of evaporation
   (iv) Equimolar concentration
6. The evaporation law, which states that the rate at which water evaporates from a surface is proportional to the difference $(p_g - p_d)$ of the partial vapour pressure at the surface (p_g) and that far from the surface (p_d), is due to
   (i) Dalton. (ii) Gibbs.
   (iii) Mollier. (iv) Onsager.
7. Which one of the following processes shown on the temperature-entropy diagram of pure substance represents correctly the adiabatic saturation process of moist air?

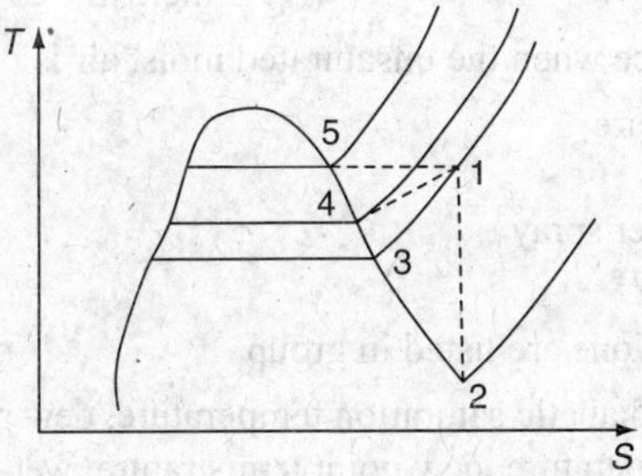

   (i) Line 1–2 (ii) Line 1–3
   (iii) Line 1–4 (iv) Line 1–5
8. The isentropic process of an incompressible substance can also be called as
   (i) isometric (ii) isobaric
   (iii) isothermal (iv) isenthalpic

9. Match List I (items) with List II (temperatures) for a natural draft cooling tower supplied with hot water sprayed at the top and cold water collected at the bottom of the tower and select the correct answer using the codes given in the following lists:

List I (Item)	*List II (Temperature)*
A. Dry-bulb temperature of atmosphere	1. 18°C
B. Wet-bulb temperature of atmosphere	2. 25°C
C. Dew point temperature of atmosphere	3. 21°C
D. Cold water temperature	4. 20°C

Codes:

	A	B	C	D		A	B	C	D
(i)	2	4	1	3	(ii)	2	4	3	1
(iii)	4	3	2	1	(iv)	3	4	1	2

10. Identify the enthalpy correction line in the outline of the following psychrometric chart.

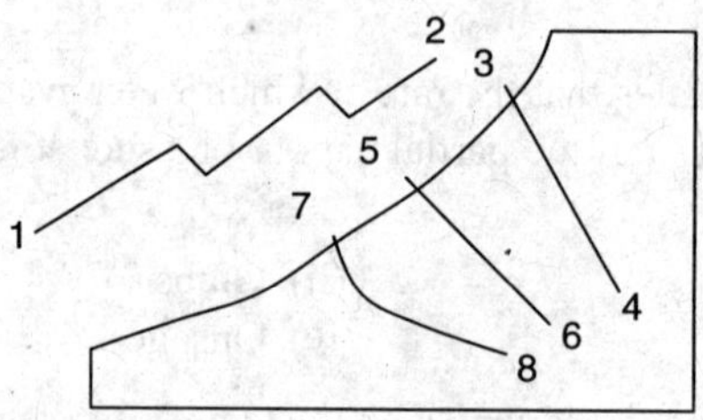

(i) Line 1–2
(ii) Line 3–4
(iii) Line 5–6
(iv) Line 7–8

11. The dew formation takes place when the unsaturated moist air is
    (i) cooled at constant pressure
    (ii) saturated adiabatically
    (iii) humidified with hot water spray
    (iv) dehumidified chemically

12. Thermodynamic properties alone are listed in group
    (i) dry-bulb temperature, adiabatic saturation temperature, dew point temperature
    (ii) adiabatic saturation temperature, dew point temperature, wet-bulb temperature
    (iii) dew point temperature, wet-bulb temperature, dry-bulb temperature
    (iv) wet-bulb temperature, dry-bulb temperature, adiabatic saturation temperature

13. If the atmospheric air were saturated, then its specific humidity and relative humidity are respectively
    (i) 1.0 and 1.0
    (ii) < 1.0 and < 1.0
    (iii) < 1.0 and 1.0
    (iv) 1.0 and < 1.0

14. Evaporative air cooling is effective when
    (i) dry-bulb temperature is very close to wet-bulb temperature
    (ii) dry-bulb temperature is high and relative humidity is high
    (iii) dry-bulb temperature is high and relative humidity is low
    (iv) dry-bulb temperature is low and relative humidity is high
15. With reference to the psychrometric processes shown on the following chart, match List I with List II and select the correct answer using the codes given in the following lists:

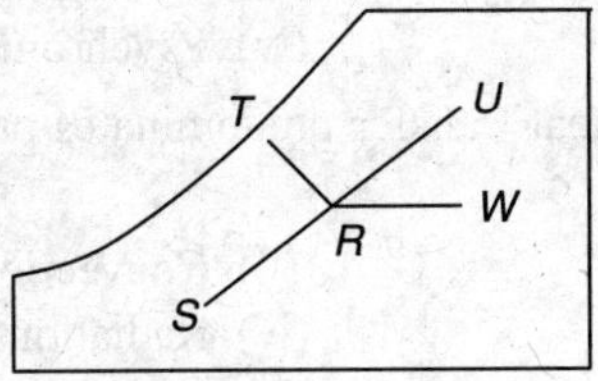

List I	*List II*
A. Process *RS*	1. Cooling and humidifying
B. Process *RT*	2. Sensible heating
C. Process *RU*	3. Cooling and dehumidifying
D. Process *RW*	4. Heating and humidifying

Codes:

	A	B	C	D
(i)	2	3	1	4
(ii)	1	4	2	3
(iii)	3	1	4	2
(iv)	3	2	1	4

16. The constant 0.622, which appears in the specific humidity expression of moist air $\left(\text{i.e., SH} = 0.622\frac{p_v}{p - p_v}\right)$, is
    (i) the ratio of molecular mass of water-vapour to molecular mass of air
    (ii) the ratio of molecular mass of air to molecular mass of water-vapour
    (iii) the gas constant of moist air
    (iv) a constant obtained from experiment
17. Which process is represented as straight line that is diagonal upwards to the right (NE direction) on the psychrometric chart?
    (i) Evaporative cooling
    (ii) Humidifying
    (iii) Heating and humidifying
    (iv) Cooling and dehumidifying

18. Assertion (A): Liquid nitrogen finds wide application as cryogenic fluid.

    Reason (R): The composition of nitrogen is the highest in the atmosphere.

    (i) Both A and R are true and R is the correct explanation of A
    (ii) Both A and R are true and R is not the correct explanation of A
    (iii) A is true but R is false
    (iv) A is false but R is true

19. Which instrument shows directly the humidity of the atmosphere?

    (i) Hydroscope (ii) Hyetograph
    (iii) Hygrograph (iv) Psychrometer

20. Which primary mode/modes of heat transfer predominates/predominate in the latent heat release from supersaturating steam?

    (i) Conduction (ii) Convection
    (iii) Conduction and convection (iv) Radiation

Answers

1. (ii)	2. (iii)	3. (ii)	4. (ii)	5. (i)	6. (i)	7. (iii)	8. (iii)	9. (i)	10. (iv)
11. (i)	12. (i)	13. (iii)	14. (iii)	15. (iii)	16. (i)	17. (iii)	18. (i)	19. (iii)	20. (iv)

9

Combustion and Flue Gas Analysis

1. Explain how chemical symbols are assigned for substances and chemical compounds.

Ans: Chemical symbols:

The elements/substances which have been discovered carry distinguishing names. Instead of writing the name of the element, only the first letter of the English or Latin names of the commoner elements is used to denote it, while less common elements, which share the same initial letter, are distinguished by an additional letter, thus carbon is denoted by C and chlorine is denoted by Cl. Chemical compounds are indicated by a combination of symbols which represent the separate molecules from which the compound was formed. Hydrogen uniting with oxygen produces steam. Symbolically the reaction is denoted by

$$2\,H_2 + O_2 \rightleftharpoons 2\,H_2O$$

Table 9.1 gives the symbols, relative atomic mass and molecular mass of some of the elements which are the constituents of commonly used fuels and the constituents of products of combustion.

Table 9.1 Atomic and Molecular Masses of Elements and Compounds

Name of Substance	Symbol	Atomic Mass	Molecular Mass	
Carbon	C	12	12	Element
Hydrogen	H_2	1	2	Element
Nitrogen	N_2	14	28	Element
Oxygen	O_2	16	32	Element
Sulphur	S	32	32	Element
Carbon monoxide	CO	28	12 + 16 = 28	Compound
Carbon dioxide	CO_2		12 + 32 = 44	Compound
Water	H_2O		2 + 16 = 18	Compound

2. Define a fuel.

Ans: A fuel is a defined as a substance, which once it has been raised to ignition temperature, continues to burn without any external support provided sufficient quantity of oxygen is available. The major constituents of a fuel are Carbon and Hydrogen and most fuels belong to the family of substances called Hydrocarbons. Fuels may be solid (coal), liquid (oil) or gaseous (natural gas).

3. What are exothermic and endothermic chemical reactions?

Ans: Combustion means burning. It is a chemical reaction, which is accompanied with the release of energy. Such a chemical reaction is called exothermic. A chemical reaction in which energy is absorbed is called endothermic.

4. What is the significance and uses of chemical equations?

Ans: A chemical equation is a symbolic way of representing chemical reaction which takes place when molecules of two or more substances combine together. In writing the chemical equations some times more than one molecule will have to be represented. In such cases the number of molecules is represented by putting the same number on the left of the symbol, for example, $2O_2$ represents two molecules of oxygen and $4CO_2$ represents 4 molecules of carbon dioxide.

How many kg of oxygen are required to burn completely one kg of hydrogen?

The equation of this reaction is

$$2H_2 + O_2 \rightleftharpoons 2\,H_2O$$

$$2(2 \times 1) + (2 \times 16) \rightleftharpoons 2(2 \times 1 + 16)$$

Hence, 4 units of H_2 combine with 32 units of O_2 to form 36 units of steam or 4 kg of H_2 will require 32 kg of O_2 to form 36 kg of steam (H_2O).

Mass of oxygen required for complete combustion of 1 kg of hydrogen $= \frac{32}{4} = 8$ kg.

Consider the following cases:

$$C + O_2 \rightleftharpoons CO_2$$

$$(12) + (2 \times 16) \rightleftharpoons (12 + 2 \times 16)$$

(i.e. 12 kg of C will require 32 kg of O_2 to form 44 kg of CO_2)

$$2C + O_2 \rightleftharpoons 2CO$$

$$2(12) + (2 \times 16) \rightleftharpoons 2(12 + 16)$$

(i.e. 24 kg of C will combine with 32 kg of O_2 to form 56 kg of CO)

$$S + O_2 \rightleftharpoons SO_2$$

$$32 + (2 \times 16) \rightleftharpoons (32 + 2 \times 16)$$

(i.e. 32 kg of SO_2 will require 32 kg of O_2 to form 64 kg of SO_2)

Formation of chemical equations: Since a chemical equation represents a redistribution of atoms on combustion, it follows that the same number of atoms of any one element must exist on both sides of the equation. Every atom of the interacting substances on the LHS of a chemical equation must also appear on the RHS of the equation, although usually in some fresh associations of atoms. This information is valuable, because it enables us to build up chemical equations.

5. Form the chemical equation that represents the complete combustion of marsh gas, CH_4.

Ans: When marsh gas is burnt completely, carbon dioxide (CO_2) and steam (H_2O) are produced; so we can write

$$CH_4 + xO_2 \rightleftharpoons CO_2 + y\,H_2O$$

The values of x and y may then be determined from the condition that the same number of atoms of each element exist on both sides of the equation.

Hence,

$$C: 1 = 1$$
$$H_2: 4 = 2\,y \rightarrow y = 2$$
$$O_2: 2\,x = 2 + y \rightarrow x = 2$$

Therefore,

$$CH_4 + 2O_2 \rightleftharpoons CO_2 + 2H_2O$$

$$(12 + 4 \times 1) + 2\,(2 \times 16) \rightleftharpoons (12 + 2 \times 16) + 2\,(2 \times 1 + 16)$$

$$16 \qquad 64 \qquad 44 \qquad 36$$

$$80 \qquad\qquad 80$$

From this example, it should be understood that a chemical equation is not only a qualitative expression, which expresses the result of reactions, but that it also has a quantitative significance, and in this respect the equation obeys mathematical laws, the total mass on the two sides being the same.

6. In what respect the combustion of gaseous fuels differs from the combustion of solid and liquid fuels?

Ans: Since the mass of a gas is difficult to measure directly, it is more usual to consider the volumes which unite rather than mass. Take for instance the molecular equation

$$2H_2 + O_2 \rightleftharpoons 2H_2O$$

and multiply this through by a constant n which represents the number of molecules in 1 m^3 of a gas at NTP conditions, then

$$2nH_2 + nO_2 \rightleftharpoons 2nH_2O$$

Two m^3 of H_2 combine with one m^3 of O_2 to produce 2 m^3 of steam at the same state as H_2 and O_2; provided the steam produced obeys the laws of perfect gases. At high temperatures, steam obeys the gas laws and at low temperatures the prefix $2n$ for H_2O has no significance since the steam would condense and its volume would then be negligible in comparison with that of the gas.

7. Explain the following: stoichiometric air, excess air.

Ans: The application of chemistry to combustion problems is in the determination of the minimum amount of air required to burn completely a solid or liquid fuel of known composition. This air is also known as stoichiometric air which is defined as the quantity of air required to burn one unit of fuel completely. If the theoretical or stoichiometric mass of air required is known, then the actual air supply can be adjusted to avoid undue heat losses due to insufficient or excess air.

Example 1: A fuel oil contains 84 percent of C and 16 percent of H_2. Determine the minimum mass of air required to burn one kg of this fuel.

The chemical equations for the reaction are

$$\underset{12\quad 32\qquad 44}{C + O_2 \rightleftharpoons CO_2} \qquad \text{and} \qquad \underset{4\qquad 32\qquad 36}{2H_2 + O_2 \rightleftharpoons 2H_2O}$$

Hence, 1 kg of carbon requires $\frac{32}{12}$ kg of O_2 and produces $\frac{44}{12}$ kg of CO_2 and 1 kg of H_2 requires 8 kg of O_2 to form 9 kg of steam.

In 1 kg of fuel, there are 0.84 kg of C and 0.16 kg of H_2, so that the total oxygen required is $\frac{32}{12} \times 0.84 + 8 \times 0.16 = 3.52\,\text{kg}$.

Air contains a large proportion of N_2, and for engineering purposes we can assume the gravimetric analysis as N_2, 77 percent; O_2, 23 percent, and the volumetric analysis as N_2, 79 percent; O_2, 21 percent. Hence, the mass of air required to supply 3.52 kg of O_2 is

$$3.52 \times \frac{100}{23} = 15.3\text{ kg.}$$

In practice, to ensure that the fuel is completely burnt, excess air to the extent of nearly 30 percent must be supplied to liquid fuels and even more to solid fuels. Of course, it is undesirable to supply more than the stoichiometric quantity of air when fuel is being burned for heating purposes, because the excess air merely lowers the gas temperature and so lowers the heat transfer rate. The quantity of excess air is usually expressed as

$$\text{Percent excess air} = \frac{x - x_{\text{stoich}}}{x_{\text{stoich}}} \times 100\,\%$$

Gas turbines, however, must operate with about 300 percent excess air in order to avoid melting the turbine blades.

8. Explain the mixture strength as applied to petrol engines.

Ans: For petrol engines, a different term is adopted for comparing x and x_{stoich}. In petrol engines, x can have values both above and below x_{stoich}. It is therefore usual to express the variation in air-fuel ratio in terms of the mixture strength defined by the relation

$$\text{Mixture strength} = \frac{x_{\text{stoich}}}{x} \times 100\%$$

A typical range of mixture strength is from a weak mixture of 90 percent to a rich mixture of 120 percent. The terms 'weak' and 'rich' express, respectively, the deficiency and the excess of the fuel in a given quantity of air entering the engine, by comparison with the stoichiometric quantity of fuel. The 'theoretical', or stoichiometric, air-fuel ratio corresponds to mixture strength of 100 percent. When the actual air-fuel ratio x is less than x_{stoich}, the concentration in the exhaust gases is negative; the assumption is therefore wrong. An approximate rule is to assume that all the carbon is burned to the monoxide, CO and thereafter any remaining oxygen is equally shared between this gas, oxidizing to CO_2 and the hydrogen, oxidizing to H_2O. The exhaust gases then comprise of CO_2, CO, H_2O, H_2 and N_2.

The minimum volume of air required for complete combustion of a gaseous fuel of known composition: Let us consider the combustion of marsh gas, CH_4. The complete reaction is

$$CH_4 + 2\,O_2 \rightleftharpoons CO_2 + 2\,H_2O$$

In terms of volume this reads: one m³ of CH_4 combines with two m³ of O_2 to form one m³ of CO_2 and two m³ of H_2O (steam); provided all these volumes are measured at the same pressure and temperature, and the constituents may be considered to obey laws of perfect gases.

Example 2: Find the volume of air required for the combustion of 1 m³ of a gas having the analysis CH_4, 39.5 percent; H_2, 46 percent; N_2, 0.5 percent; CO, 7.5 percent; H_2O, 2 percent; CO_2, 4.5 percent. What will be the volumetric analysis of the products of combustion at 100°C?

Table 9.2 The Computation of the Air Required is Effected by the Tabular Method

Constituent	Volume (m³)	Combustion Equation	O_2 Reqd (m³)	Products CO_2 (m³)	Products H_2O (m³)	N_2 (m³)
CH_4	0.395	$CH_4 + 2O_2 \rightleftharpoons CO_2 + 2H_2O$	0.7900	0.395	0.79	–
H_2	0.460	$2H_2 + O_2 \rightleftharpoons 2H_2O$	0.2300	–	0.46	–
N_2	0.005	–	–	–	–	0.005
CO	0.075	$2CO + O_2 \rightleftharpoons 2CO_2$	0.0375	0.075	–	–
H_2O	0.020	–	–	–	0.02	–
CO_2	0.045	–	–	0.045	–	–
	1.000	Total O_2 required	1.0575	0.515	1.27	0.005

Air required at the supply temperature and pressure of the gas $= \dfrac{1.0575}{0.21} = 5.04\text{ m}^3$.

N_2 associated with the supply air $= 5.04 \times 0.79 = 3.9816\text{ m}^3$

Total volume of N_2 in the products $= 3.9816 + 0.005 = 3.9866\text{ m}^3$

Total volume of CO_2 in the products $= 0.515\text{ m}^3$

Total volume of H_2O (steam) in the products $= 1.27\text{ m}^3$

Total volume of the flue gas $= 3.9866 + 0.515 + 1.27 = 5.7716\text{ m}^3$

The analysis of the products of combustion will be H_2O (steam), 1.27; CO_2, 0.515; and N_2, 3.9866 m³ or expressed as percentage, H_2O, 22 percent; CO_2, 8.92 percent and N_2, 69.07 percent.

9. Draw a neat sketch of Orsat apparatus which is used for flue gas analysis and explain. Mention the different chemicals used for absorbing different constituents.

Ans: The Orsat apparatus is very convenient for analyzing flue gases containing CO_2, O_2 and CO by volume. If the remainder of flue gas is nitrogen, it will also be known by difference. The unit consists of three pairs of flasks containing chemicals to absorb different gases. In its simplest form, the apparatus is arranged for the absorption of CO_2, O_2 and CO, the nitrogen content of the gas being obtained by difference.

Referring to Figure 9.1, *C* is the water-jacketed eudiometer graduated up to 100 cubic centimetres. The base of the eudiometer is connected to an aspirator bottle *A*, the purpose of which is to charge or discharge the eudiometer. Flasks 1, 2 and 3, with the pair flasks behind them, contain respectively solutions of caustic soda (NaOH) for the absorption of CO_2, pyrogallic acid in alkali for the absorption of

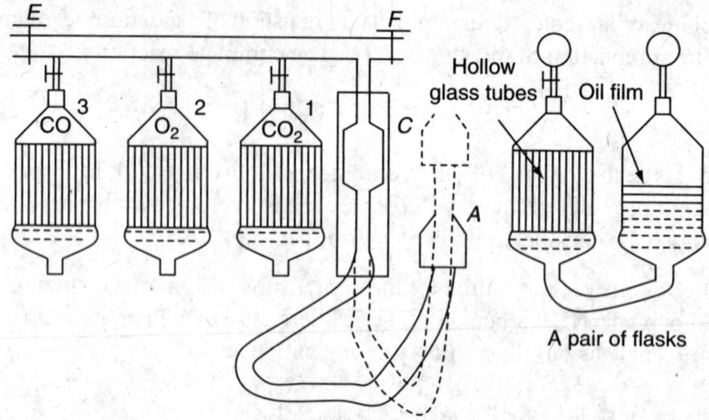

Figure 9.1 Orsat Apparatus

O_2 and cuprous chloride (which is made by dissolving copper oxide (CuO) in about twenty times its mass of concentrated hydrochloric acid (HCl)) for the absorption of CO. To accelerate absorption of the gases, the flasks 1, 2 and 3 are packed with small glass tubes, the wetted surfaces of which is bared when, by raising the aspirator bottle and opening the tap communicating with the particular flask, the gas sample is introduced into the flask. The expelled reagent is accommodated in the open pair flask, and is protected from reaction with the atmosphere by a film of oil.

On opening the cock *F* the aspirator bottle *A* is raised to expel any residual gas in the eudiometer. Care should be taken to prevent the salt water (salt water is used to avoid absorption of the gases), discharged from *A* from entering the horizontal tube between *EF* of the apparatus. On closing *F*, the level of the reagents in flasks 1, 2 and 3 must be raised to the level of tight-fitting cork by opening the respective cocks above the flasks and lowering gently the aspirator bottle. Expel the exhausted gas through *F* and on closing this cock open *E* and lower bottle *A* to induce gas sample into the eudiometer. Close *E* and test for gas leaks in the apparatus by placing *A* on top of the apparatus and note if saline solution rises in the eudiometer, or if the reagents fall in the flasks. If all is satisfactory, expel the trial charge and induce a fresh one of 100 cubic centimetres. To ensure that the test is conducted at constant pressure the water level in *A* should be brought horizontal with that in the eudiometer, which is protected from temperature changes by a water jacket.

The CO_2 will be the first to be absorbed, and this is effected by opening the cock communicating with flask 1 and raising the aspirator bottle *A* until the saline solution reaches 100 mark on the eudiometer. To accelerate the absorption, the aspirator bottle should be raised and lowered slowly. After a few trials, the aspirator bottle should be lowered to bring the reagent back to its original level, when the cock on flask 1 should be closed. After leveling, the liquid *A* and *C* read the eudiometer which gives the percentage of CO_2 directly. The procedure is repeated with other flasks to find the content of O_2 and CO. If the remainder of the flue gas is nitrogen, it will be known by difference. It should be noted that this apparatus is not sufficiently sensitive to give accurate values of the CO content, because of the small amount to be measured, but it will detect it. The apparatus is handy, easy to use and fairly accurate. In order to ensure that only dry gas is being analyzed, the flue gas is passed through U-tube, containing calcium chloride or some other drying agent, before it enters the apparatus (cock *E*).

Solved Problems

1. A solid fuel, having a composition by mass of 84 percent carbon, 14 percent hydrogen and 2 percent sulphur, is burned completely with oxygen. Determine the minimum quantity of oxygen required for combustion. If the fuel is burned completely with air, find the quantity of air required for combustion.

Table 9.3 Combustion of 1 kg of Fuel

Constituent	Mass (kg)	Chemical Equation	Mass of O_2 Required (kg)
C	0.84	$C + O_2 \rightleftharpoons CO_2$	2.24
H_2	0.14	$2H_2 + O_2 \rightleftharpoons 2H_2O$	1.12
S	0.02	$S + O_2 \rightleftharpoons SO_2$	0.02

The minimum quantity of O_2 required = 2.24 + 1.12 + 0.02 = 3.38 kg/kg of fuel.

The minimum quantity of air required $= \dfrac{3.38}{0.23} = 14.69$ kg/kg of fuel

2. The percentage composition of a sample of coal was found to be: carbon = 89.3 percent, hydrogen = 5.0 percent, oxygen = 4.2 percent and nitrogen = 1.5 percent. Determine the stoichiometric air-fuel ratio by mass. If 30 percent excess air is supplied, find the percentage composition of dry flue gas by volume.

Table 9.4 Combustion of 1 kg of Fuel

Substance	Mass (kg)	Chemical Equation	Mass of O_2 Required (kg)	CO_2	H_2O	N_2
C	0.893	$C + O_2 \rightleftharpoons CO_2$	2.381	3.274	–	–
H_2	0.050	$2H_2 + O_2 \rightleftharpoons 2H_2O$	0.400	–	0.45	–
N_2	0.015	–	–	–	–	0.015
O_2	0.042	–	–	–	–	–

Mass of O_2 required for combustion = 2.381 + 0.400 = 2.781 kg

Mass of O_2 to be supplied by air = 2.781 – 0.042 = 2.739 kg

Hence, stoichiometric air-fuel ratio $= \dfrac{2.739}{0.23} = 11.91$ kg/kg of fuel

Mass of air supplied = 1.3 × 11.91 = 15.483 kg

Mass of O_2 supplied = 0.23 × 15.483 = 3.561 kg

Mass of N_2 supplied = 0.77 × 15.483 = 11.922 kg

Mass of O_2 in the flue gas = 3.561 + 0.042 – 2.781 = 0.822 kg

Mass of N_2 in the flue gas = 11.922 + 0.015 = 11.937 kg

Mass of CO_2 in the flue gas = 3.274 kg

Table 9.5 Volumetric Analysis of Dry Flue Gas

Substance	Mass (kg)	Molecular Mass	Proportionate Volume	Percentage Composition
CO_2	3.274	44	0.07441	14.14%
O_2	0.822	32	0.02569	4.88%
N_2	11.937	28	0.42632	80.98%
		Total	0.52642	

3. A gas has the following composition by volume: hydrogen, 45 percent; nitrogen, 4 percent; marsh gas (CH_4), 36 percent; carbon monoxide, 15 percent. Find the volume of air required for the combustion of 1 m³ of the gas (oxygen in air is 21 percent by volume).

Table 9.6 Combustion of 1 m³ of Gas

Constituents	Volume (m³)	Combustion Equation	O_2 Reqd. (m³)	CO_2	H_2O	N_2
H_2	0.45	$H_2 + \frac{1}{2} O_2 \rightleftharpoons H_2O$	0.225	–	0.45	–
CH_4	0.36	$CH_4 + 2 O_2 \rightleftharpoons CO_2 + 2 H_2O$	0.72	0.36	0.72	–
CO	0.15	$CO + \frac{1}{2} O_2 \rightleftharpoons CO_2$	0.075	0.15	–	–
N_2	0.04	–	–	–	–	0.04
		Total volume of O_2 reqd.	1.020			

$$\text{Volume of air required for combustion} = \frac{1.02}{0.21} = 4.857 \text{ m}^3 \text{ of air per m}^3 \text{ of gas.}$$

4. Determine the composition (on a dry basis) of the products formed by complete combustion (with 20 percent excess air) of a casinghead gas with the following volumetric composition: CH_4, 60 percent; C_2H_6, 30 percent; N_2, 10 percent.

Table 9.7 Combustion of 1 m³ of Casinghead Gas

Constituent	Volume (m³)	Combustion Equation	O_2 Reqd. (m³)	CO_2 (m³)	H_2O (m³)	N_2 (m³)
CH_4	0.60	$CH_4 + 2 O_2 \rightleftharpoons CO_2 + 2 H_2O$	1.20	0.60	1.20	–
C_2H_6	0.30	$C_2H_6 + 3\frac{1}{2} O_2 \rightleftharpoons 2 CO_2 + 3 H_2O$	1.05	0.60	0.90	–
N_2	0.10	–	–	–	–	0.10
		Total O_2 required	2.25	1.20	2.10	0.10

$$\text{Air required at the supply temperature and pressure of the gas} = \frac{2.25}{0.21} = 10.71 \text{ m}^3$$

Actual quantity of air supplied = 1.2 × 10.71 = 12.852 m³ per m³ of the gas.

Volume of N_2 in the air = 0.79 × 12.852 = 10.153 m³

Total volume of N_2 in the products = 10.153 + 0.10 = 10.253 m³

Volume of O_2 in the products = 0.21 × 12.852 − 2.25 = 2.699 − 2.25 = 0.449 m³

Volume of CO_2 in the products = 1.20 m^3

Total volume of dry flue gases = 10.253 + 0.449 + 1.20 = 11.902 m^3

Table 9.8 Composition of Products on Dry Basis

Constituent	Volume (m^3)	Percentage
CO_2	1.200	10.08%
O_2	0.449	3.77%
N_2	10.253	86.15%
	11.902	

5. A producer has the following analysis by volume: hydrogen, 18.23 percent; carbon monoxide, 25.07 percent; carbon dioxide, 5.2 percent; nitrogen, 51 percent. Estimate the minimum quantity of air required for the complete combustion of 1 m^3 of the gas and the percentage contraction in volume after combustion and the composition of the products of combustion.

Table 9.9 Combustion of 1 m^3 of the Producer Gas

Constituent	Volume (m^3)	Combustion Equation	O_2 Reqd. (m^3)	CO_2 (m^3)	H_2O (m^3)	N_2 (m^3)
H_2	0.1823	$H_2 + \frac{1}{2} O_2 \rightleftharpoons H_2O$	0.09115	–	0.1823	–
CO	0.2507	$CO + \frac{1}{2} O_2 \rightleftharpoons CO_2$	0.12535	0.2507	–	–
CO_2	0.052	–	–	0.0520	–	–
N_2	0.51	–	–	–	–	0.51
		Total O_2 required	0.2165	0.3027	0.1823	0.51

$$\text{Minimum quantity of air required to burn 1 m}^3 \text{ of gas} = \frac{0.2165}{0.21} = 1.0309 \text{ m}^3$$

Volume of CO_2 = 0.3027 m^3

Volume of H_2O = 0.1823 m^3

Total volume of N_2 = 0.51 + 0.79 × 1.0309 = 1.3244 m^3

Table 9.10 Composition of Products of Combustion

Constituent	Volume (m^3)	Percentage
CO_2	0.3027	16.7%
H_2O	0.1823	10.1%
N_2	1.3244	73.2%
	1.8094	

$$\text{Contraction in Volume} = \frac{0.1823}{1.8094} \times 100$$

$$= 10.07\%$$

6. A stoichiometric mixture of *n*-octane (C_8H_{18}) and air burns completely. Determine the stoichiometric air-fuel ratio by mass and the molar composition of the products of combustion.

The combustion equation is

$$C_8H_{18} + 12\frac{1}{2}\, O_2 = 8\, CO_2 + 9\, H_2O$$

$$(8 \times 12 + 18 \times 1) + (12\frac{1}{2} \times 32) = [8(12 + 32)] + [9(2 \times 1 + 16)]$$

$$(114) + (400) = (352) + (162)$$

Mass of oxygen required for complete combustion is $\frac{400}{114} = 3.509$ kg

Mass of air required for complete combustion is $\frac{3.509}{0.23} = 15.256$ kg

Stoichiometric air-fuel ratio is 15.256 kg of air per kg of fuel

Mass of CO_2: $\frac{352}{114} = 3.0877$ kg; Mass of H_2O: $\frac{162}{114} = 1.421$ kg;

Mass of N_2: $15.256 - 3.509 = 11.747$ kg

Table 9.11 Products of Combustion

Constituent	Mass (m)	Molecular Mass (M)	Proportional Volume (m/M)	Molar Composition
CO_2	3.0877	44	0.07017	12.34%
H_2O	1.4210	18	0.07894	13.88%
N_2	11.7470	28	0.41953	73.78%
			Σ 0.56864	

7. A liquid fuel has a composition, by mass of 84 percent carbon and 16 percent of hydrogen. The fuel is used in an engine at mixture strength of 115 percent. On the assumption that all the hydrogen is burned completely and that there is no 'free' oxygen in the exhaust gas estimate (i) the molar composition of the exhaust gas and (ii) the 'Orsat analysis' of the exhaust gas.

Table 9.12 Combustion of 1 kg of Fuel

Constituent	Mass (kg)	Chemical Equation	O_2 Reqd.	Products CO_2	Products H_2O
C	0.84	$C + O_2 \rightleftharpoons CO_2$ 12 32 44	2.24	3.08	–
H_2	0.16	$2H_2 + O_2 \rightleftharpoons 2H_2O$ 4 32 36	1.28	–	1.44
			Total O_2 required, Σ 3.52		

Stoichiometric/minimum air required $= \frac{3.52}{0.23} = 15.304$ kg

$$\text{Mixture strength} = \frac{\text{Stoichiometric air fuel ratio}}{\text{Actual air fuel ratio}} \times 100$$

$$\text{Actual air fuel ratio} = \frac{15.304}{115} \times 100 = 13.308 \text{ kg/kg of fuel}$$

Since the actual air supplied is less than the stoichiometric air and all the hydrogen is burned completely, there will be incomplete combustion of carbon. There will be CO_2 and CO in the exhaust gases.

Mass of O_2 supplied $= 0.23 \times 13.308 = 3.0608$ kg

Mass of O_2 required for complete combustion of $H_2 = 1.28$ kg

Mass of O_2 available for combustion of carbon $= 3.0608 - 1.28 = 1.7808$ kg

Let α be the mass of carbon which will give CO_2 on combustion and $(0.84 - \alpha)$ will be the mass of carbon which burns to give CO.

Mass of O_2 required for combustion of carbon into $CO_2 = \left(\frac{32}{12}\right)\alpha$

Mass of O_2 required for combustion of carbon into $CO = \left(\frac{32}{24}\right)(0.84-\alpha)$

Hence, $\left(\frac{32}{12}\right)\alpha + \left(\frac{32}{24}\right)(0.84-\alpha) = 1.7808$ and on simplification we get $\alpha = 0.4956$

Combustion of carbon: 0.4956 kg burns to CO_2 and 0.3444 kg burns to CO

Therefore, mass of $CO_2 = \left(\frac{44}{12}\right) \times 0.4956 = 1.8172$ kg; mass of $CO = \left(\frac{56}{24}\right) \times 0.3444 = 0.8036$ kg;

mass of $H_2O = 1.44$ kg (from the table) and mass of $N_2 = 0.77 \times 13.308 = 10.247$ kg.

Table 9.13 Molar Composition of the Exhaust Gas

Constituent	Mass (m, kg)	Molecular Mass (M)	Molar Composition Kmol (m/M)
CO_2	1.8172	44	0.041
CO	0.8036	28	0.029
H_2O	1.4400	18	0.080
N_2	10.2470	28	0.366

Table 9.14 Orsat Analysis of the Exhaust Gas

Constituent	Molar Composition	Percentage
CO_2	0.041	9.4%
CO	0.029	6.7%
N_2	0.366	83.9%
	Σ 0.436	

8. A boiler uses coal having the following percentage composition by mass: C = 80 percent, H_2 = 6.5 percent, O_2 = 4.5 percent, Moisture = 2 percent and the remainder 7 percent being ash. Calculate the stoichiometric air required for complete combustion of 1 kg of this fuel. If the actual air supplied is 17 kg per kg of fuel, hydrogen is completely burnt, 80 percent of carbon burns to CO_2 and the remaining 20 percent to CO determine the volumetric analysis of the dry products of combustion.

Table 9.15 Combustion of 1 kg of Coal

Constituent	Mass (kg)	Combustion Equation	O_2 Reqd. (kg)	CO_2	H_2O	N_2
C	0.800	$C + O_2 \rightleftharpoons CO_2$	2.133	2.933	–	–
H_2	0.065	$2H_2 + O_2 \rightleftharpoons 2H_2O$	0.520	–	0.585	–
O_2	0.045	–	–	–	–	–
H_2O	0.020	–	–	–	0.020	–
Ash	0.070	–	–	–	–	–
			$\Sigma O_2 = 2.653$ kg			

Total mass of O_2 required $= 2.653 - 0.045 = 2.608$ kg

Stoichiometric air required $= \dfrac{2.608}{0.23} = 11.34$ kg/kg of fuel

Actual quantity of air supplied = 17 kg which contains 3.91 kg of O_2 and 13.09 kg of N_2.

Mass of carbon burns to $CO_2 = 0.8 \times 0.8 = 0.64$ kg

Mass of carbon burns to $CO = 0.2 \times 0.8 = 0.16$ kg

Table 9.16 Combustion of 1 kg of Fuel with 17 kg of Air

Const.	Mass (kg)	Combustion Equation	O_2 Reqd. (kg)	CO_2	CO^*	H_2O	O_2	N_2
C	0.64	$C + O_2 \rightleftharpoons CO_2$	1.707	2.347	–	–		
C	0.16	$2C + O_2 \rightleftharpoons 2CO$	0.213	–	0.373	–		
H_2	0.065	$2H_2 + O_2 \rightleftharpoons 2H_2O$	0.520	–	–	0.585		
H_2O	0.02	–	–	–	–	0.020		
O_2	0.045	–	*	–	–	–	*	
	3.91	–	*	–	–	–	*	
N_2	13.09	–	–	–	–	–		13.09

*Mass of O_2 in the products of combustion $= 0.045 + 3.91 - (1.707 + 0.213 + 0.52)$

$$= 3.955 - 2.440 = 1.515 \text{ kg}$$

Table 9.17 Volumetric Composition of Dry Products of Combustion

Constituent	Mass (m)	Molecular Mass (M)	Proportionate Volume (m/M)	Percentage
CO_2	2.347	44	0.05334	9.17%
CO	0.373	28	0.01332	2.29%
O_2	1.515	32	0.04734	8.14%
N_2	13.090	28	0.46750	80.40%
			Σ dry Products = 0.5815	

9. A boiler burns fuel having a composition by mass of 82 percent carbon, 5 percent hydrogen, 6 percent oxygen and 7 percent ash. The composition by volume of dry flue gas is 13.2 percent carbon dioxide, 1.5 percent carbon monoxide, 6.8 percent oxygen and 78.5 percent nitrogen. Evaluate (i) the air-fuel ratio, (ii) the percentage excess air supplied and (iii) the volume of air handled by the boiler fan in m^3/s, given that the hourly fuel consumption of the boiler is 1000 kg. Assume that the air enters the fan at a pressure of 100×10^3 N/m^2 and a temperature of 40°C.

   (i) 1 mole of exhaust (flue) gas contains 0.132 moles of (monatomic) carbon in CO_2, 0.015 moles of (monatomic) carbon in CO and 0.785 moles of (diatomic) nitrogen.

   The mass ratio of carbon to nitrogen in the flue gas is therefore,

$$\frac{12(0.132+0.015)}{28(0.785)} = 0.08025 \quad (1)$$

   In 1 kg of fuel, the mass of carbon present is 0.82 kg and mass of oxygen is 0.06 kg.

   If the air fuel ratio is AFR, then the mass of nitrogen present in the combustion process will be equal to $0.77\left(\text{AFR} - \frac{0.06}{0.23}\right) = 0.77(\text{AFR} - 0.260869)$.

$$\text{The mass ratio of carbon to nitrogen in the reactants} = \frac{0.82}{0.77(\text{AFR} - 0.260869)} \quad (2)$$

$$\text{Equating (1) and (2),} \quad \frac{0.82}{0.77(\text{AFR} - 0.26089)} = 0.08025.$$

   On simplification, we get AFR = 13.53 kg/kg of fuel

   (ii) Total mass of oxygen required for complete combustion of 0.82 kg of carbon and 0.05 kg of hydrogen:

$$0.82 \times \frac{32}{12} + 0.05 \times \frac{32}{4} = 2.1867 + 0.4 = 2.5867 \text{ kg}$$

   Oxygen needed from air = 2.5867 – 0.06 = 2.5267 kg

$$\text{Air needed for combustion} = \frac{2.5267}{0.23} = 10.9856 \text{ kg}$$

$$\text{Excess air supplied} = \frac{(13.53-10.9856)}{10.9856} \times 100 = 23.16\%$$

   (iii) Mass of air supplied for combustion of 1000 kg of coal/h = 13530 kg of air/h

$$\text{Volume of air handled by the fan} = \frac{13530 \times 287 \times 313}{3600 \times 100000} = 3.376 \text{ m}^3/\text{s}$$

10. A fuel gas has a composition, by volume, of 30 percent carbon monoxide, 10 percent hydrogen, 3 percent methane, 5 percent carbon dioxide and 52 percent nitrogen. The gas is burned in a furnace, the air fuel ratio by volume being 2. Assuming combustion to be complete, determine the volume analysis of the dry exhaust gas.

   Volume of air supplied = 2 m^3 (O_2 = 0.42 m^3 and N_2 = 1.58 m^3)

Table 9.18 Combustion of 1 m³ of the Fuel Gas

Const.	Volume (m³)	Combustion Equation	O_2 Reqd. (m³)	CO_2	H_2O	O_2	N_2
CO	0.30	$CO + \frac{1}{2} O_2 \rightleftharpoons CO_2$	0.15	0.30	–	–	–
H_2	0.10	$H_2 + \frac{1}{2} O_2 \rightleftharpoons H_2O$	0.05	–	0.10	–	–
CH_4	0.03	$CH_4 + 2\,O_2 \rightleftharpoons CO_2 + 2\,H_2O$	0.06	0.03	0.06	–	–
CO_2	0.05	–	–	0.05	–	–	–
N_2	0.52	–	–	–	–	–	0.52
			Total O_2 reqd. $\Sigma O_2 = 0.26$				
(Air)							
O_2	0.42	$O_2 = 0.42 - 0.26 = 0.16$				0.16	–
N_2	1.58						1.58
				0.38	0.16	0.16	2.10

Table 9.19 Volume of Dry Exhaust Gas

Constituent	Volume	Percentage
CO_2	0.38	14.39
O_2	0.16	6.06
N_2	2.10	79.55
	Σ 2.64	

Practice Problems

1. A fuel oil contains 85 percent carbon and 15 percent of hydrogen by mass. Determine the minimum mass of air required for the complete combustion of 1 kg of this oil. If the air actually supplied is 20 percent in excess of the minimum required, find the percentage composition of the dry products of combustion by volume.

   (15.07; CO_2, 12.01%; O_2, 3.67%; N_2, 84.32%)

2. The percentage composition of a sample of anthracite was found by analysis to be C, 90; H_2, 3.3; O_2, 3.0; N_2, 0.8; Ash, 2.0. Calculate the minimum mass of air for complete combustion of 1 kg of this fuel. If 50 percent excess air is supplied, find the percentage composition of the dry flue gases by volume.

   (11.49 kg; CO_2, 12.69%; SO_2, 0.05%; O_2, 7.0%; N_2, 80.26%)

3. Estimate the volume of air required for the combustion of 1 m³ of gas having the analysis CH_4, 39.5 percent; H_2, 46.0 percent; N_2, 0.5 percent; CO, 7.5 percent; H_2O, 2.0 percent; CO_2, 4.5 percent. What is the volumetric analysis of the products of combustion at 100°C?

   (5.04 m³; CO_2, 8.93%; H_2O, 22.01%; N_2, 69.06%)

4. Find, for benzene, C_6H_6, the air/fuel ratio, by mass, giving chemically correct combustion. In an engine test, using this fuel, the air/fuel ratio by mass was found to be 11/1. Estimate the percentage composition of the exhaust gases by mass.

   (13.37; H_2O, 5.78%; CO, 7.98%; CO_2, 15.66%; N_2, 70.58%)

5. The volumetric analysis of producer gas supplied to an engine is carbon dioxide, 7.66 percent; carbon monoxide, 22.27 percent; hydrogen, 20.19 percent; marsh gas (CH_4), 2.778 percent; nitrogen, 47.1 percent. The exhaust gases contained 10 percent of oxygen by volume. Estimate the quantity of air actually supplied per m^3 of gas.

(2.917 m^3)

Objective Type Questions

1. The volumetric composition of carbon monoxide is more in the fuel
   (i) Blast furnace gas
   (ii) Coal gas
   (iii) Natural gas
   (iv) Producer gas

2. All chemical reactions must satisfy
   (i) the conservation of mass and the first law of thermodynamics
   (ii) the conservation of mass and the second law of thermodynamics
   (iii) the conservation of energy and the second law of thermodynamics
   (iv) the first law of thermodynamics and the second law of thermodynamics

3. Match List I (gadgets undergoing a thermodynamic process) with the List II (property of the system that remains unchanged) and select the correct answer using the codes given in the following lists:

List I	*List II*
A. Bomb calorimeter	1. Pressure
B. Exhaust gas calorimeter	2. Enthalpy
C. Junker gas calorimeter	3. Volume
D. Throttling calorimeter	4. Specific heat

Codes:

	A	B	C	D
(i)	3	4	1	2
(ii)	3	1	4	2
(iii)	2	4	1	3
(iv)	4	3	2	1

4. Which one of the following combustion products/constituents contributes significantly to global warming?
   (i) Carbon dioxide (CO_2)
   (ii) Carbon monoxide (CO)
   (iii) Oxides of nitrogen (NOX)
   (iv) Unburned hydrocarbon (UHC)

5. List I gives different terms related to combustion while List II furnishes the outcome of the events that follow. Match List I with List II and select the correct answer using the codes given below the lists:

List I	*List II*
A. Association	1. Pseudo shock
B. Dissociation	2. Knock
C. Flame front	3. Endothermic
D. Abnormal combustion	4. Exothermic

Codes:

	A	B	C	D
(i)	1	4	3	2
(ii)	4	3	1	2
(iii)	2	4	1	3
(iv)	3	2	4	1

6. Match List I (measuring appliances) with List II (properties/composition of fuel) and select the correct answer using the codes given below the lists:

List I	*List II*
A. Hydrometer	1. Vapour pressure
B. Bomb caloriemeter	2. Products of combustion
C. Reid bomb	3. Specific gravity
D. Orsat apparatus	4. Calorific value

Codes:

	A	B	C	D
(i)	2	4	1	3
(ii)	3	4	1	2
(ii)	3	1	4	2
(iv)	2	1	4	3

7. Which one of the following parameters is significant to ascertain chemical equilibrium of a system?
   (i) Maxwell relation
   (ii) Helmholtz function
   (iii) Gibbs function
   (iv) Onsager relation
8. The chemical formula of normal cetane, a primary fuel of diesel engines, is
   (i) $C_6 H_{14}$
   (ii) $C_7 H_{16}$
   (iii) $C_8 H_{18}$
   (iv) $C_{16} H_{34}$
9. Which emission let out by automobiles and other IC engines contributes significantly to the formation of photochemical smog in the atmosphere?
   (i) Carbon dioxide
   (ii) Carbon monoxide
   (iii) Oxides of nitrogen
   (iv) Unburned hydrocarbon
10. The unburned hydrocarbon in the exhaust gases from an IC engine can be traced by
    (i) an Orsat apparatus
    (ii) an exhaust gas calorie meter
    (iii) a flame ionization detector
    (iv) a catalytic converter

11. The Orsat apparatus, which gives volumetric percentage of four constituents of dry flue gases, is arranged for absorption of three gases and the fourth content being obtained by difference. Match List I (Gases) with List II (Solutions/difference) and select the correct answer using the codes given in the following lists:

List I (Gases)	*List II (Solutions/difference)*
A. Carbon dioxide	1. By difference
B. Carbon monoxide	2. Caustic soda
C. Oxygen	3. Pyrogallic acid
D. Nitrogen	4. Cuprous chloride

Codes:

	A	B	C	D
(i)	2	4	1	3
(ii)	4	1	2	3
(iii)	2	4	3	1
(iv)	3	2	4	1

12. The minimum amount of oxygen needed to burn completely one kg of methane, CH_4 (taking atomic masses of H, C and O as 1, 12 and 16, respectively) is

(i) 2 (ii) 4
(iii) 6 (iv) 8

13. Identify the fuel whose constant volume heat of combustion is equal to the constant pressure heat of combustion.

(i) Acetylene (C_2H_2) (ii) Ethene (C_2H_4)
(iii) Ethane (C_2H_6) (iv) Propene (C_3H_6)

14. The solution kept in Orsat apparatus for absorbing carbon monoxide while analyzing the flue gases is

(i) caustic soda (ii) pyrogallic acid
(iii) cuprous chloride (iv) hydrogen sulphide

15. The mass of air required for complete combustion of unit mass of fuel can always be calculated from the formula

(i) 0.1152 (C) + 0.3456 (H)
(ii) 0.1152 (C) + 0.3456 [(H) – 0.125 (O)]
(iii) 0.1152 (C) + 0.3456 [(H) – 0.125 (O)] + 0.0432 (S)
(iv) 0.1152 (C) + 0.3456 [(H) + 0.125 (O)] + 0.0432 (S)

16. Which thermodynamic property variance furnishes the calorific value of fuel on combustion in a bomb calorimeter?

(i) Internal energy (ii) Enthalpy
(iii) Entropy (iv) Pressure

17. Which hydro-carbon fuel has got its hydrogen carbon ratio based on atomic masses a minimum?

(i) Butane (ii) Ethane
(iii) Methane (iv) Propane

18. An analysis of the products of combustion through the Orsat apparatus furnishes
    (i) the percent mass on wet basis
    (ii) the percent mass on dry basis
    (iii) the percent volume on wet basis
    (iv) the percent volume on dry basis
19. The molecular structure of the paraffin family pictured below represents

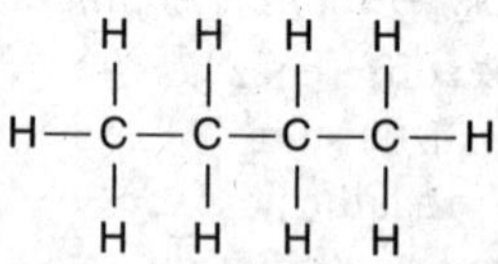

    (i) ethane
    (ii) propane
    (iii) butane
    (iv) hexane
20. The two basic processes involved in the theoretical analysis of chemical reactions by the application of the first law of thermodynamics are
    (i) isothermal and isometric
    (ii) isometric and isobaric
    (iii) isothermal and isobaric
    (iv) adiabatic and non-adiabatic

Answers

1. (iv) 2. (iv) 3. (i) 4. (i) 5. (ii) 6. (ii) 7. (iii) 8. (iv) 9. (i) 10. (iii)
11. (iii) 12. (ii) 13. (ii) 14. (iii) 15. (iii) 16. (i) 17. (i) 18. (iv) 19. (iii) 20. (iv)

10

Heat Pump and Refrigeration Cycles

1. Explain the gadgets Refrigerator and Heat pump.

Ans: In power cycle, heat is received by the working fluid at a high temperature and rejected at a low temperature, while a net amount of work is done by the fluid. Cycles can be conceived in which the reverse happens, heat being received at a low temperature and rejected at a high temperature, while a net amount of work is done on the fluid. The latter are called heat pump or refrigeration cycles. The term **heat pump** is usually applied to a machine whose principal purpose is to supply heat at an elevated temperature, and the term **refrigerator** to one whose purpose is the extraction of heat

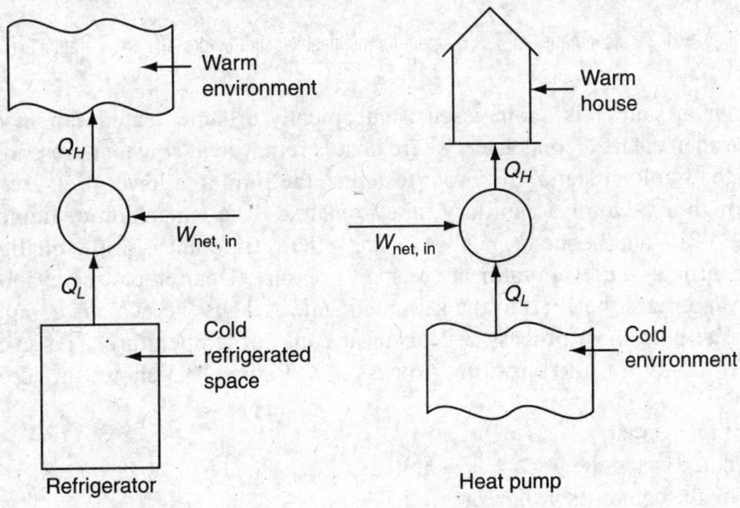

Figure 10.1 Refrigeration Cycle for a Refrigerator and a Heat Pump

from a cold space. This distinction in terminology is arbitrary because a heat pump and refrigerator are identical in principle and it is possible to use one machine to fulfil the function of a heat pump and refrigerator simultaneously. When no distinction is necessary we shall refer to the cycle as a refrigeration cycle (see Figure 10.1).

Practical refrigeration cycles are composed of flow processes, each process being carried out in separate component. We shall assume that changes in potential and kinetic energy between the inlet and outlet of each component are negligible. The working fluids are almost always in the liquid or vapour phase. Air is occasionally used, but only in air-conditioning plant where the air fulfils the dual function of the refrigerating fluid and air-conditioning medium.

2. Describe the vapour-compression refrigeration system with the help of flow and *T-s* diagrams.

Ans: Although the reversed Carnot cycle is a standard one, it is not a practical device for refrigeration purposes. The heat absorption and heat rejection processes are allowed to take place at constant temperatures to get the maximum coefficient of performance possible. This is accomplished by operating a refrigeration device on a vapour-compression cycle. The schematic of the equipment for the cycle, along with *T-s* diagram of the ideal cycle, is shown in Figure 10.2.

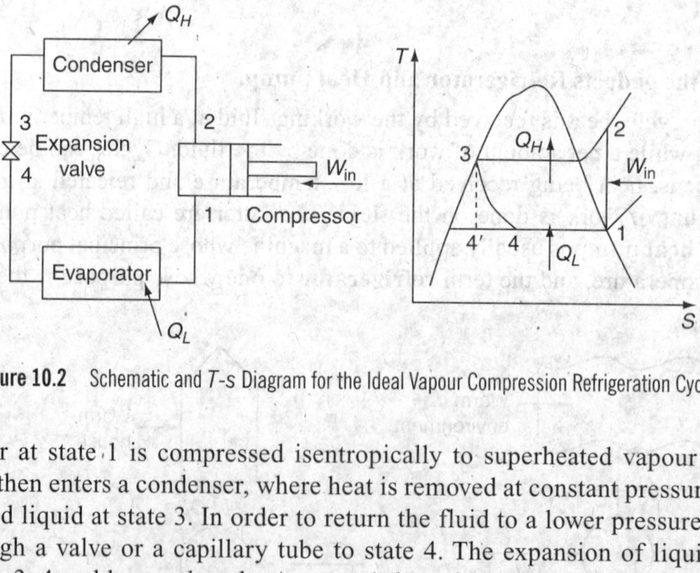

Figure 10.2 Schematic and *T*-s Diagram for the Ideal Vapour Compression Refrigeration Cycle

Saturated vapour at state 1 is compressed isentropically to superheated vapour at state 2. The refrigerant vapour then enters a condenser, where heat is removed at constant pressure until the fluid becomes a saturated liquid at state 3. In order to return the fluid to a lower pressure, it is expanded adiabatically through a valve or a capillary tube to state 4. The expansion of liquid refrigerant is a throttling process 3–4 and hence, $h_3 = h_4$. At state 4 the refrigerant is a low-quality wet mixture. Finally, it passes through the evaporator at constant pressure. Heat enters the evaporator from the cold source and vapourizes the fluid to the saturated state 1. Thus the cycle is completed. The process 4–1 and a large portion of process 2–3 occur at constant temperature. This cycle contains an irreversible process within it, the throttling process 3–4. Figure 10.3 shows the ideal cycle on the *p-h* diagram.

For unit mass of refrigerant,

Heat absorbed in the evaporator, $Q_2 = h_1 - h_4$ kJ

Work required by the compressor, $W = h_2 - h_1$ kJ

Heat rejected in the condenser, $Q_1 = h_2 - h_3$ kJ

$$(\text{C.O.P})_{\text{refrigerator}} = \frac{Q_2}{W} = \frac{h_1 - h_4}{h_2 - h_1},$$

$$(\text{C.O.P})_{\text{heat pump}} = \frac{Q_1}{W} = \frac{h_2 - h_3}{h_2 - h_1}$$

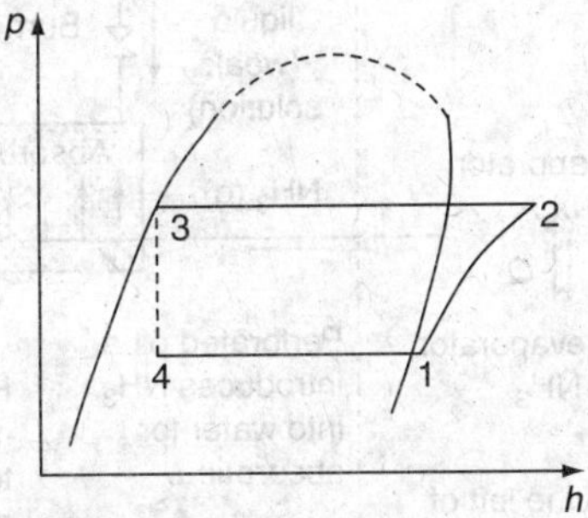

Figure 10.3 *p-h* Diagram of Ideal Vapour Compression Refrigeration Cycle

The standard unit of refrigeration is given on the basis of tons of refrigeration or denoted by the symbol *TR*. It is equivalent to the production of cold at the rate at which heat is removed from1 tonne of water to freeze it to ice at 0°C in 24 h. In SI unit 1 $TR = 3.5$ kJ/s = 3.5 kW.

3. Explain with a neat sketch the working of vapour absorption system.

Ans: Any vapour compression refrigeration system requires a power input to compress the refrigerant vapour from the evaporator pressure to the condenser pressure. Pumping a liquid through the same pressure difference at the same mass flow rate would require less power. To condense the pure refrigerant at the evaporator pressure would defeat the purpose of the refrigeration system. Hence, a workable arrangement is the absorption refrigeration cycle in which the refrigerant is dissolved in a liquid at the evaporator pressure, the liquid is pumped to the condenser pressure and the refrigerant is then separated at the condenser pressure from the liquid.

Figure 10.4 shows a flow diagram of an elementary absorption cycle that uses ammonia as the refrigerant and water (or a water–ammonia mixture) as the liquid in which the refrigerant is carried from the low pressure to the high pressure. Ammonia enters the condenser as a vapour, is condensed and enters the expansion valve as a liquid. It partially flashes to vapour when it expands through the valve, and it is further vapourized as it absorbs heat in the evaporator. Thus, the part of the flow diagram to the left of line *A–A* in the figure is identical with that of a vapour compression cycle explained already. However, to the right of the line *A–A* we see that in place of a compressor is an absorber, a pump, a generator and a return line for liquid. Ammonia vapour from the evaporator is dissolved in water in the absorber. This process is exothermic, just as a condensation process of a pure substance is, so heat must be removed from the absorber to keep its temperature constant. Also, at a given pressure the amount of ammonia that can be dissolved in water increases as the temperature is decreased; so the absorber temperature must be kept as low as possible. The strong ammonia–water solution is then pumped to the generator, which is at the condenser pressure. Heat is

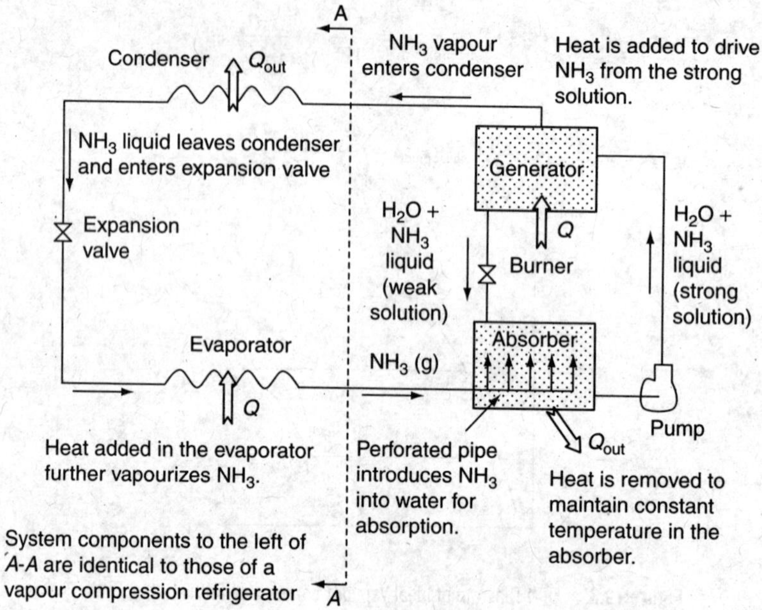

Figure 10.4 Basic Aqua Ammonia Absorption Refrigeration Cycle

added to the solution in the generator to drive much of the ammonia out of solution (an endothermic process, just as an evaporation process of a pure substance is). Ammonia vapour, or a vapour mixture that is very rich in ammonia, then goes to the condenser, and the weak ammonia–water solution left in the generator passes through a valve back to the absorber. The liquid that travels through the absorber–pump–generator–valve cycle is simply a transport medium for the refrigerant, carrying it from the evaporator pressure to the condenser pressure in the liquid phase. Notice that the strong solution flows from the absorber that needs to be kept cold to the generator that needs to be kept hot, while weak solution flows in the opposite direction. This situation suggests that a heat exchanger that preheats the fluid flowing toward the generator by taking heat from the fluid flowing toward the absorber would certainly be desirable. In fact, such heat exchangers are usually used, although one is not shown in Figure 10.4.

Actual ammonia absorption systems always involve some modifications not shown in Figure. The pump delivers cold liquid from the absorber to the generator where the temperature must be high, and hot liquid flows from the generator back to the absorber, which must be kept cold; so a heat exchanger is always used to transfer heat from the weak solution to the strong solution. Also, carrying water vapour into the condenser must be avoided because it will freeze in the expansion valve and evaporator; therefore, a rectifier is placed between the generator and the condenser. Its function is to remove traces of water from the refrigerant before it reaches the condenser. Figure 10.5 shows an aqua ammonia refrigeration system with a heat exchanger and rectifier.

Another pair of substances commonly used for absorption refrigeration systems is lithium bromide and water. An aqueous solution of lithium bromide is the carrier and water is the refrigerant; therefore, lithium bromide absorption systems are used only for applications where the minimum temperature is above 0°C.

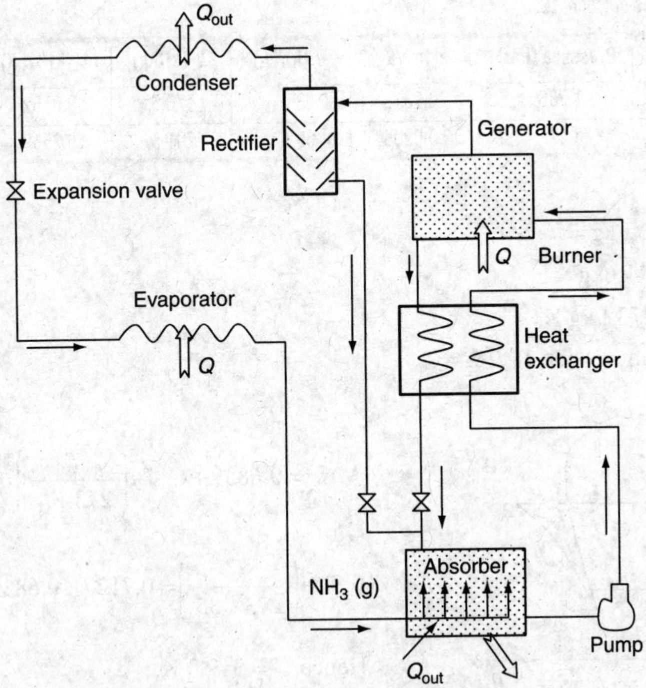

Figure 10.5 Aqua Ammonia System with Heat Exchanger and Rectifier

Solved Problems

1. A Carnot refrigerator operates in a room in which the temperature is 25°C and consumes 2 kW of power when operating. If the food compartment of the refrigerator is to be maintained at 3°C, determine the rate of heat removal from the food compartment.
Given data: $T_1 = 273 + 25 = 298$ K and $T_2 = 273 + 3 = 276$ K

$$\text{C.O.P of the Carnot refrigerator} = \frac{T_2}{T_1 - T_2} = \frac{276}{298 - 276} = 12.54.$$

$$\text{C.O.P of the Carnot refrigerator} = \frac{Q_2}{W} = \frac{\text{heat removed}}{\text{work input}} = 12.54.$$

The rate of heat removal from the compartment = 12.54 × 2 = 25.08 kJ/s.

2. An ideal refrigeration cycle with Freon-12 as the working fluid operates with an evaporator temperature of –25°C and a condenser temperature of 35°C. The mass flow rate of refrigerant is 3 kg/min through the cycle. Compute the coefficient of performance, the tons of refrigeration, the power input per ton of refrigeration and the coefficient of performance of a Carnot refrigerator operating under the same maximum and minimum temperatures. Assume Cp for Freon-12 vapour = 0.8 kJ/ kg K. Use the following data (Table 10.1) for Freon-12:

Table 10.1 Properties of Freon-12

Temperature (°C)	Pressure (bar)	v_g (m³/kg)	h_f (kJ/kg)	h_g (kJ/kg)	s_f (kJ/kg K)	s_g (kJ/kg K)
−25	1.236	0.1313	13.3	176.5	0.0552	0.7127
35	8.47	0.0206	69.5	201.5	0.2559	0.6839

Given data:

$h_3 = 69.5$ kJ/kg $= h_4$

$h_1 = 176.5$ kJ/kg

$s_1 = 0.7127$ kJ/kg K

For isentropic compression 1–2, $s_2 = s_1$

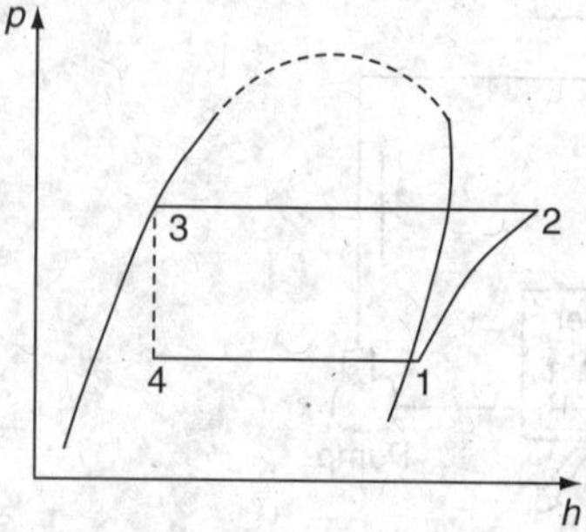

$$s_2 = 0.6839 + C_p\,\ell n\left(\frac{T}{273+t_s}\right) = 0.7127$$

$$0.8\,\ell n\left(\frac{T}{273+35}\right) = 0.7127 - 0.6839 = 0.0288$$

Hence, $T = 319.3$ K,

$h_2 = 201.5 + 0.8\,(319.3 - 308) = 210.5$ kJ/kg.

$$\text{C.O.P of the refrigerator} = \frac{h_1 - h_4}{h_2 - h_1} = \frac{(176.5-69.5)}{(210.5-176.5)} = 3.14.$$

Tons of refrigeration:

Amount of heat removed = (3/60) × (176.5 − 69.5) = 5.35 kJ/s.

$$TR = \frac{5.35}{3.5} = 1.53.$$

$$(\text{C.O.P})_{\text{Carnot}} = \frac{T_2}{T_1 - T_2} = \frac{(273-25)}{(273+35)-(273-25)} = 4.13$$

3. A 15 *TR* R12 vapour compression system operates between a condenser temperature of 40°C and an evaporator temperature of 5°C. Determine the compressor discharge temperature assuming the specific heat of its vapour as 0.8 kJ/kg K. Calculate the theoretical piston displacement and power consumption per ton of refrigeration. Use the following data (Table 10.2) for Freon-12:

Table 10.2 Properties of Freon-12

Temperature (°C)	Pressure (bar)	v_g (m3/kg)	h_f (kJ/kg)	h_g (kJ/kg)	s_f (kJ/kg K)	s_g (kJ/kg K)
5	3.62	0.0475	40.7	189.7	0.1587	0.6942
40	9.60	0.0182	74.6	203.2	0.2718	0.6825

Given data:

$h_3 = 74.6$

$s_1 = 0.6942$

$s_2 = s_1$

$$0.6942 = 0.6825 + 0.8\,\ell n\left(\frac{T}{273+40}\right)$$

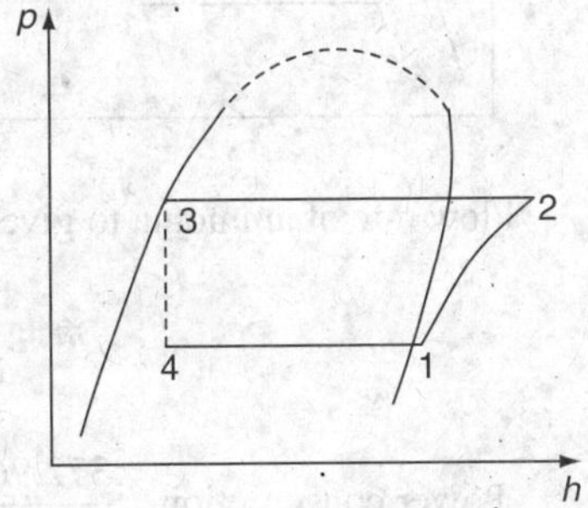

Hence, $T = 317.6$ K,

$h_2 = 203.2 + 0.8\ (317.6 - 313) = 206.88$ kJ/kg.

Temperature of refrigerant at compressor discharge line = 317.6 K or 44.6°C.

Total volume of refrigerant at compressor inlet:

$$\text{Flow rate of refrigerant} = \frac{15 \times 3.5}{(h_1 - h_4)}$$

$$= \frac{15 \times 3.5}{(189.7 - 74.6)} = 0.456 \text{ kg/s.}$$

Theoretical piston displacement:

$P_{dis} = 0.456 \times 0.0475 = 0.02166$ m³/s.

Power consumption:

Power input $= 0.456 \times (h_2 - h_1) = 0.456 \times (206.88 - 189.7) = 7.83$ kW.

$$\text{Power consumption} = \frac{7.83}{15} = 0.522 \text{ kW/ton of refrigeration.}$$

4. A simple saturation ammonia compression system has a high pressure of 1.35 MN/m² and a low pressure of 0.19 MN/m². Find per 4,00,000 kJ/h of refrigerating capacity, the power consumption of the compressor and C.O.P of the cycle. The properties of Ammonia are given in Table 10.3.

Table 10.3 Properties of Ammonia

p (bar)	t (°C)	v_g (m³/kg)	h_f (kJ/kg)	h_g (kJ/kg)	s_f (kJ/kg K)	s_g (kJ/kg K)	Superheated Vapour			
							h (50°C)	s	h (100°C)	s
1.9	−20	0.624	89.8	1420	0.363	5.624	1535	6.039	1646	6.379
13.5	35	0.096	347.5	1471	1.282	4.930	1616	5.368	1744	5.702

Given data:

$h_1 = 1420$ kJ/kg

$s_1 = 5.624$ kJ/kg K

$h_3 = 347.5$ kJ/kg K

$s_2 = s_1$

Since s_2 is greater than entropy of superheated vapour at 50°C, the temperature lies betwee 50°C and 100°C.

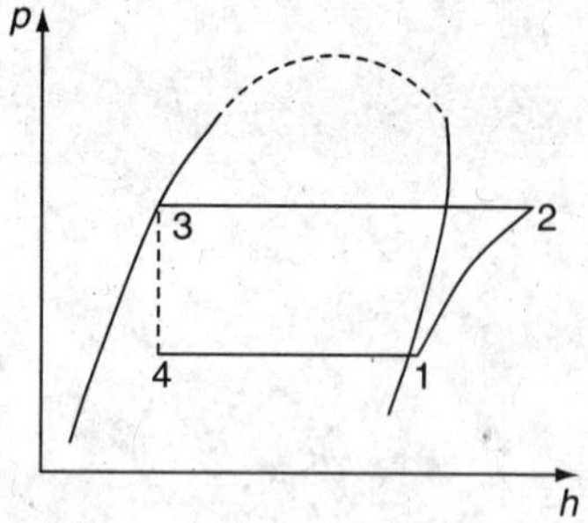

By interpolation, we get

$$\frac{\Delta t}{50} = \frac{5.624 - 5.368}{5.702 - 5.368} \quad \text{or} \quad \Delta t = 38.3,$$

$$t_2 = 50 + \Delta t = 88.3°\text{C and}$$

$$h_2 = 1616 + \frac{38.3}{50}(1744 - 1616) = 1714 \text{ kJ/kg}.$$

Flow rate of ammonia to give 4,00,000 kJ/h of refrigerating capacity:

$$m = \frac{400000}{(h_1 - h_4)} = \frac{400000}{(1420 - 347.5)} = 372.96 \text{ kg/h}.$$

$$\text{Power consumption} = \frac{372.96 \times (h_2 - h_1)}{3600} = \frac{372.96 \times (1714 - 1420)}{3600} = 30.458 \text{ kW}.$$

$$\text{C.O.P} = \frac{(h_1 - h_4)}{(h_2 - h_1)} = \frac{(1420 - 347.5)}{(1714 - 1420)} = 3.647.$$

5. A R12 refrigerating machine operates between a condenser temperature of 40°C and an evaporator temperature of 5°C. Calculate the increase (percent) in the theoretical piston displacement and power consumption of the cycle if the evaporator temperature is reduced to 0°C. Assume C_p for the refrigerant as 0.8 kJ/kg K. Use the following data (Table 10.4) for R12:

Table 10.4 Properties of R12

p (bar)	t (°C)	v_g (m³/kg)	h_f (kJ/kg)	h_g (kJ/kg)	s_f (kJ/kg K)	s_g (kJ/kg K)	Superheated Vapour			
							h (50°C)	s	h (100°C)	s
3.08	0	0.0554	36.1	187.5	0.1420	0.6966	200.5	0.7423	213.5	0.7853
3.62	5	0.0475	40.7	189.7	0.1587	0.6942	202.9	0,7401	216.1	0.7830
9.60	40	0.0182	74.6	203.2	0.2718	0.6825	218.5	0.7300	233.4	0.7732

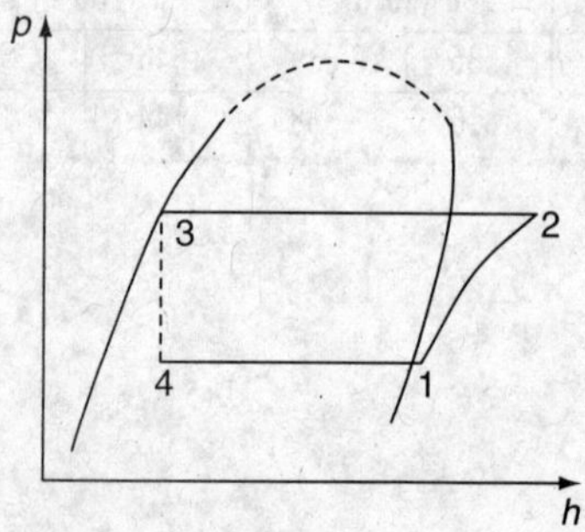

Condenser temp. = 40°C

Evaporator temp. = 5°C

$$h_1 = 189.7,\ h_3 = 74.6,$$

For isentropic compression $s_2 = s_1$,

$$s_1 = 0.6942. \text{ Hence, } s_2 = 0.6942.$$

For throttling process $h_4 = h_3$

$$h_4 = 74.6.$$

Since s_2 is less than 0.7300, the temperature of the refrigerant will be more than 40°C and less than 50°C.

$$0.6942 = 0.6825 + 0.8\,\ell n\left(\frac{T}{T_g}\right) = 0.6825 + 0.8\,\ell n\left(\frac{T}{313}\right)$$

Solving the above equation, we get $T = 317.6$ K.

Enthalpy of the refrigerant delivered by the compressor:

$$h_2 = 203.2 + 0.8(317.6 - 313) = 206.88 \text{ kJ/kg}.$$

Assume 1 kg of refrigerant is flowing through the cycle.

The quantity of heat absorbed, $Q_2 = h_1 - h_4 = 189.7 - 74.6 = 115.1$.

Power needed by the compressor, $W = h_2 - h_1 = 206.88 - 189.7 = 17.18$ kJ.

Volume of gas displaced by the piston $= 1 \times 0.0475 = 0.0475$ m^3.

The evaporator temperature is now reduced to 0°C.

Hence, $h_6 = 187.5$, $s_6 = 0.6966$, $h_3 = 74.6$ and $s_7 = s_6$.

$$0.6966 = 0.6825 + 0.8\,\ell n\left(\frac{T}{313}\right)$$

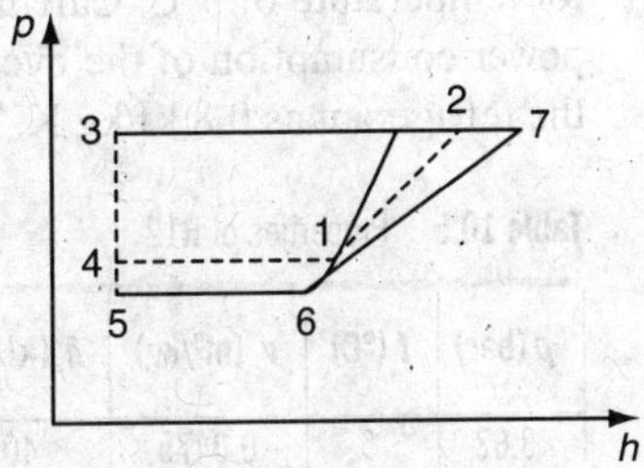

Solving the above equation, we get

$$T = 318.56 \text{ K},$$

$$h_7 = 203.2 + 0.8(318.56 - 313) = 207.65 \text{ and}$$

$$h_5 = h_3 = 74.6.$$

Amount of refrigerant required to absorb 115.1 kJ of heat from the evaporator,

$$m_1 = \frac{115.1}{h_6 - h_5} = \frac{115.1}{187.5 - 74.6} = 1.019 \text{ kg}.$$

Volume of gas to be handled by the compressor $= 1.019 \times 0.0554 = 0.056$ m^3.

Power required by the compressor $= 1.019 \times (h_7 - h_6) = 1.019 \times (207.65 - 187.5)$

$$= 20.53 \text{ kJ}.$$

The increase in theoretical piston displacement $= \dfrac{(0.056 - 0.0475)}{0.0475}100 = 17.89\%$.

The increase in power consumption $= \dfrac{(20.53 - 17.18)}{17.18}100 = 19.5\%$.

Practice Problems

1. A Carnot refrigerator operates in a room in which the temperature is 30°C and consumes 3 kW of power when operating. If the food compartment of the refrigerator is to be maintained at 3°C, determine the rate of heat removal from the food compartment.

(30.67 kJ/s)

2. A refrigerator is to remove heat from the cooled space at a rate of 300 kJ/min to maintain its temperature at −8°C. If the air surrounding the refrigerator is at 25°C, determine the minimum power input required for this refrigerator.

(622.6 W)

3. An air-conditioning system operating on the reversed Carnot cycle is required to transfer heat from a house at a rate of 750 kJ/min, to maintain its temperature at 20°C. If the outdoor air temperature is 35°C, determine the power required to operate this air-conditioning system.

(0.741 kW)

4. The performance of a heat pump degrades (i.e. its C.O.P decreases) as the temperature of the heat source decreases. This makes using heat pumps at locations with severe weather conditions unattractive. Consider a house that is heated and maintained at 20°C by a heat pump during the winter. What is the maximum C.O.P for this heat pump, if heat is extracted from the outdoor air at (i) 10°C, (ii) −5°C and (iii) −30°C?

(29.3; 11.72; 5.86)

5. A R12 refrigerating machine operates between a condenser temperature of 40°C and an evaporator temperature of 5°C. Calculate the increase (percent) in the theoretical piston displacement and power consumption of the cycle if the condenser temperature is increased to 45°C. Assume C_p for the refrigerant as 0.8 kJ/kg K. The refrigerant has the properties given below:

Table 10.5 Properties of R12

p (bar)	t (°C)	v_g (m³/kg)	h_f (kJ/kg)	h_g (kJ/kg)	s_f (kJ/kg K)	s_g (kJ/kg K)	Superheated Vapour			
							h (50°C)	s	h (100°C)	s
3.62	5	0.0475	40.7	189.7	0.1587	0.6942	202.9	0.7401	216.1	0.7830
9.60	40	0.0182	74.6	203.2	0.2718	0.6825	218.5	0.7300	233.4	0.7732
10.84	45	0.0160	79.7	204.9	0.2877	0.6812	220.6	0.7291	235.7	0.7724

(4.63%; 21.59%)

Objective Type Questions

1. Freon-12 is a compound consisting of atoms
   (i) carbon, hydrogen and fluorine
   (ii) carbon, chlorine and fluorine
   (iii) hydrogen, carbon and chlorine
   (iv) fluorine, chlorine and hydrogen
2. To meet the varying load requirements in a refrigerant plant, the appropriate expansion device used is
   (i) automatic expansion valve
   (ii) thermostatic expansion valve
   (iii) low-side float valve
   (iv) high-side float valve
3. Which basic cycle has throttling process associated with it?
   (i) Diesel cycle
   (ii) Gas turbine cycle
   (iii) Otto cycle
   (iv) Refrigeration cycle

4. Under cooling in a refrigeration cycle
   (i) decreases the work input
   (ii) increases the work input
   (iii) increases the refrigerating effect
   (iv) decreases the refrigerating effect
5. If an engine of 40 percent thermal efficiency drives a refrigerator having a C.O.P of 5, then the heat input to the engine for 1 kJ of heat removed from the cold body by the refrigerator, in kJ, is
   (i) 0.50 (ii) 0.75
   (iii) 1.00 (iv) 1.25
6. If ammonia absorption refrigeration cycle is analysed from the stand point of the second law of thermodynamics, then the largest irreversibility occurs in
   (i) the absorber (ii) the condenser
   (iii) the evaporator (iv) the rectifier
7. Refrigerant flowing through the capillary tubes in a domestic refrigerator undergoes
   (i) isothermal process (ii) isentropic process
   (iii) isenthalpic process (iv) isobaric process
8. Identify the irreversible process in the vapour compression refrigeration cycle shown on the pressure–enthalpy diagram given below.

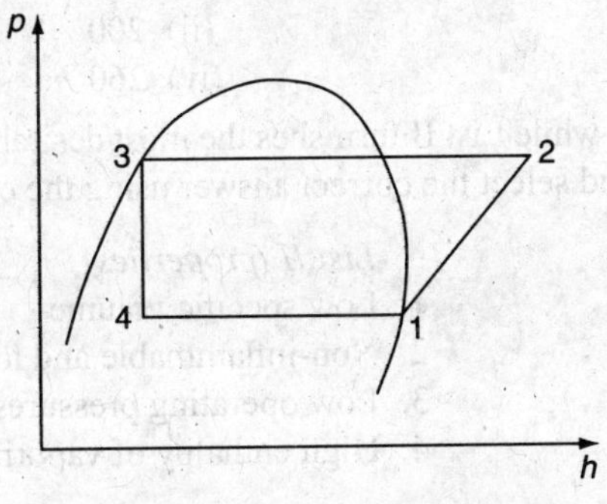

   (i) Process 1–2 (ii) Process 2–3
   (iii) Process 3–4 (iv) Process 4–1
9. Which one of the following axioms forms the basis for thermoelectric refrigeration?
   (i) The Compton effect (ii) The Peltier effect
   (iii) The Seebeck effect (iv) The Thomson effect
10. Three out of the four processes undergone by the refrigerant in an ideal vapour compression refrigerating machine delivering dry and saturated vapour to the condenser appear as straight lines when represented on
   (i) the pressure–volume diagram
   (ii) the temperature–entropy diagram
   (iii) the pressure–temperature diagram
   (iv) the enthalpy–entropy diagram

11. The three most commonly used halogens to produce synthetic refrigerants from methane, ethane or propane are

(i) astatine, bromine and chlorine
(ii) bromine, chlorine and fluorine
(iii) chlorine, fluorine and iodine
(iv) fluorine, iodine and astatine

12. The axiom that the passage of electricity through a junction of two different metals (antimony and copper) produces a rise in temperature at the junction when passing in one direction and a drop in temperature when passed in the reverse direction is known as

(i) the Compton effect
(ii) the Thomson effect
(iii) the Peltier effect
(iv) the Seebeck effect

13. Which one of the following gadgets is known as vacuum refrigeration?

(i) Vapour absorption system
(ii) Vapour compression system
(iii) Cascade system
(iv) Steam jet system

14. Which property diagram is more suited for the refrigeration cycle analysis?

(i) Pressure–temperature
(ii) Temperature–entropy
(iii) Enthalpy–entropy
(iv) Pressure–enthalpy

15. If the vapour pressure of liquid ammonia and solid ammonia can be approximated as $\ln p = 15.0 - \frac{3063}{T}$ and $\ln p = 18.0 - \frac{3753}{T}$ in the vicinity of the triple point, then the triple point temperature for ammonia, in K, is

(i) 170
(ii) 200
(iii) 230
(iv) 260

16. List I gives the refrigerants while List II furnishes the most desirable properties possessed by them. Match List I with List II and select the correct answer using the codes given below the lists.

List I (refrigerants)	*List II (properties)*
A. Ammonia	1. Low specific volume
B. Carbon dioxide	2. Non-inflammable and low toxicity
C. Freons	3. Low operating pressures
D. Sulphour dioxide	4. High enthalpy of vaporization

Codes:

	A	B	C	D
(i)	4	1	2	3
(ii)	2	1	4	3
(iii)	4	1	3	2
(iv)	4	3	1	2

17. An ideal vapour compression refrigeration cycle operates with dry saturated vapour coming out of the compressor and saturated liquid leaving the condenser. When this cycle is represented on pressure-enthalpy co-ordinates of the ideal refrigerant, the diagram assumes the shape of

(i) a rectangle
(ii) a parallelogram
(iii) a quadrilateral
(iv) a trapezium

18. The pressure of vapour in the evaporator of a refrigerating cycle is more than one atmosphere in order
    (i) to minimize the possibility of air infiltration into the system
    (ii) to increase the COP of the plant
    (iii) to increase the refrigerating capacity of the plant
    (iv) to minimize the compressor work input

19. Which one of the following changes/sets of changes in the higher and lower temperatures (T_1 and T_2) of a Carnot refrigerator would result in the maximum improvement in its COP?
    (i) $(T_2 + \Delta T)$
    (ii) $(T_1 - \Delta T)$
    (iii) $(T_2 + \Delta T)$ and $(T_1 - \Delta T)$
    (iv) $(T_2 + \Delta T)$ and $(T_1 + \Delta T)$

20. The Brayton cycle is reversed and operated as a refrigerator between the high pressure p_2 and low pressure p_1. The ideal COP, in terms of the pressure ratio r_p ($r_p = p_2/p_1$) and the adiabatic index γ ($\gamma = C_p / C_v$), is
    (i) $\dfrac{1}{(r_p)^\gamma - 1}$
    (ii) $\dfrac{(r_p)^\gamma}{(r_p)^\gamma - 1}$
    (iii) $\dfrac{(r_p)^{(\gamma-1)/\gamma}}{(r_p)^{(\gamma-1)/\gamma} - 1}$
    (iv) $\dfrac{1}{(r_p)^{(\gamma-1)/\gamma} - 1}$

Answers

1. (ii) 2. (ii) 3. (iv) 4. (iii) 5. (i) 6. (i) 7. (iii) 8. (iii) 9. (ii) 10. (ii)
11. (ii) 12. (iii) 13. (iv) 14. (iv) 15. (iii) 16. (i) 17. (iv) 18. (i) 19. (iii) 20. (iv)

11

Additional Solved Problems

1. A certain gas with molecular mass of 28 occupies 16 m³ per kg. What volume will 1 kg of another gas occupy if its molecular mass is 32?

Ans: We know

$$\rho = \frac{1}{v} = n \times \text{molecular mass, where } n \text{ is the number of molecules in unit volume}$$

$$\frac{1}{v_1} = n \times 28 \text{ and } \frac{1}{v_2} = n \times 32\text{, Hence, } \frac{v_2}{v_1} = \frac{28}{32}$$

Volume of the other gas, $v_2 = \frac{28}{32} \times 16 = 14 \text{ m}^3 \text{per kg.}$

2. A certain gas has initial pressure, volume and temperature as 7 bar, 0.08 m³ and 150°C, respectively. The gas expands in such a manner that the final volume and the temperature become 0.25 m³ and 10°C, respectively. Find the final pressure.

Ans:

$$\frac{p_1 V_1}{T_1} = \frac{p_2 V_2}{T_2}; \quad p_2 = \frac{V_1}{V_2} \times \frac{T_2}{T_1} \times p_1 = \frac{0.08}{0.25} \times \frac{283}{423} \times 7 = 1.498 \text{ bar.}$$

Final pressure = 1.498 bar.

3. Nitrogen is stored at 140 bar, 27°C in a steel flask of 0.05 m³ volume. The flask is to be protected against excessive pressure by a fusible plug which will melt and allows the gas to escape if the temperature rises too high. (i) At what temperature must the fusible plug melt in order to limit the pressure of the full flask to a maximum of 160 bar?

(ii) How many kg of nitrogen will the flask hold at the designed condition? Assume gas constant for nitrogen as 297 N.m/kg K.

Ans:

$$p_1 V_1 = m_1 R T_1; \; m = \frac{140\times10^5\times0.05}{297\times300} = 7.856 \text{ kg.}$$

For a maximum pressure, $p_2 = 160$ bar; $p_2 V_2 = m_2 R T_2$

$$T = \frac{160\times10^5\times0.05}{7.856\times297} = 342.8\,\text{K}$$

(i) The temperature at which the fusible plug will melt is 342.8 K or 69.8°C.
(ii) The mass of nitrogen the flask holds is 7.856 kg.

4. The temperature of the charge of gas at the beginning of fuel admission in a diesel engine cylinder is 300°C. The volume of the gas charge at the end of heat supply is 1.65 times the volume at the beginning of heat supply. Assuming that the heat supply is at constant pressure, calculate the temperature at the end of constant pressure heating.

Ans: For constant pressure process,

$$\frac{V_3}{T_3} = \frac{V_2}{T_2}; \; T_3 = \frac{V_3}{V_2}\times T_2 = 1.65\times573 = 945.45 \text{ K.}$$

The temperature at the end of heat supply is 945.45 K or 672.45°C.

5. The initial pressure and volume of air are 1 bar and 0.7 m³, respectively. After being compressed at 15°C (constant temperature) its volume became 0.14 m³. What is its pressure then? The air is then heated at constant pressure until its temperature is 55°C. What is its volume then?

Ans: For constant temperature process,

$$p_1V_1 = p_2V_2; \; p_2 = p_1\frac{V_1}{V_2} = 1\times10^5\times\frac{0.7}{0.14} = 5\times10^5 \text{ or 5 bar.}$$

For constant pressure process, $V_3 = V_2\times\frac{T_3}{T_2} = 0.14\times\frac{328}{288} = 0.159 \text{ m}^3.$

6. A vessel contains 1.5 m³ of air at a pressure of 7 bar and temperature 40°C. If one-fifth of this air be removed, what will be the pressure of air, the temperature remains the same.

Ans:

$$p_1V_1 = m_1RT_1; \; m_1 = \frac{7\times10^5\times1.5}{287\times313} = 11.69 \text{ kg.}$$

Mass of air to be removed = (1/5) × m_1 = (1/5) × 11.69 = 2.34 kg.
Mass of air now in the vessel, m_2 = 11.69 − 2.34 = 9.35 kg.

Pressure of air in the vessel, $p_2 = \frac{m_2RT_2}{V_2} = \left(\frac{9.35\times287\times313}{1.5}\right)\frac{1}{10^5} = 5.599$ bar.

7. A vacuum gauge connected to a tank reads 30 kPa at a location where the barometric reading is 755 mmHg. Determine the absolute pressure in the tank. Assume specific gravity of mercury as 13.59.

Ans: Atmospheric pressure, $p = \rho g h = 13.59 \times 1000 \times 9.81 \times 0.755 = 1.006 \times 10^5$ N/m².
Pressure in the tank = 100.6 – 30 = 70.6 kPa (abs).

8. 2.27 kg of air at a pressure of 3.5 bar and at a temperature of 25°C occupies a volume of 0.566 m³. Calculate the specific heats of air given that the adiabatic index is 1.4.

Ans:

$$pV = mRT; \quad R = \frac{pV}{mT} = \frac{3.5\times10^5\times0.566}{2.27\times298} = 292.8\frac{\text{Nm}}{\text{kgK}} \quad \text{or} \quad \frac{\text{J}}{\text{kg K}}.$$

$$C_p - C_v = R$$

$$C_v = \frac{R}{\gamma-1} \text{ and } C_p = \frac{\gamma R}{\gamma-1}; \quad C_v = \frac{292.8}{0.4} = 732\frac{\text{J}}{\text{kg K}} \text{ and } C_p = \frac{1.4\times292.8}{0.4} = 1024.8\frac{\text{J}}{\text{kg K}}.$$

9. An air receiver has a volume of 0.4245 m³ and contains air at a pressure of 2.067 bar and temperature of 15°C. Find the pressure in the receiver when an additional 0.908 kg of air has been pumped into it at 15°C.

Ans:

$$m_1 = \frac{p_1 V_1}{RT_1} = \frac{2.067\times10^5\times0.4245}{287\times288} = 1.06 \text{ kg.}$$

$$m_2 = 1.06 + 0.908 = 1.968 \text{ kg.}$$

$$p_2 = \frac{m_2 R T_2}{V_2} = \frac{1.968\times287\times288}{0.4245} = 3.83\times10^5\frac{\text{N}}{\text{m}^2} \quad \text{or} \quad 3.83 \text{ bar.}$$

10. A 20 m³ capacity tank contains nitrogen at 25°C and 800 kPa. The gas is allowed to escape until the pressure in the tank drops to 600 kPa. If the temperature at this point is 20°C, determine the amount of nitrogen that has escaped.

Ans:

$$R = \frac{8.314}{28}\times1000 = 296.9\frac{\text{N.m}}{\text{kg K}}$$

$$m_1 = \frac{800\times10^3\times20}{296.9\times298} = 180.8 \text{ kg;} \quad m_2 = \frac{600\times10^3\times20}{296.9\times293} = 137.9 \text{ kg.}$$

The quantity of nitrogen escaped from the tank = $m_1 - m_2$ = 180.8 – 137.9 = 42.9 kg.

11. A quantity of certain gas is compressed from the initial state of 0.085 m³ and 1.034 bar to a final condition of 0.034 m³ and 3.996 bar, respectively. If the specific heat of the gas at constant volume is 0.718 and at constant pressure is 1.005 and the observed temperature rise is 146°C, calculate the change of internal energy.

Ans:

$$p_1 V_1 = mRT_1 \quad \text{and } p_2 V_2 = mRT_2; \quad C_p - C_v = R$$

$$m(C_p - C_v)(T_2 - T_1) = (p_2 V_2 - p_1 V_1)$$

Mass of the gas, $m = \dfrac{(3.996 \times 0.034 - 1.034 \times 0.085) \times 10^5}{(1.005 - 0.718) \times 10^3 \times (146)} = 0.114$ kg.

The change in internal energy, $\Delta U = m\, C_v\, (T_2 - T_1) = 0.114 \times 0.718 \times 146$

$= 11.95$ kJ.

12. Five kg of air is expanded isothermally from 7 bar and 35°C to 1.5 bar. Calculate the work done, change in internal energy and heat transfer.

Ans:

$$p_1 V_1 = m R T_1;\; V_1 = \frac{m R T_1}{p_1} = \frac{5 \times 287 \times 308}{7 \times 10^5} = 0.6314 \text{ m}^3.$$

For isothermal process $p_1 V_1 = p_2 V_2$; $V_2 = \dfrac{7 \times 10^5 \times 0.6314}{1.5 \times 10^5} = 2.9465 \text{ m}^3.$

Work done, ${}_1W_2 = p_1 V_1 \ln\left(\dfrac{V_2}{V_1}\right) = \dfrac{7 \times 10^5 \times 0.6314}{10^3} \ln\left(\dfrac{2.9465}{0.6314}\right) = 680.8$ kN.m.

Change in internal energy = 0

$${}_1Q_2 = U_2 - U_1 + {}_1W_2 = 0 + 680.8$$

Heat transfer = 680.8 kJ.

13. Air at a pressure of 1.25 bar and 25°C is compressed with an index of 1.3. Due to compression, its volume is decreased from 0.028 m³ to 0.0042 m³. Determine the final temperature and work of compression.

Ans:

$$p_1 V_1^{1.3} = p_2 V_2^{1.3};\; p_2 = p_1 \left(\frac{V_1}{V_2}\right)^{1.3} = 1.25 \times \left(\frac{0.028}{0.0042}\right)^{1.3} = 14.72 \text{ bar.}$$

$$\frac{T_2}{T_1} = \left(\frac{V_1}{V_2}\right)^{(1.3-1)};\; T_2 = 298 \times \left(\frac{0.028}{0.0042}\right)^{0.3} = 526.5 \text{ K.}$$

Final Temperature = 526.5 K or 253.5°C.

Work of compression $= \dfrac{(p_1 V_1 - p_2 V_2)}{(n-1)} = \dfrac{(1.25 \times 0.028 - 14.72 \times 0.0042) \times 10^5}{(1.3 - 1) \times 10^3} = -8.94$ kN.m

(–ve sign indicates that work is done on the system).

14. 0.25 m³ of a gas under a pressure of 1.4 MPa and temperature 60°C is expanded to a pressure of 350 kPa along a curve, the equation of which is $pV^{1.3}$ = constant. If the specific heats at constant pressure and constant volume are 2.18 and 1.55 kJ/kgK respectively, find (i) the heat transfer and (ii) the entropy change during the process.

Ans:

$$p_1 = 14 \text{ bar}; V_1 = 0.25 \text{ m}^3; T_1 = 333 \text{ K}; p_2 = 3.5 \text{ bar}.$$

$$R = C_p - C_v = 2.18 - 1.55 = 0.63 \text{ kJ/kg K or } 630 \text{ N.m/kg K}.$$

$$p_1 V_1^{1.3} = p_2 V_2^{1.3};\ V_2 = V_1\left(\frac{p_1}{p_2}\right)^{(1/1.3)} = 0.25 \times \left(\frac{14}{3.5}\right)^{(1/1.3)} = 0.726 \text{ m}^3.$$

$$T_2 = T_1\left(\frac{V_1}{V_2}\right)^{(n-1)} = 333 \times \left(\frac{0.25}{0.726}\right)^{0.3} = 241.8 \text{ K};\ m = \frac{p_1 V_1}{R T_1} = \frac{14 \times 10^5 \times 0.25}{630 \times 333} = 1.668 \text{ kg}.$$

$$\text{Work done} = \frac{p_1 V_1 - p_2 V_2}{n-1} = \frac{(14 \times 0.25 - 3.5 \times 0.726) \times 10^5}{(1.3-1) \times 1000} = 319.667 \text{ kN.m.}$$

$$\Delta U = U_2 - U_1 = m\,C_v(T_2 - T_1) = 1.668 \times 1.55 \times (241.8 - 333) = -235.788 \text{ kJ}.$$

Heat transfer, ${}_1Q_2 = U_2 - U_1 + {}_1W_2 = -235.788 + 319.667 = 83.879$ kJ.

$$\text{Entropy change}, S_2 - S_1 = m\left[C_v \ln\left(\frac{T_2}{T_1}\right) + R\ln\left(\frac{V_2}{V_1}\right)\right]$$

$$= 1.668\left[1.55\ln\left(\frac{241.8}{333}\right) + 0.63\ln\left(\frac{0.726}{0.25}\right)\right] = 0.2929\ \frac{\text{kJ}}{\text{K}}.$$

15. A mixture of gases expands at constant pressure from 1 MPa, 0.03 m³ to 0.06 m³ with 84 kJ of heat transfer from the surroundings. There is no work interaction other than done on the piston. Find the change in internal energy for the gaseous mixture. The same mixture expands through the same state path while a stirring device does 21 kJ of work on the system. Find the heat transfer, work transfer and change in internal energy.

Ans: (i) Work done $= p(V_2 - V_1) = 10 \times 10^5 \times (0.06 - 0.03)/1000 = 30$ kN.m or kJ.

${}_1Q_2 = 84$ kJ.

${}_1Q_2 = U_2 - U_1 + {}_1W_2$; $84 = U_2 - U_1 + 30$.

Change in internal energy = 84 – 30 = 54 kJ.

(ii) When work is done on the system through a stirrer, additional 21 kJ of heat is added to the system.

${}_1Q_2 = U_2 - U_1 + {}_1W_2$; $84 + 21 = U_2 - U_1 + 30$.

Change in internal energy = 84 + 21 – 30 = 75 kJ.

16. A system is subjected to a cycle consisting of three processes namely constant volume, constant pressure and adiabatic. 185 kJ of heat is supplied to the system at constant volume. The system rejects 190 kJ of heat and 120 kJ of work is done on it at constant pressure. The system is brought to its initial state by an adiabatic process. Determine the adiabatic work. Determine also the values of internal energy at all end states, its initial value being 100 kJ.

Ans:

1–2 Constant volume process

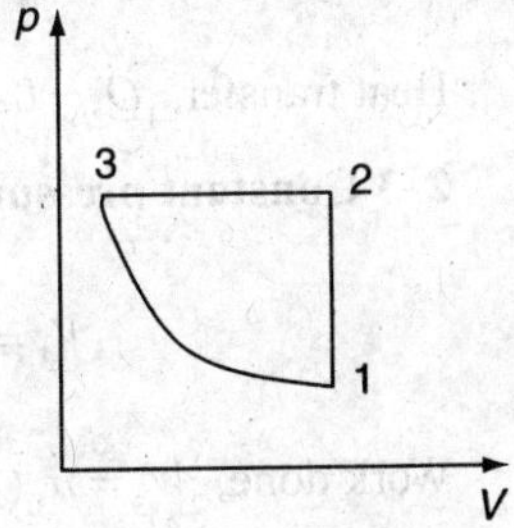

$$_1Q_2 = 185 \text{ kJ};\ _1W_2 = 0;\ U_1 = 100 \text{ kJ}.$$

$$_1Q_2 = U_2 - U_1 + {_1W_2}$$

$$185 = U_2 - 100 + 0.$$

$$U_2 = 285 \text{ kJ}.$$

2–3 Constant pressure process

$$_2Q_3 = -190 \text{ kJ};\ _2W_3 = -120 \text{ kJ};$$

$$_2Q_3 = U_3 - U_2 + {_2W_3}$$

$$-190 = U_3 - 285 - 120$$

$$U_3 = 215 \text{ kJ}.$$

3–1 Adiabatic process

$$_3Q_1 = 0;$$

$$_3Q_1 = U_1 - U_3 + {_3W_1}$$

$$0 = 100 - 215 + {_3W_1}$$

$$_3W_1 = 115 \text{ kJ}.$$

17. 0.5 kg of air is compressed reversibly and adiabatically from 80 kPa, 60°C to 0.4 MPa and is then expanded at constant pressure to the original volume. Sketch the processes on $p - V$ and $T - S$ planes. Compute the heat transfer and work transfer for the whole path.

Ans:

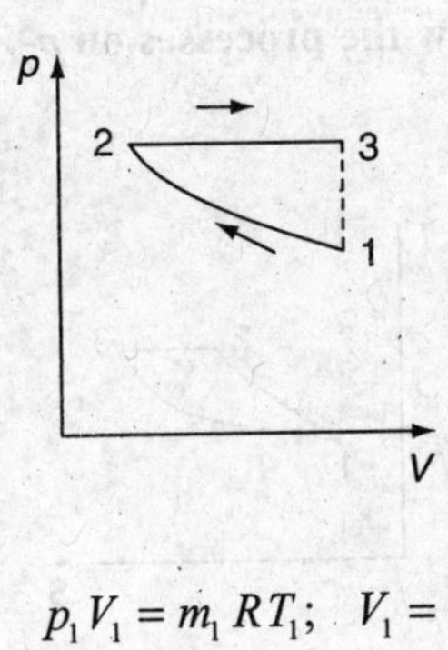

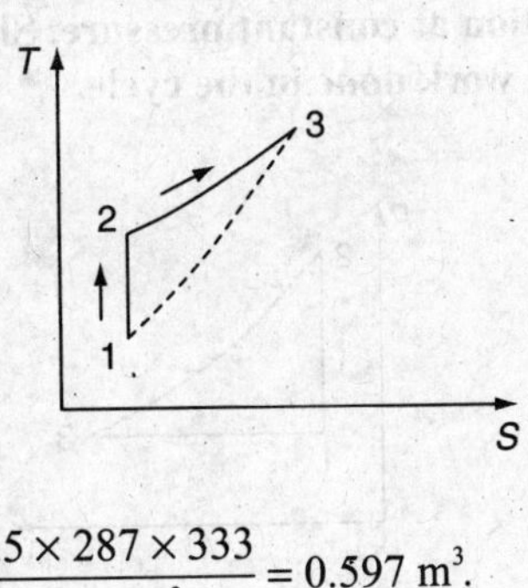

$$p_1V_1 = m_1RT_1;\quad V_1 = \frac{0.5 \times 287 \times 333}{80 \times 10^3} = 0.597 \text{ m}^3.$$

1–2 Adiabatic process

$$T_2 = T_1\left(\frac{p_2}{p_1}\right)^{\frac{\gamma-1}{\gamma}} = 333\left(\frac{400}{80}\right)^{\frac{0.4}{1.4}} = 527.4 \text{ K};\quad V_2 = V_1\left(\frac{p_1}{p_2}\right)^{\frac{1}{\gamma}} = 0.597\left(\frac{80}{400}\right)^{\frac{1}{1.4}} = 0.189 \text{ m}^3.$$

$$\text{Work done, } {}_1W_2 = \frac{p_1V_2 - p_2V_2}{\gamma - 1} = \frac{(0.8\times0.597 - 4\times0.189)\times10^5}{(1.4-1)\times10^3} = -69.6 \text{ kJ.}$$

Heat transfer, ${}_1Q_2 = 0$.

2–3 Constant pressure process

$$V_3 = V_1 = 0.597 \text{ m}^3;\ T_3 = T_2\left(\frac{V_3}{V_2}\right) = 527.4\left(\frac{0.597}{0.189}\right) = 1665.9 \text{ K.}$$

Work done, ${}_2W_3 = p_3\,(V_3 - V_2) = 400000 \times (0.597 - 0.189) = 163200$ N.m or 163.2 kJ.

$${}_2Q_3 = U_3 - U_2 + {}_3W_2 = 0.5 \times 0.718 \times (1665.9 - 527.4) + 163.2 = 571.92 \text{ kJ.}$$

Total work done = –69.6 + 163.2 = 93.6 kJ or kN.m.
Total heat transfer = 0 + 571.92 = 571.92 kJ.

18. The internal energy of a certain system is given by $U = (25 + T/4)$ kJ, where T = temperature in °C. When this system executes a certain process, work done per degree temperature change is given by $(dW/dT) = 37(T)^{0.5}$ N.m. Find the heat transfer when the system temperature increases from 100° to 260°C.

Ans:

$$U_2 - U_1 = \left(25 + \frac{260}{4}\right) - \left(25 + \frac{100}{4}\right) = \frac{160}{4} = 40 \text{ kJ.}$$

$${}_1W_2 = \int dW = \int 37\sqrt{T}\,dT = \int_{100}^{260} 37\sqrt{T}\,dT = 37\left(\frac{260^{1.5} - 100^{1.5}}{1.5}\right) = 78{,}745 \text{ N.m.}$$

Heat transfer during the process, ${}_1Q_2 = U_2 - U_1 + {}_1W_2 = 40 + 78.745 = 118.745$ kJ.

19. Air at a pressure of 1 bar and 25°C is heated at constant volume till the pressure is doubled. It is then expanded isothermally to the original pressure and then cooled to the initial condition at constant pressure. Show the processes on *p-V* and *T-S* diagrams and calculate the work done in the cycle.

Ans:

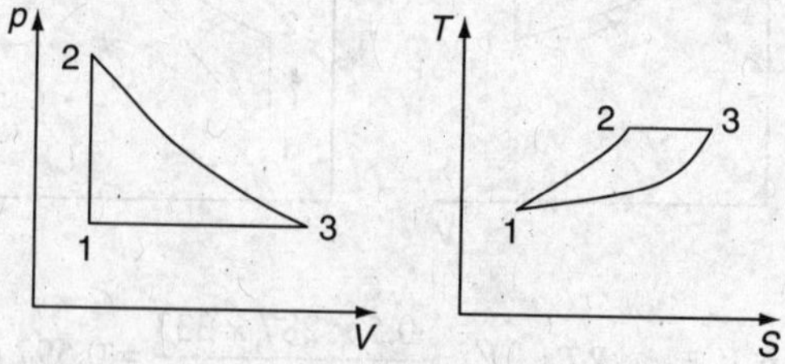

$$p_1 = 1 \text{ bar},\ T_1 = 298 \text{ K};\ p_2 = 2p_1 = 2 \text{ bar.}$$

Assume 1 kg of air is present in the system.

$$V_1 = \frac{RT_1}{p_1} = \frac{287\times298}{1{,}00{,}000} = 0.85526 \text{ m}^3.$$

1–2 Constant volume process

$$V_2 = V_1 = 0.85526 \text{ m}^3,\ T_2 = T_1\left(\frac{p_2}{p_1}\right) = 298 \times 2 = 596 \text{ K}.$$

$$_1W_2 = 0.$$

2–3 Isothermal expansion process

$p_3 = p_1 = 1$ bar; $p_2 V_2 = p_3 V_3$.

$$V_3 = V_2\left(\frac{p_2}{p_3}\right) = 0.85526 \times 2 = 1.71052 \text{ m}^3.$$

$$_2W_3 = p_2 V_2 \ln\left(\frac{V_3}{V_2}\right) = \frac{2\times10^5 \times 0.85526}{10^3}\ln\left(\frac{1.71052}{0.85526}\right) = 118.56 \text{ kN.m.}$$

3–1 Constant pressure process

$$_3W_1 = p_1(V_1 - V_3),$$

$$_3W_1 = \frac{1\times10^5(0.85526 - 1.71052)}{1000} = -85.526 \text{ kN.m.}$$

Work done in the cycle = 0 + 118.56 – 85.526 = 33.034 kN.m or kJ.

20. Compare the work done in expanding 1 kg of air at 10 bar and 300°C to double its volume by the following processes:

(i) isothermal,
(ii) isentropic and
(iii) $pV^{1.3}$ = constant.

Also compare the exit pressures. Show the processes on *p-V* and *T-S* diagrams.

Ans: (i) *Isothermal process*

$$p_1 V_1 = p_2 V_2,\quad p_2 = p_1\frac{V_1}{V_2} = 10 \times \frac{V_1}{2V_1} = 5 \text{ bar.}$$

$$\text{Work done} = p_1 V_1 \ln\left(\frac{V_2}{V_1}\right) = m_1 R T_1 \ln\left(\frac{2V_1}{V_1}\right) = 0.287 \times 573 \times \ln(2) = 113.98 \text{ kN.m.}$$

(ii) *Isentropic process p V^γ = constant*

$$p_2 = p_1\left(\frac{V_1}{V_2}\right)^\gamma = 10 \times \left(\frac{1}{2}\right)^{1.4} = 3.789 \text{ bar};\ T_2 = T_1\left(\frac{V_1}{V_2}\right)^{(\gamma-1)} = 573\left(\frac{1}{2}\right)^{0.4} = 434.2 \text{ K.}$$

$$\text{Work done} = \frac{p_1 V_1 - p_2 V_2}{\gamma - 1} = \frac{mR(T_1 - T_2)}{\gamma - 1} = \frac{287(573 - 434.2)}{0.4 \times 1000} = 99.589 \text{ kN.m.}$$

(iii) *Polytropic process, p $V^{1.3}$ = constant*

$$p_2 = p_1\left(\frac{V_1}{V_2}\right)^n = 10\times\left(\frac{1}{2}\right)^{1.3} = 4.061 \text{ bar};\ T_2 = T_1\left(\frac{V_1}{V_2}\right)^{(n-1)} = 573\left(\frac{1}{2}\right)^{0.3} = 465.42 \text{ K.}$$

$$\text{Work done} = \frac{p_1V_1 - p_2V_2}{n-1} = \frac{mR(T_1 - T_2)}{n-1} = \frac{287(573-465.42)}{0.3\times1000} = 102.918 \text{ kN.m.}$$

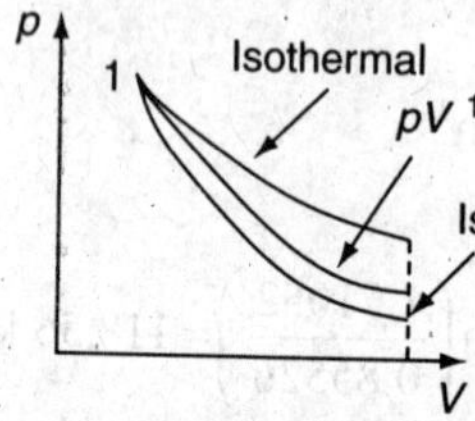

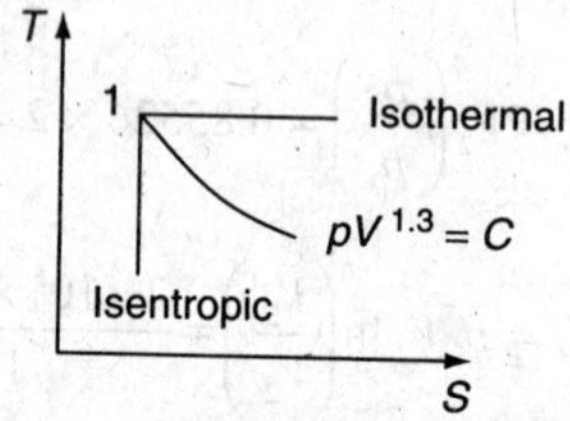

21. A cycle is completed by taking air at 1 bar and 27°C through isothermal compression to 10 bar and then constant pressure heating to a condition such that isentropic expansion will take it to the initial condition. Determine the work output of the cycle and also the heat input during the constant pressure process.

Ans: Assume 1 kg of air is undergoing the cycle.

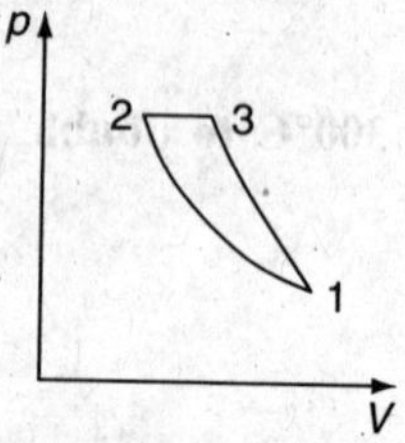

$$p_1 = 1 \text{ bar}, p_2 = 10 \text{ bar}, T_1 = 300 \text{ K.}$$

$$p_1V_1 = RT_1,\quad V_1 = \frac{287\times300}{100000} = 0.861 \text{ m}^3.$$

1–2 Isothermal compression process

$$T_2 = T_1 = 300 \text{ K};$$

$$V_2 = \frac{p_1}{p_2}V_1 = \frac{1}{10}\times0.861 = 0.0861 \text{ m}^3$$

$$_1W_2 = p_1V_1\ln\left(\frac{V_2}{V_1}\right) = \frac{1\times10^5\times0.861}{10^3}\ln\left(\frac{0.0861}{0.861}\right) = -198.25 \text{ kN.m}$$

$$_1Q_2 = {_1W_2} = -198.25 \text{ kJ.}$$

Since the point 3 is to be obtained from the process 3–1, we shall use the expression for the temperature ratio during the adiabatic expansion process.

$$\frac{T_3}{T_1} = \left(\frac{p_3}{p_1}\right)^{(\gamma-1)/\gamma};\quad T_3 = 300\left(\frac{10}{1}\right)^{0.4/1.4} = 579.2 \text{ K}$$

2–3 Constant pressure heating process

$$\frac{V_2}{T_2} = \frac{V_3}{T_3};\quad V_3 = V_2\frac{T_3}{T_2} = 0.0861\times\frac{579.2}{300} = 0.1662 \text{ m}^3$$

$$_2W_3 = p_2(V_3 - V_2) = 10 \times (10)^5 \times (0.1662 - 0.0861) = 80100 \text{ N.m or } 80.1 \text{ kN.m.}$$

$$_2Q_3 = C_p(T_3 - T_2) = 1.005 \times (579.2 - 300) = 280.596 \text{ kJ.}$$

3–1 Isentropic expansion

$$_3W_1 = \frac{(p_3V_3 - p_1V_1)}{(\gamma - 1)} = \frac{(10 \times 0.1662 - 1 \times 0.861) \times 10^5}{(1.4 - 1) \times 10^3} = 200.25 \text{ kN.m.}$$

$$_3Q_1 = 0$$

Net work output = –198.25 + 80.1 + 200.25 = 82.1 kN.m.
Heat input during constant pressure heating = 280.596 kJ.

22. A certain mass of air undergoes a cyclic process as detailed below: 0.01 m³ of air initially at a pressure of 5 bar is expanded at constant pressure. At the end of this process the air is further expanded reversibly and adiabatically to a pressure of 2 bar and volume of 0.025 m³. From this state the air is returned to the initial state through a reversible isothermal compression process. Calculate the work interaction in each of the above processes. Show the processes on *p-V* diagram and calculate the net work output for the cycle.

Ans:

$$p_1 = 5 \text{ bar}, p_2 = 5 \text{ bar}, p_3 = 2 \text{ bar}, V_1 = 0.01 \text{ m}^3 \text{ and } V_3 = 0.025 \text{ m}^3.$$

We shall find the volume of air at point 2 using the relation for adiabatic process $pV^\gamma = C$.

$$\frac{V_2}{V_3} = \left(\frac{p_3}{p_2}\right)^{1/\gamma}; \quad V_2 = 0.025 \times \left(\frac{2}{5}\right)^{1/1.4} = 0.01299 \text{ m}^3.$$

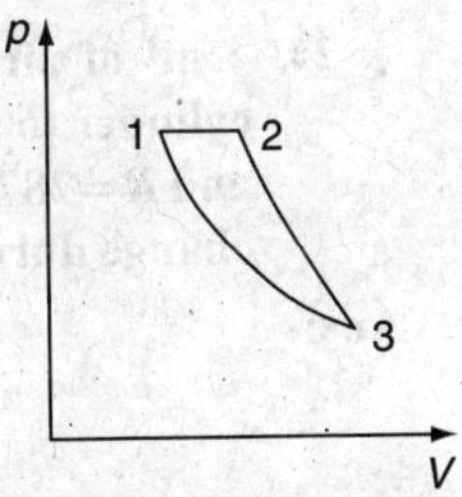

1–2 Constant pressure process

$$_1W_2 = p_1(V_2 - V_3) = 500000 \times (0.01299 - 0.01)$$

$$= 1495 \text{ N.m or } 1.495 \text{ kN.m.}$$

2–3 Isentropic process

$$_2W_3 = \frac{(p_2V_2 - p_3V_3)}{(\gamma - 1)} = \frac{(5 \times 0.01299 - 2 \times 0.025) \times 10^5}{(1.4 - 1) \times 10^3}$$

$$= 3.74 \text{ kN.m.}$$

3–1 Isothermal process

$$_3W_1 = p_3V_3 \ln\left(\frac{V_1}{V_3}\right) = \frac{2 \times 10^5 \times 0.025}{10^3} \ln\left(\frac{0.01}{0.025}\right) = -4.581 \text{ kN.m.}$$

Net work output for the cycle = 1.495 +3.74 – 4.581 = 0.654 kN.m.

23. 0.4 m³ of air at 10 bar and 100°C expands to 1 bar adiabatically and then compressed isothermally to the original volume. Find the final temperature and pressure and change in internal energy. Assume $C_p = 1.005$ and $C_v = 0.718$.

Ans:

1–2 Adiabatic process

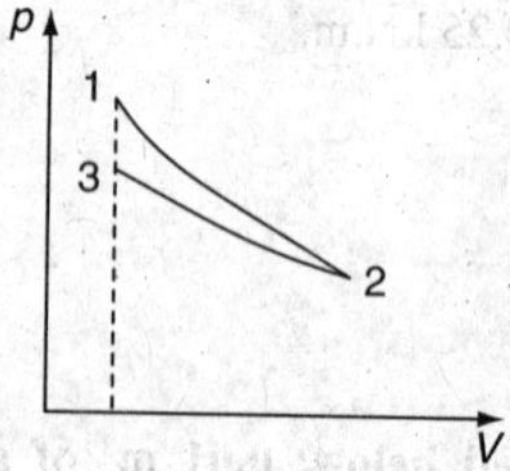

$$p_1 V_1^{\gamma} = p_2 V_2^{\gamma};\quad V_2 = V_1\left(\frac{p_1}{p_2}\right)^{1/\gamma} = 0.4\left(\frac{10}{1}\right)^{1/1.4} = 2.07 \text{ m}^3$$

$$T_2 = T_1\left(\frac{p_2}{p_1}\right)^{\gamma-1/\gamma} = 373\left(\frac{1}{10}\right)^{0.4/1.4} = 193.2 \text{ K.}$$

2–3 Isothermal process

$$p_3V_3 = p_2V_2;\ V_3 = V_1 = 0.4 \text{ m}^3;\ T_3 = T_2 = 193.2 \text{ K.}$$

$$p_3 = \frac{p_2V_2}{V_3} = \frac{1\times 2.07}{0.4} = 5.175 \text{ bar};\quad m = \frac{p_1V_1}{RT_1} = \frac{10\times10^5\times0.4}{287\times373} = 3.736 \text{ kg.}$$

Final temperature, $T_3 = 193.2$ K

Final pressure, $p_3 = 5.175$ bar

The change in internal energy, $\Delta U = m\, C_v\,(T_3 - T_1)$

$$= 3.736\times0.718\times(193.2-373)$$
$$= -482.3 \text{ kJ.}$$

24. 4 m³ of air at a pressure of 5 bar and temperature of 27°C expands reversibly in a cylinder to a pressure of 1 bar. The final volume is 20 m³. Assume constant specific heats and $R = 287$ J/kg K. Calculate (i) the heat transferred during the process; (ii) the entropy change during the process.

Ans:

$$V_1 = 4 \text{ m}^3;\ T_1 = 300 \text{ K};\ p_1 = 5 \text{ bar};\ V_2 = 20 \text{ m}^3;\ p_2 = 1 \text{ bar};$$

$$m = \frac{5\times10^5\times4}{287\times300} = 23.228 \text{ kg.}$$

$$p_1V_1^n = p_2V_2^n;\quad n = \frac{\ln(5/1)}{\ln(20/4)} = 1.$$ Since the index of expansion is 1, the process is isothermal.

$${}_1W_2 = p_1V_1\ln\left(\frac{V_2}{V_1}\right) = \frac{5\times10^5\times4}{10^3}\ln\left(\frac{20}{4}\right) = 3218.87 \text{ kN.m}$$

${}_1Q_2 = {}_1W_2$ for an isothermal process.

Heat transferred = 3218.87 kJ

For isothermal process, $S_2 - S_1 = m\,R\ln\left(\frac{V_2}{V_1}\right) = 23.228\times0.287\times\ln\left(\frac{20}{4}\right) = 10.729\ \frac{\text{kJ}}{\text{K}}$.

25. In a steady-flow system a substance flows at the rate of 5 kg/sec. It enters the system at a pressure of 600 kPa, velocity 300 m/sec, internal energy 2 MJ/kg and a specific volume 0.38 m³/kg. It leaves the system at a pressure of 150 kPa, velocity 150 m/sec, internal energy 1.8 MJ/kg and specific volume 1.26 m³/kg. During its passage through the system, the substance loses 100 kJ/kg of heat to the surroundings. Determine the power of the system stating whether it is from or to the system.

Ans:

$$\dot{Q}+m\left[\frac{V_1^2}{2}+p_1 v_1+u_1\right]=\dot{W}_s+m\left[\frac{V_2^2}{2}+p_2 v_2+u_2\right]$$

$$-100\times 5+5\left[\frac{300^2}{2000}+600\times 0.38+2000\right]=\dot{W}_s+5\left[\frac{150^2}{2000}+150\times 1.26+1800\right]$$

$$\dot{W}_s=-500+5\left[\frac{300^2-150^2}{2000}+(600\times 0.38-150\times 1.26)+(2000-1800)\right]$$

$$\dot{W}_s=-500+5[33.75+39.0+200]=863.75\frac{\text{kJ}}{\text{sec}}$$

Power of the system = 863.75 kW (from the system).

26. 15 kg/sec of gases flow through a gas turbine power plant developing a power of 12,000 kW. The gases enter with an enthalpy of 1260 kJ/kg at 50 m/sec and leave with an enthalpy of 400 kJ/kg at 110 m/sec. Determine the heat rejection rate from the turbine.

Ans:

$$Q+m\left[\frac{V_1^2}{2}+h_1+..\right]=W_s+m\left[\frac{V_2^2}{2}+h_2+...\right]$$

$$Q=12000+15\left[\frac{110^2-50^2}{2\times 1000}+(400-1260)\right]=12000+15(4.8-860)=-828\frac{\text{kJ}}{\text{sec}}$$

Heat rejection rate from the turbine = 828 kJ/sec.

27. Air flows steadily at the rate of 20 kg/min through an air compressor, entering at 6.5 m/sec velocity, 1.02 bar pressure and 0.845 m³/kg volume and leaving at 4.5 m/sec, 7.03 bar and 0.165 m³/kg. The internal energy of the fluid leaving is 88 kJ/kg greater than that of the air entering. Cooling water in the compressor jacket absorbs heat from the air at the rate of 3700 kJ/min. Calculate the power input required in kW and find the ratio of inlet pipe to outlet pipe diameter.

Ans:

$$Q+m\left(\frac{V_1^2}{2}+u_1+p_1 v_1+g z_1\right)=W+m\left(\frac{V_2^2}{2}+u_2+p_2 v_2+g z_2\right)$$

$$-\frac{3700}{60}+\frac{20}{60}\left(\frac{6.5^2}{2000}+u_1+102\times 0.845+0\right)=W_s+\frac{20}{60}\left[\frac{4.5^2}{2000}+u_2+703\times 0.165+0\right]$$

$$-61.67 = W_s + \frac{20}{60}\left[\frac{4.5^2 - 6.5^2}{2000} + 88 + (703 \times 0.165 - 102 \times 0.845)\right]$$

$$W_s = -100.93 \text{ kJ/sec.}$$

Power input = 100.93 kW

$$\text{Area of inlet pipe, } a_1 = \frac{m v_1}{V_1} = \left(\frac{20}{60}\right)\frac{0.845}{6.5} = 0.04333 \text{ m}^2$$

$$\text{Area of outlet pipe, } a_2 = \frac{m v_2}{V_2} = \left(\frac{20}{60}\right)\frac{0.165}{4.5} = 0.01222 \text{ m}^2$$

$$\text{The ratio of inlet pipe to outlet pipe diameter} = \sqrt{\frac{a_1}{a_2}} = \sqrt{\frac{0.04333}{0.01222}} = 1.88.$$

28. Steam enters an adiabatic turbine steadily at 3 MPa and 400°C and leaves at 50 kPa and 100°C. If the power output of the turbine is 2 MW and the kinetic energy changes are negligible, determine the mass rate of steam flowing through the turbine.

Ans: The following values are obtained from the steam tables:
Enthalpy of steam at 3 MPa and 400°C, $h_1 = 3230.9$ kJ/kg
Enthalpy of steam at 50 kPa and 100°C, $h_2 = 2682.5$ kJ/kg

$$Q + m(h_1 + ...) = W_s + m(h_2 + ...)$$

$$0 + m(3230.9) = \frac{2 \times 10^6}{10^3} + m(2682.5)$$

$$m = \frac{2000}{(3230.9 - 2682.5)} = 3.6469 \frac{\text{kg}}{\text{sec}}$$

Mass rate of steam flowing through the turbine = 3.6469 kg/sec.

29. In a gas turbine, the flow rate of air is 4 kg/sec. At entrance, the velocity is 250 m/sec and enthalpy is 6420 kJ/kg. At exit, the velocity is 170 m/sec and enthalpy is 4800 kJ/kg. As the air passes through the turbine, a loss of heat equal to 40 kJ/kg occurs. Find the power developed in the turbine.

Ans:

$$Q + m\left[\frac{V_1^2}{2} + h_1 + ...\right] = W_s + m\left[\frac{V_2^2}{2} + h_2 + ...\right]$$

$$-4 \times 40 + 4\left[\frac{250^2}{2000} + 6420\right] = W_s + 4\left[\frac{170^2}{2000} + 4800\right]$$

$$W_s = -160 + 4\left[\frac{(250^2 - 170^2)}{2000} + (6420 - 4800)\right]$$

$$= 6387.2 \text{ kW.}$$

30. A nozzle is a device for increasing the velocity of a steady flowing stream. At the inlet to a certain nozzle, the enthalpy of the fluid passing is 2990 kJ/kg and the velocity is 60 m/sec. At the discharge end, the enthalpy is 2760 kJ/kg. The orientation of nozzle is horizontal and there is negligible heat loss from it. (i) Find the velocity at exit of the nozzle. (ii) If the inlet area is 1000 cm² and the specific volume at inlet is 0.187 m³/kg, find the mass flow rate. (iii) If the specific volume at the nozzle exit is 0.498 m³/kg, find the exit area of the nozzle.

Ans: (i) The steady flow energy equation (SFEE) is modified for nozzle as

$$Q+m\left(\frac{V_1^2}{2}+h_1+0\right)=0+m\left(\frac{V_2^2}{2}+h_2+0\right)$$

The aforesaid equation is rewritten taking the heat loss as negligible (i.e. Q = 0)

$$V_2=\sqrt{2\times 1000(h_1-h_2)+V_1^2}=\sqrt{2000(2990-2760)+60^2}=680.8\,\frac{\text{m}}{\text{sec}}$$

The velocity of the stream at exit is 680.8 m/sec.

(ii) The mass rate of flow is obtained from the continuity equation

$$m=\frac{a_1 V_1}{v_1}=\frac{1000\times 10^{-4}\times 60}{0.187}=32.08\,\frac{\text{kg}}{\text{sec}}$$

(iii) The exit area is also obtained from the continuity equation

$$a_2=\frac{m\,v_2}{V_2}=\frac{32.08\times 0.498}{680.8}=0.02346\ \text{m}^2 \text{ or } 234.6\ \text{cm}^2.$$

31. Steam at 10 bar and 0.95 dryness is throttled to 0.5 bar. Determine the end condition. Use steam tables only.

Ans: At 10 bar, 0.95 dryness,

$$h_1 = 762.6 + 0.95 \times 2013.6 = 2675.5$$

During throttling process, the enthalpy remains the same.

$$h_2 = h_1,\ h_2 = 2675.5.$$

At 0.5 bar (81.3°C),

$$h_f = 340.6;\ h_g = 2646;$$

Since h_2 is greater than h_g, the steam after throttling is superheated.

$$C_p=\frac{(2877.7-2780.1)}{50}=1.952,\quad h_2=h_g+C_p\left(T_2-T_{g2}\right)$$

$$2675.5 = 2646 + 1.952[T_2 - (273+81.3)],\ T_2 = 369.4\ \text{K or } 96.4°\text{C}.$$

The condition of steam after throttling is superheated and the temperature of steam is 96.4°C or the degree of superheat is 15.1°C.

32. Steam at 10 bar and 300°C is cooled at constant volume till the pressure is 0.5 bar. Find the end condition using steam tables. Also find the change in internal energy and entropy.

Ans:

At 10 bar, 300°C

Specific volume, $v_1 = 0.258$ m³/kg

Specific enthalpy, $h_1 = 3051.2$

Specific entropy, $s_1 = 7.123$

Specific internal energy, $u_1 = 2793.2$

At 0.5 bar

$h_f = 340.6$

$h_{fg} = 2305.4$

$h_g = 2646.0$

$s_f = 1.0912$

$S_{fg} = 6.5035$

$v_g = 3.24$ m³/kg

During constant volume process,

$$v_1 = v_2 = 0.258 \text{ m}^3/\text{kg}$$

Since v_2 is less than v_g (3.24), the steam is wet after the cooling process.

$$\text{Dryness fraction, } x_2 = \frac{0.258}{3.24} = 0.0796$$

$$h_2 = 340.6 + 0.0796 \times 2305.4 = 524.1$$

$$u_2 = 524.1 - 0.5 \times 100 \times 0.258 = 511.2$$

$$s_2 = 1.0912 + 0.0796 \times 6.5035 = 1.6088$$

The end condition of steam is wet with a dryness fraction of 0.0796.

The change in internal energy, $u_2 - u_1 = 511.2 - 2793.2 = -2282$ kJ/kg.

The change in entropy, $s_2 - s_1 = 1.6088 - 7.123 = -5.5142$ kJ/kg K.

33. A pressure cooker contains 1.5 kg of saturated steam at 5 bar. Find the quantity of heat which must be rejected so as to reduce the quality at 60% dry. Determine the pressure and temperature of the steam at the new state.

Ans:

At 5 bar pressure, $h_g = 2747.5$; $v_g = 0.375$ m³/kg.

At constant volume process, $V_2 = V_1$ or $v_2 = v_1 = 0.375$.

Since the dryness fraction is given as 0.6, the final specific volume of dry steam is calculated from the relation, $v_2 = x_2 v_g$, where v_g is the specific volume of dry steam at the end condition.

$$v_g = \frac{v_2}{x_2} = \frac{0.375}{0.6} = 0.625 \frac{\text{m}^3}{\text{kg}}$$

From steam tables, the pressure is found as 2.9 bar corresponding to the specific volume of dry and saturated steam 0.625 m³/kg.

At 2.9 bar pressure, $h_f = 556.5$; $h_{fg} = 2166.6$; $v_2 = 0.375$

$$h_1 = 2747.5;\ u_1 = h_1 - p_1 v_1 = 2747.5 - 5 \times 100 \times 0.375 = 2560;$$

$$h_2 = 556.5 + 0.6 \times 2166.6 = 1856.46;$$

$$u_2 = h_2 - p_2 v_2 = 1856.46 - 2.9 \times 100 \times 0.375 = 1747.71.$$

For a constant volume process,

$$_1q_2 = u_2 - u_1 + 0 = 1747.71 - 2560 = -812.29 \text{ kJ/kg}$$

Heat rejection rate, $_1Q_2 = 1.5 \times (-812.29) = -1218$ kJ.

Final pressure, $p_2 = 2.9$ bar and final temperature, $t_2 = 132.4°C$ (t_s for 2.9 bar).

34. Steam at a pressure of 10 bar with a degree of super heat of 100°C is expanded isentropically in an open system to a pressure of 0.1 bar. Calculate (i) the work done by steam/kg and (ii) specific volume of steam at the end of expansion.

Ans: At 10 bar, $t_s = 179.9°C$, $h_g = 2776.2$, $s_g = 6.5828$

The specific heat of superheated steam at 10 bar is found from the enthalpy of superheated steam table as given below

$$C_p = \frac{h_{\text{sup}1} - h_{\text{sup}2}}{t_{\text{sup}1} - t_{\text{sup}2}} = \frac{2942.6 - 2827.9}{250 - 200} = 2.294 \frac{\text{kJ}}{\text{kg K}}$$

$$h_1 = h_g + C_p(t_2 - t_s) = 2776.2 + 2.294(100) = 3005.6$$

$$s_1 = s_g + C_p \ln\frac{(273+279.9)}{(273+179.9)} = 6.5828 + 2.294\ln\left(\frac{552.9}{452.9}\right) = 7.0405 \frac{\text{kJ}}{\text{kg K}}$$

At 0.1 bar pressure,

$$h_f = 191.8,\ h_{fg} = 2392.8,\ h_g = 2584.7,\ s_f = 0.649,\ s_{fg} = 7.501,\ s_g = 8.150,\ v_g = 14.67$$

During the isentropic expansion the entropy remains the same. $s_2 = s_1 = 7.0405$

Since s_2 (7.0405) is less than s_g (8.15), the condition of steam is wet.

$$x_2 = \frac{7.0405 - 0.649}{7.501} = 0.852$$

$$h_2 = 191.8 + 0.852 \times 2392.8 = 2230.5$$

Work done $= h_1 - h_2 = 3005.6 - 2230.5 = 775.1$ kJ/kg.

Specific volume of steam at the end of expansion, $v_2 = 0.852 \times 14.67 = 12.4988 \text{ m}^3/\text{kg}$.

35. A pressure cooker has a volume of 4 litres and contains 0.75 kg of water at 20°C. The lid is secured and heat is added with the air being vented in the process, until the pressure is 200 kPa. Determine the heat added and final quality.

Ans:

Enthalpy of water at 20°C, $h_1 = u_1 = 84$ kJ/kg

Volume of steam at the end of the process $= 0.004 \text{ m}^3$

Specific volume of steam at the end of the process, $v_2 = \frac{0.004}{0.75} = 0.0053 \frac{\text{m}^3}{\text{kg}}$

At 2 bar pressure,

$$h_f = 504.7$$

$$h_{fg} = 2201.6$$

$$v_g = 0.885$$

Since v_2 is less than v_g, the condition of steam is wet at the end of process.

$$x_2 v_g = 0.0053; x_2 = (0.0053)/(0.885) = 0.006.$$

$$h_2 = 504.7 + 0.006 \times 2201.6 = 517.9; u_2 = 517.9 - 200 \times 0.0053 = 516.8 \text{ kJ/kg}$$

$$_1q_2 = u_2 - u_1 + {}_1w_2 = 516.8 - 84 + 0 = 432.8 \text{ kJ/kg}$$

$$\text{Heat added, } {}_1Q_2 = 0.75 \times 432.8 = 324.6 \text{ kJ.}$$

Final quality is 0.6% dry or dryness fraction 0.006.

36. A vessel of capacity 30 litres contains steam at a pressure of 10 bar and 95% quality. Determine its internal energy. The pressure of steam falls to 3 bar due to radiation loss. What will be the quality of steam at this stage? Neglecting the volume of water present, calculate the heat lost due to radiation.

Ans:

$$V_1 = 0.03 \text{ m}^3$$

As the steam undergoes a constant volume process $V_2 = V_1 = 0.03$ m³
At 10 bar pressure,

$$h_f = 762.6; h_{fg} = 2013.6; v_g = 0.194$$

At 3 bar pressure,

$$h_f = 561.4; h_{fg} = 2163.2; v_g = 0.606$$

Specific volume steam at the initial state, $v_1 = 0.95 \times 0.194 = 0.1843$

Mass of steam present in the vessel, $m = \dfrac{0.03}{0.1843} = 0.1628$ kg.

During a constant volume process, $V_2 = V_1$ or $v_1 = v_2$

Final specific volume, $v_2 = 0.1843 = x_2 \times 0.606$,

Final dryness fraction, $x_2 = \dfrac{0.1843}{0.606} = 0.304$

Quality of steam after radiation loss = 30.4% dry.

$$h_1 = 762.6 + 0.95 \times 2013.6 = 2675.5; u_1 = 2675.5 - 1000 \times 0.1843 = 2491.2$$

$$h_2 = 561.4 + 0.304 \times 2163.2 = 1219.0; u_2 = 1219.0 - 300 \times 0.1843 = 1163.7$$

$$_1q_2 = u_2 - u_1 + {}_1w_2 = 1163.7 - 2491.2 + 0 = -1327.5.$$

Heat lost due to radiation, $_1Q_2 = m \times {}_1q_2 = 0.1628 \times (-1327.5) = -216.117$ kJ.

37. Steam at 10 bar and 0.92 dryness expands hyperbolically in a piston-cylinder arrangement to 4 bar. Determine the end condition, work done and the change in internal energy.

Ans: At 10 bar pressure,

$$h_f = 762.6; h_{fg} = 2013.6; v_g = 0.194.$$

At 4 bar pressure,

$$h_f = 604.6;\ h_{fg} = 2133;\ v_g = 0.462$$

$$v_1 = 0.92 \times 0.194 = 0.1785 \text{ m}^3\text{/kg.}$$

For hyperbolic process, $p\,v$ = constant,

$$p_1 v_1 = p_2 v_2;\ 10 \times 10^5 \times (0.1785) = 4 \times 10^5 \times (v_2)$$

$$v_2 = \frac{10 \times 10^5 \times 0.1785}{4 \times 10^5} = 0.4463.$$

Since v_2 is less than v_g (0.462), the condition of the steam is wet.

$$x_2 \times 0.462 = 0.4463;$$

$$x_2 = \frac{0.4463}{0.462} = 0.966.$$

Dryness fraction of steam after expansion is 0.966.

$$h_1 = 762.6 + 0.92 \times 2013.6 = 2615.1;\ u_1 = 2615.1 - 1000 \times 0.1785 = 2436.6.$$

$$h_2 = 604.6 + 0.966 \times 2133 = 2665.078;\ u_2 = 2665.078 - 400 \times 0.4463 = 2486.558.$$

$$\text{Work done} = p_1 v_1 \ln\left(\frac{v_2}{v_1}\right) = 1000 \times 0.1785 \times \ln\left(\frac{0.4463}{0.1785}\right) = 163.577 \frac{\text{kJ}}{\text{kg}}$$

Change in internal energy $= u_2 - u_1 = 2486.558 - 2436.6 = 49.958$ kJ/kg.

38. Steam is taken from a boiler at a pressure of 15 bar and 250°C. It is expanded with an index of 1.25 to a pressure of 1.5 bar. Determine the final condition of steam if the mass of the steam is 0.9 kg. Also compute the work done and heat transferred.

Ans: At 15 bar, 250°C,

$$h_{sup} = 2923.3;\ u_{sup} = 2695.3;\ v_{sup} = 0.152;$$

At 1.5 bar pressure,

$$h_f = 467.1;\ h_{fg} = 2226.2;\ v_g = 1.159;$$

$$V_1 = 0.9 \times v_{sup} = 0.9 \times 0.152 = 0.1368;$$

$$\text{p}_1 V_1^n = p_2 V_2^n;\ \ V_2 = \left(\frac{p_1}{p_2}\right)^{1/n} V_1 = \left(\frac{15}{1.5}\right)^{1/1.25} \times 0.1368 = 0.863 \text{ m}^3$$

Specific volume of steam after expansion, $v_2 = V_2/m = (0.863) / (0.9) = 0.959$ m³/kg.
Since v_2 is less than v_g (1.159), the condition of steam after expansion is wet.
Dryness fraction, $x_2 = (v_2)/(v_g) = (0.959)/(1.159) = 0.83$.

$$h_1 = h_{sup} = 2923.3;\ u_1 = 2695.3;\ h_2 = h_f + x_2\, h_{fg} = 467.1 + 0.83 \times 2226.2 = 2314.8;$$

$$u_2 = h_2 - p_2 v_2 = 2314.8 - 150 \times 0.959 = 2170.95.$$

$$\text{Work done, } {}_1W_2 = \frac{p_1V_1 - p_2V_2}{n-1} = \frac{1500 \times 0.1368 - 150 \times 0.863}{1.25-1} = 303 \text{ kN.m.}$$

$$U_1 = 0.9 \times 2695.3 = 2425.8;\ U_2 = 0.9 \times 2170.95 = 1953.9;$$

$$\text{Heat transferred, } {}_1Q_2 = U_2 - U_1 + {}_1W_2 = (1953.9 - 2425.8) + 303 = -168.9 \text{ kJ.}$$

39. Steam at 15 bar and 300°C is throttled till its pressure is 10 bar and then it is expanded isentropically passing through a turbine till its pressure drops to 1 bar. Find the condition of the steam leaving the turbine and the isentropic heat drop.

Ans: At 15 bar and 300°C,

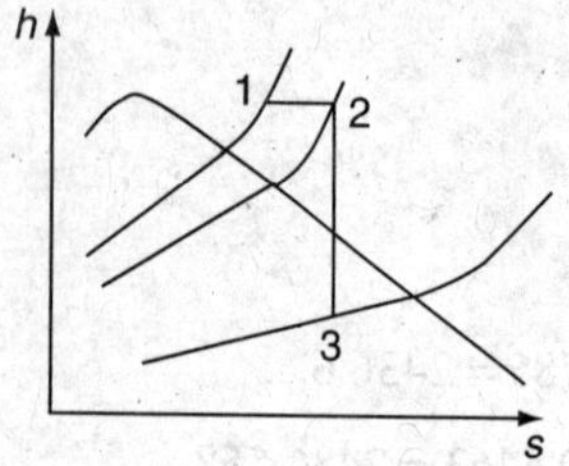

$$h_{sup} = 3037.6$$

At 10 bar pressure,

$$h_g = 2776.2;\ t_s = 179.9°\text{C};\ s_g = 6.5828$$

At 1 bar pressure,

$$h_f = 417.5;\ h_{fg} = 2257.9;\ s_f = 1.3027;$$

$$s_{fg} = 6.0571;\ s_g = 7.3598.$$

1–2 Throttling process

$$h_1 = h_2 = 3037.6$$

Since h_2 is more than h_g (2776.2), the steam is superheated after throttling.

The specific heat at constant pressure for superheated steam at 10 bar is obtained from the enthalpy table given as follows:

$$C_p = \frac{3051.2 - 2942.6}{300 - 250} = 2.17$$

$$3037.6 = 2776.2 + C_p[T_2 - (273 + 179.9)];\ T_2 = 573.3 \text{ K.}$$

$$s_2 = s_g + C_p \ln\left(\frac{T_2}{T_s}\right) = 6.5828 + 2.17 \ln\left(\frac{573.3}{452.9}\right) = 7.094 \frac{\text{kJ}}{\text{kg K}}$$

2–3 Isentropic process

$$s_2 = s_3,\ s_2 = 7.094 = s_3$$

Since s_3 is less than s_g (7.3598), the condition of steam is wet.

$$s_3 = s_f + x_3 s_{fg};\ 1.3027 + x_3 \times 6.0571 = 7.094;\ \text{hence, } x_3 = (7.094 - 1.3027)/6.0571$$

$$x_3 = 0.956$$

$$h_3 = 417.5 + 0.956 \times 2257.9 = 2576.0.$$

The condition of steam leaving the turbine is wet with dryness fraction 0.956.

The isentropic heat drop, $\Delta h = h_2 - h_3 = 3037.6 - 2576.0 = 461.6$ kJ/kg.

40. 10 kg of steam at 50 bar and 0.9 dryness fraction is heated isothermally till its entropy becomes 72.87 kJ/K. Find the change in (a) enthalpy, (b) internal energy and (c) entropy. Also find the work accompanying the process. Represent the process on *T-s* and *h-s* diagrams.

Ans: At 50 bar pressure,

$$t_s = 263.9°C$$

$$s_f = 2.9206;\ s_{fg} = 3.0529$$

$$h_f = 1154.5;\ h_{fg} = 1639.7$$

$$v_g = 0.0394$$

$$s_1 = 2.9206 + 0.9 \times 3.0529$$
$$= 5.6682 \text{ kJ/kg K.}$$

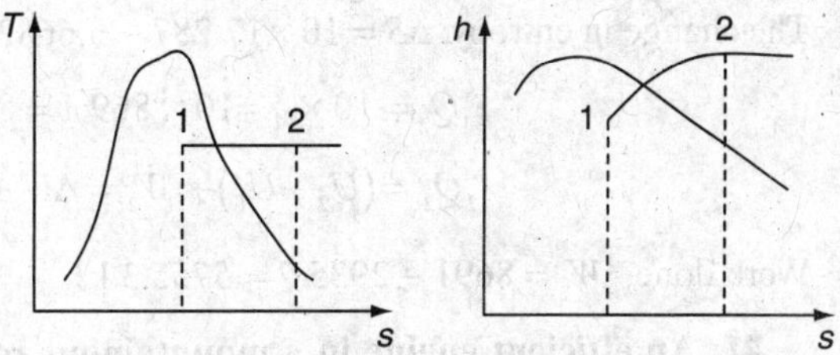

Final entropy, $S_2 = 72.87$ kJ/K.

$$s_2 = \frac{72.87}{10} = 7.287 \text{ kJ/kg K.}$$

$$q = T(s_2 - s_1) = 536.9 \times (7.287 - 5.6682) = 869.1 \text{ kJ/kg.}$$

The point 2 is to be located on the Mollier chart along the isotherm (263.9°C) and the pressure can be read. The enthalpy can be read directly from the chart.

The problem can also be solved using the steam tables.

The following entropy values are taken from the super heated steam tables:

Pr	250°C	300°C
3.0	7.517	7.702
4.0	7.379	7.566
5.0	7.271	7.460
6.0	7.182	7.372

From the entropy values, we can conclude that the final pressure is 5 bar.

At 5 bar pressure and 250°C,

$$h_{sup} = 2960.7;\ v_{sup} = 0.474.$$

The specific heat of superheated steam at 5 bar is found from the enthalpy of superheated steam as follows:

$$C_p = \frac{h_{sup1} - h_{sup2}}{t_{sup1} - t_{sup2}} = \frac{3064.2 - 2960.7}{300 - 250} = 2.07 \frac{\text{kJ}}{\text{kg k}}$$

The specific volume of steam is also found as follows:

$$\frac{v_2}{T_2} = \frac{v_{sup}}{T_{sup}};\quad \frac{v_2}{273 + 263.9} = \frac{0.474}{273 + 250};\quad v_2 = \frac{536.9}{523} \times 0.474 = 0.486 \frac{\text{m}^3}{\text{kg}}$$

$$h_2 = 2960.7 + 2.07(536.9 - 523) = 2989.5 \text{ kJ/kg.}$$

$$u_2 = h_2 - p_2 v_2 = 2989.5 - 5 \times 100 \times 0.486 = 2746.5 \text{ kJ/kg.}$$

$$h_1 = 1154.5 + 0.9 \times 1639.7 = 2630.23 \text{ kJ/kg.}$$

$$u_1 = 2630.23 - 50 \times 100 \times (0.9 \times 0.0394) = 2452.93 \text{ kJ/kg.}$$

The change in enthalpy, $\Delta H = 10 \times (2989.5 - 2630.23) = 3592.7$ kJ.
The change in internal energy, $\Delta U = 10 \times (2746.5 - 2452.93) = 2935.7$ kJ.
The change in entropy, $\Delta S = 10 \times (7.287 - 5.6682) = 16.188$ kJ/K.

$${}_1Q_2 = 10 \times q = 10 \times 869.1 = 8691 \text{ kJ.}$$

$${}_1Q_2 = (U_2 - U_1) + {}_1W_2 = \Delta U + {}_1W_2; \quad 8691 = 2935.7 + {}_1W_2.$$

Work done, ${}_1W_2 = 8691 - 2935.7 = 5755.3$ kJ.

41. An efficient engine in a mountainous region using hot spring water at 57°C(as source) receiving 418 kJ from the source is said to reject 325 kJ to melting ice at 0°C(as sink). Comment on the engine efficiency.

Ans:

$$T_1 = 273 + 57 = 330 \text{ K}; \; T_2 = 273 + 0 = 273 \text{ K}$$

$$\eta_{\text{Carnot}} = \frac{T_1 - T_2}{T_1} = \frac{330 - 273}{330} = 0.1727$$

$$\eta_{\text{engine}} = \frac{418 - 325}{418} = 0.2224.$$

Since the engine efficiency is more than the Carnot efficiency, the engine cannot function.

42. An engineer claims his engine develops 3.68 kW. On testing, the engine consumes 0.44 kg of fuel per hour. The calorific value of the fuel is 41,833 kJ/kg. The maximum temperature recorded in the cycle is 1400°C and the minimum is 350°C. Find whether the engineer is justified in his claim.

Ans:

$$T_1 = 273 + 1400 = 1673 \text{ K}; \; T_2 = 273 + 350 = 623 \text{ K.}$$

$$\eta_{\text{carnot}} = \frac{T_1 - T_2}{T_1} = \frac{1673 - 623}{1673} = 0.6276 \text{ or } 62.76\%$$

$$\eta_{\text{engine}} = \frac{3.68}{\frac{0.44}{3600} \times 41833} = 0.7197 \text{ or } 71.97\%$$

Since the engine efficiency is more than the Carnot efficiency, the claim is incorrect.

43. A series combination of three Carnot engines operate between 1000 K and 300 K. Obtain the intermediate temperatures if the work done by the engines are in the ratio 5:4:3.

Ans: For engine 1,

$$\eta_1 = \frac{1000 - T_1}{1000}; \quad W_1 = \eta_1 Q_1; \quad W_1 = Q_1 - Q_2$$

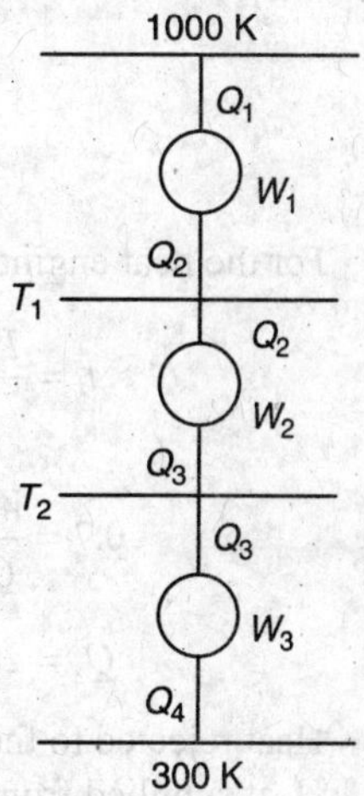

For engine 2,

$$\eta_2 = \frac{T_1 - T_2}{T_1}; \quad W_2 = \eta_2 Q_2; \quad W_2 = Q_2 - Q_3$$

For engine 3,

$$\eta_3 = \frac{T_2 - 300}{T_2}; \quad W_3 = \eta_3 Q_3; \quad W_3 = Q_3 - Q_4$$

We know the following relations for the Carnot engines

$$\frac{Q_1}{1000} = \frac{Q_2}{T_1} = \frac{Q_3}{T_2} = \frac{Q_4}{300}$$

$$\frac{W_1}{W_2} = \frac{5}{4} = \frac{Q_1 - Q_2}{Q_2 - Q_3} \quad (1)$$

$$\frac{W_2}{W_3} = \frac{4}{3} = \frac{Q_2 - Q_3}{Q_3 - Q_4} \quad (2)$$

We can express Q_1, Q_2, and Q_3 in terms of Q_4, T_1, and T_2 and simplify as follows:

$$\frac{5}{4} = \frac{\frac{1000\,Q_4}{300} - \frac{T_1 Q_4}{300}}{\frac{T_1 Q_4}{300} - \frac{T_2 Q_4}{300}} = \frac{1000 - T_1}{T_1 - T_2}; \quad 5T_1 - 5T_2 = 4000 - 4T_1; \quad 9T_1 - 5T_2 = 4000 \quad (3)$$

$$\frac{4}{3} = \frac{\frac{T_1 Q_4}{300} - \frac{T_2 Q_4}{300}}{\frac{T_2 Q_4}{300} - Q_4} = \frac{T_1 - T_2}{T_2 - 300}; \quad 4T_2 - 1200 = 3T_1 - 3T_2; \quad 3T_1 - 7T_2 = -1200 \quad (4)$$

Solving equations (3) and (4), we get the intermediate temperatures as follows:

$$T_1 = 708.33 \text{ K and } T_2 = 475 \text{ K.}$$

44. A heat pump working on reversed Carnot cycle takes heat from a reservoir at 5°C and delivers heat to reservoir at 60°C. The heat pump is driven by a reversible heat engine which takes heat from a reservoir at 840°C and rejects heat to the reservoir at 60°C. The reversible heat engine also drives another machine that consumes 30 kW. The heat pump extracts 17 kJ/sec from the 5°C reservoir. Determine (i) the rate of heat supplied from the 840°C reservoir and (ii) the rate of heat rejected to 60°C reservoir.

Ans:

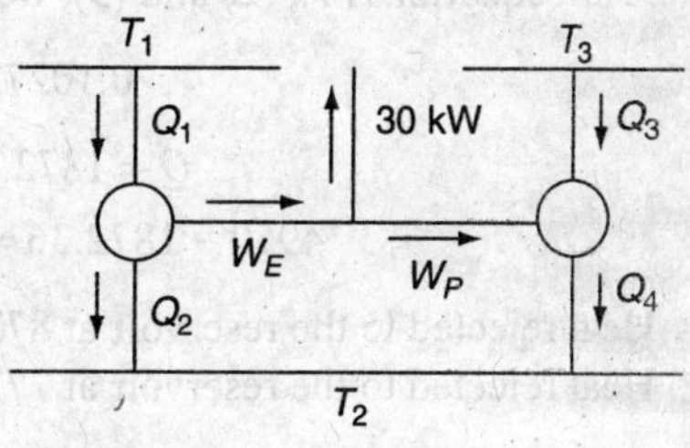

$$T_1 = 273 + 840 = 1113 \text{ K};$$

$$T_2 = 273 + 60 = 333 \text{ K};$$

$$T_3 = 273 + 5 = 278 \text{ K};$$

$$(\text{C.O.P})_{\text{H.P}} = \frac{T_2}{T_2 - T_3} = \frac{333}{333 - 278} = 6.05$$

For a reversible Carnot cycle (heat pump),

$$\frac{Q_4}{T_2} = \frac{Q_3}{T_3}; Q_4 = \frac{333}{278} \times 17 = 20.36 \frac{\text{kJ}}{\text{sec}}.$$

$$W_P = 20.36 - 17 = 3.36 \text{ kJ/sec.}$$

For the heat engine,

$$\eta = \frac{T_1 - T_2}{T_1} = \frac{1113 - 333}{1113} = 0.7; W_E = 3.36 + 30 = 33.36 \text{ kJ/sec;}$$

$$0.7 = \frac{W_E}{Q_1}; \quad Q_1 = \frac{33.36}{0.7} = 47.657 \frac{\text{kJ}}{\text{sec}}; Q_2 = 47.657 - 33.36 = 14.297 \text{ kJ/sec;}$$

$$Q_4 = 20.36 \text{ kJ/sec.}$$

Heat rejected to the reservoir at 60°C = $Q_4 + Q_2$ = 20.36 + 14.297 = 34.657 kJ/sec.
Heat supplied from the reservoir at 840°C = Q_1 = 47.657 kJ/sec.

45. A reversible engine takes 4950 kJ of heat per minute from a reservoir at 700°C and develops 825 kJ of work per minute when executing complete cycles. The engine rejects heat to two reservoirs at 600°C and 500°C. Find the heat rejected to each sink.

Ans:

$$T_1 = 273 + 700 = 973 \text{ K}; T_2 = 273 + 600 = 873 \text{ K}; T_3 = 273 + 500 = 773 \text{ K}.$$

Let us assume that one cycle of the engine receives Q amount of heat from the reservoir at 973 K and rejects the heat to the reservoir at 873 K and another cycle of the engine receives (4950 – Q) of heat from the reservoir at 973 K and rejects heat to the reservoir at 773 K.

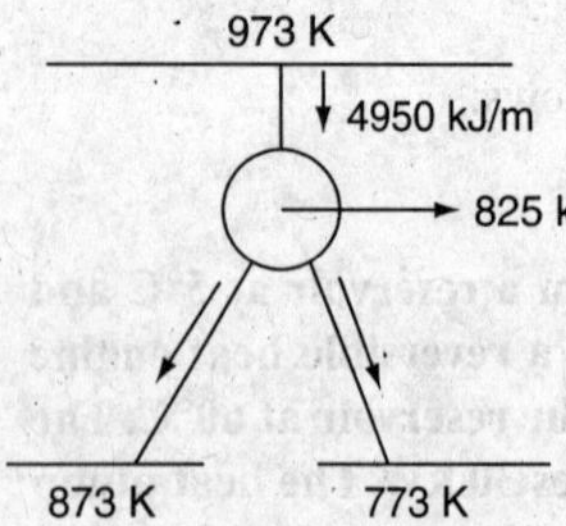

$$\eta_1 = \frac{973 - 873}{973} = 0.10277;$$

$$W_1 = 0.10277\, Q \qquad (1)$$

$$\eta_2 = \frac{973 - 773}{973} = 0.20554;$$

$$W_2 = 0.20554\,(4950 - Q) = 1017.423 - 0.20554\, Q \qquad (2)$$

$$W_1 + W_2 = 825 \qquad (3)$$

From equations (1), (2) and (3), we can solve for Q as given below:

$$0.10277\, Q + 1017.423 - 0.20554\, Q = 825$$

$$Q = 1872.36; W_1 = 0.10277 \times 1872.36 = 192.42.$$

$$4950 - 1872.36 = 3077.64; W_2 = 0.20554 \times 3077.64 = 632.58.$$

Heat rejected to the reservoir at 873 K = 1872.36 – 192.42 = 1679.94 kJ/min.
Heat rejected to the reservoir at 773 K = 3077.64 – 632.58 = 2445.06 kJ /min.

46. Two reversible heat engines are connected in series with heat rejected by the first engine being the heat input to the second engine. The temperature at which heat is added to the first engine is 1000 K and the temperature at which the second engine rejects heat is 300 K. If the first engine produces 2/3 of the work done by the second engine, find the temperature at which heat is received by the second engine.

Ans: For the reversible engines we have

$$\frac{Q_1}{1000} = \frac{Q_2}{\text{T}} = \frac{Q_3}{300}$$

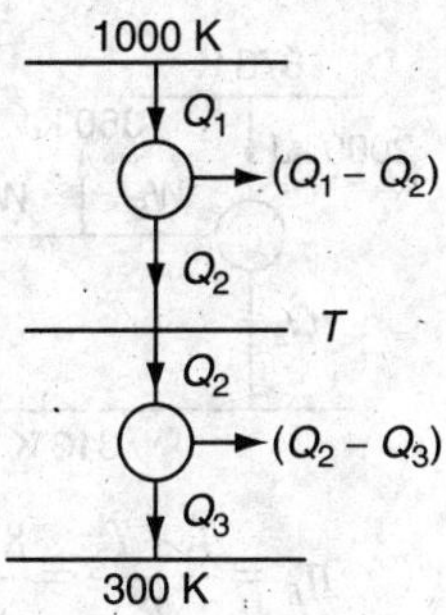

Given

$$(Q_1 - Q_2) = \frac{2}{3}(Q_2 - Q_3)$$

Substituting for Q_1 and Q_3 in terms of Q_2 and T in the aforesaid equation, we have

$$\left(\frac{1000\,Q_2}{T} - Q_2\right) = \frac{2}{3}\left(Q_2 - \frac{300\,Q_2}{T}\right)$$

$$\text{i.e.}\quad \frac{1000}{T} - 1 = \frac{2}{3}\left(1 - \frac{300}{T}\right)$$

$$\frac{1000}{T} + \frac{200}{T} = 1 + \frac{2}{3}.$$

Hence, $T = 720$ K.

The temperature at which the second engine receives heat is at 720 K.

47. A Carnot refrigerator removes heat from a sink at –20°C and rejects heat to the atmosphere at 15°C. The refrigerator is powered by the output of a Carnot engine which receives heat from a source at 600°C and rejects to the atmosphere. Determine the ratio of heat supplied to the engine to the heat removed by the refrigerator.

Ans:

$$T_1 = 273 + 600 = 873\text{ K};\ T_2 = 273 + 15 = 288\text{ K};\ T_3 = 273 - 20 = 253\text{ K}$$

Carnot engine operates between T_1 and T_2 and refrigerator operates between T_3 and T_2

$\eta_E = \dfrac{873 - 288}{873} = 0.67$, $W_E = 0.67\,Q_1$, where Q_1 is the heat supplied to the engine.

$$(\text{C.O.P})_R = \frac{253}{288 - 253} = 7.228.$$

$W_R = \dfrac{Q_3}{(\text{C.O.P})_R} = \dfrac{Q_3}{7.228}$, where Q_3 is the heat removed by the refrigerator.

Given $W_E = W_R$; (the engine drives the refrigerator)

$$0.67\,Q_1 = \frac{Q_3}{7.228};\quad \frac{Q_1}{Q_3} = \frac{1}{0.67 \times 7.228} = 0.20649.$$

The ratio of heat supplied to the engine to the heat removed by the refrigerator is 0.2065.

48. A reversible heat engine operates between two reservoirs at temperatures of 600°C and 40°C. The engine drives a reversible refrigerator which operates between reservoirs at temperatures of 40°C and −20°C. The heat transfer to the engine is 2 MJ and the net work output of the combined plant is 360 kJ. (i) Evaluate the heat transfer to the refrigerant and the net heat transfer to the reservoir at 40°C. (ii) Reconsider (i), given that the efficiency of heat engine and COP of the refrigerator are each 40% of their respective maximum possible values.

Ans:

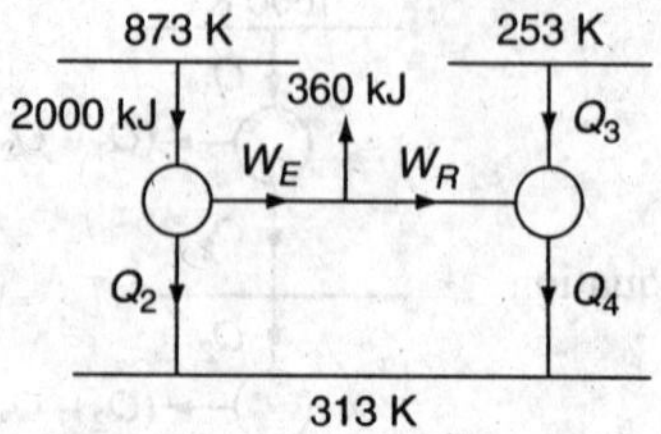

$$T_1 = 273 + 600 = 873\text{ K};$$

$$T_2 = 273 + 40 = 313\text{ K};$$

$$T_3 = 273 - 20 = 253\text{ K};$$

$$Q_1 = 2000\text{ kJ};$$

$$W_E - W_R = 360\text{ kJ};$$

(i) $$\eta_E = \frac{T_1 - T_2}{T_1} = \frac{873-313}{873} = 0.64$$

$$(\text{C.O.P})_R = \frac{T_3}{T_2 - T_3} = \frac{253}{313-253} = 4.216.$$

$$W_E = 0.64 \times 2000 = 1280\text{ kJ.}$$

$$Q_2 = 2000 - 1280 = 720\text{ kJ.}$$

$$W_R = 1280 - 360 = 920\text{ kJ.}$$

$$(\text{C.O.P})_R = \frac{Q_3}{W_R};\quad Q_3 = 4.216 \times 920 = 3878.7\text{ kJ.}$$

$$Q_4 = W_R + Q_3 = 920 + 3878.7 = 4798.7\text{ kJ.}$$

Heat transfer to the refrigerant, $Q_3 = 3878.7$ kJ.
Heat transfer to the reservoir at 40°C $= Q_2 + Q_4 = 720 + 4798.7 = 5518.7$ kJ.

(ii) $Q_1 = 2000$ kJ

$$\eta_E = 0.4 \times 0.64 = 0.256.$$

$$(\text{C.O.P})_R = 0.4 \times 4.216 = 1.686.$$

$$W_E = 0.256 \times 2000 = 512\text{ kJ.}$$

$$Q_2 = 2000 - 512 = 1488\text{ kJ.}$$

$$W_R = W_E - 360 = 512 - 360 = 152\text{ kJ.}$$

$$Q_3 = (\text{C.O.P})_R\, W_R = 1.686 \times 152 = 256.3\text{ kJ.}$$

$$Q_4 = Q_3 + W_R = 256.3 + 152 = 408.3\text{ kJ.}$$

Heat transfer to the refrigerant, $Q_3 = 256.3$ kJ.
Heat transfer to the reservoir at 40°C $= Q_2 + Q_4 = 1488 + 408.3 = 1896.3$ kJ.

49. One MJ of heat is transferred from a high temperature reservoir at 1000 K to a heat engine. From there the heat is rejected to a low temperature reservoir which is at 500 K. Determine by Clausius inequality whether the cycle is impossible, reversible or irreversible for the following cases:

(i) 500 kJ of heat rejected
(ii) 600 kJ of heat rejected
(iii) 400 kJ of heat rejected

Also calculate the efficiency of the engine.

Ans:

(i) $\oint \frac{dQ}{T} = \frac{1000}{1000} - \frac{500}{500} = 0$; Hence, the cycle is reversible.

Efficiency, $\eta = \frac{1000 - 500}{1000} = 0.5$ or 50%.

(ii) $\oint \frac{dQ}{T} = \frac{1000}{1000} - \frac{600}{500} = -0.2$; Hence, the cycle is irreversible.

Efficiency, $\eta = \frac{1000 - 600}{1000} = 0.4$ or 40%.

(iii) $\oint \frac{dQ}{T} = \frac{1000}{1000} - \frac{400}{500} = 0.2$; Hence, the cycle is impossible.

Efficiency is not defined in this case.

50. A heat engine works between two temperature limits. The higher temperature limit is increased by 200°C retaining the lower temperature the same. The efficiency increases by 12% of the original. Retaining the original higher temperature the same, the lower temperature is increased by 100°C. Now the efficiency is reduced by 20% of the original. Find the original temperatures and the efficiency.

Ans: Let T_1 and T_2 be the original temperatures in Kelvin.

$$\frac{T_1 - T_2}{T_1} = \eta, \text{ where } \eta \text{ is the original efficiency}$$

When T_1 is increased by 200 and T_2 is the same, the efficiency increases by 12%.

$$\frac{(T_1 + 200) - T_2}{(T_1 + 200)} = 1.12\eta \qquad (1)$$

When T_1 is the same and T_2 is increased by 100, the efficiency reduces by 20%.

$$\frac{T_1 - (T_2 + 100)}{T_1} = 0.8\eta \qquad (2)$$

From equation (2) $\frac{T_1 - T_2}{T_1} - \frac{100}{T_1} = 0.8\eta$ i.e. $\eta - \frac{100}{T_1} = 0.8\eta$ or $T_1 = \frac{500}{\eta}$.

$$\frac{T_1 - T_2}{T_1} = \eta; \quad 1 - \frac{T_2}{T_1} = \eta; \quad 1 - \frac{\eta}{500} T_2 = \eta; \quad T_2 = \frac{(1-\eta)}{\eta} 500.$$

Substituting for T_1 and T_2 in terms of η in equation (1), we get

$$\frac{\left(\frac{500}{\eta} + 200\right) - \frac{(1-\eta)500}{\eta}}{\left(\frac{500}{\eta} + 200\right)} = 1.12\eta.$$

The aforesaid equation simplifies to

$$\frac{(500 + 200\eta) - (1-\eta)500}{(500 + 200\eta)} = 1.12\eta \quad \text{i.e.} \ (500\eta + 200\eta) = 1.12\eta(500 + 200\eta).$$

$$\eta = 0.625; \quad T_1 = \frac{500}{0.625} = 800 \text{ K}; \quad T_2 = \frac{(1-0.625)}{0.625} 500 = 300 \text{ K}.$$

The original temperatures are 800 K and 300 K and the efficiency is 62.5%.

51. An engine working on ideal Carnot cycle is supplied with dry saturated steam at 14 bar. The steam expands in the engine cylinder isentropically to a pressure of 0.05 bar. Assume the liquid to be saturated at entry to the boiler. Find (i) work ratio, (ii) thermal efficiency and (iii) specific steam consumption.

Ans: At 14 bar,

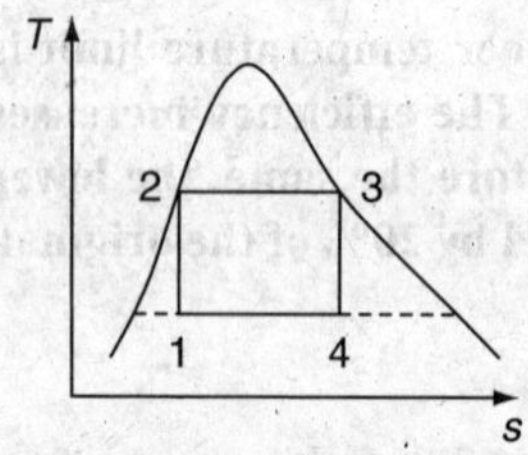

$h_3 = h_g = 2787.8$; $h_2 = h_f = 830.1$;

$s_3 = s_g = 6.4651$; $s_2 = s_f = 2.2837$;

At 0.05 bar,

$h_f = 137.8$; $h_{fg} = 2423.7$; $h_g = 2561.5$;

$s_f = 0.476$; $s_g = 8.395$;

For Carnot cycle during isentropic expansion (3–4) and isentropic compression (1–2),

$$s_3 = s_4, \text{ and } s_2 = s_1.$$

Hence,

$$x_4 = \frac{6.4651 - s_f}{s_g - s_f} = \frac{6.4651 - 0.476}{8.395 - 0.476} = 0.756.$$

$$x_1 = \frac{2.2837 - s_f}{s_g - s_f} = \frac{2.2837 - 0.476}{8.395 - 0.476} = 0.228$$

$$h_4 = h_f + x_4 h_{fg} = 137.8 + 0.756 \times 2423.7 = 1970.1$$

$$h_1 = h_f + x_1 h_{fg} = 137.8 + 0.228 \times 2423.7 = 690.4$$

Engine output $= h_3 - h_4 = 2787.8 - 1970.1 = 817.7.$
Pump work $= h_2 - h_1 = 830.1 - 690.4 = 139.7.$

$$\text{Work ratio} = \frac{(817.7 - 139.7)}{817.7} = 0.82.$$

$$\text{Thermal efficiency} = \frac{(h_3 - h_4) - (h_2 - h_1)}{(h_3 - h_2)} = \frac{817.7 - 139.7}{(2787.8 - 830.1)} = 0.3463 \text{ or } 34.63\%.$$

$$\text{Specific steam consumption} = \frac{3600}{(817.7 - 139.7)} = 5.3097 \frac{\text{kg}}{\text{kWhr}}.$$

52. A steam power plant working on the Rankine cycle is supplied with dry saturated steam at a pressure of 40 bar. The steam expands in the engine cylinder to a pressure of 0.05 bar. Assuming the feed water supplied to the boiler as saturated, find (i) cycle efficiency and (ii) specific steam consumption.

Ans: At 40 bar pressure,

$$h_g = 2800.3;\ s_g = 6.0685;$$

At 0.05 bar pressure,

$$h_f = 137.8;\ h_{fg} = 2423.7;\ s_f = 0.476;\ s_{fg} = 7.919.$$

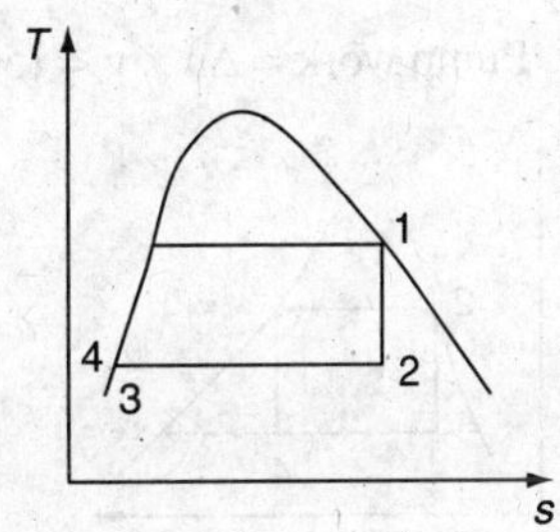

During isentropic expansion 1–2,

$$s_1 = s_2 = 6.0685$$

Since s_2 is less than $(s_f + s_{fg})$, the condition of steam is wet.

$$6.0685 = s_f + x_2\, s_{fg} = 0.476 + x_2 \times 7.919;$$

$$x_2 = \frac{6.0685 - 0.476}{7.919} = 0.7062.$$

$$h_2 = 137.8 + 0.7062 \times 2423.7 = 1849.4.$$

$$h_3 = 137.8.$$

$$h_4 = h_3 = 137.8 \text{ (pump work is neglected).}$$

$$\text{Cycle efficiency} = \frac{h_1 - h_2}{h_1 - h_4} = \frac{2800.3 - 1849.4}{2800.3 - 137.8} = 0.3571 \text{ or } 35.71\%.$$

$$\text{Specific steam consumption} = \frac{3600}{(h_1 - h_2)} = \frac{3600}{2800.3 - 1849.4} = 3.78 \frac{\text{kg}}{\text{kWhr}}.$$

53. A power generating plant uses steam as working fluid and operates on the Rankine cycle at a boiler pressure of 5 MPa, dry saturated and a condenser pressure of 50 kPa. Calculate the cycle efficiency. What will be the cycle efficiency if the plant operates on the Carnot cycle?

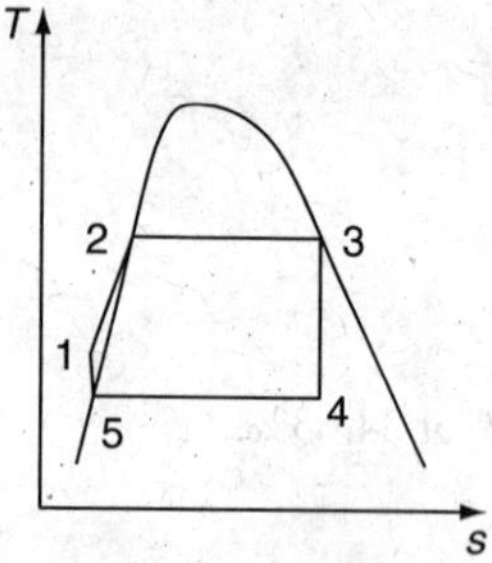

Rankine Cycle

Ans: At 50 bar pressure,

$h_g = 2794.2$; $h_f = 1154.5$; $s_g = 5.9735$; $s_f = 2.9206$.

At 0.5 bar pressure,

$h_f = 340.6$; $h_{fg} = 2305.4$; $s_f = 1.0912$; $s_{fg} = 6.5035$

$v_f = 0.00103 \text{ m}^3/\text{kg}$.

$$h_3 = 2794.2$$

During isentropic expansion 3 – 4, $s_4 = s_3 = 5.9735$.
Since s_4 is less than s_g (1.0912 + 6.5035), the steam is wet after expansion.

$$x_4 = \frac{5.9735 - s_f}{s_{fg}} = \frac{5.9735 - 1.0912}{6.5035} = 0.75.$$

$$h_4 = 340.6 + 0.75 \times 2305.4 = 2069.65.$$

$$h_5 = 340.6.$$

Pump work = $\Delta p \times v_f = (50 - 0.5) \times 100 \times 0.00103 = 5.09$ kJ/kg

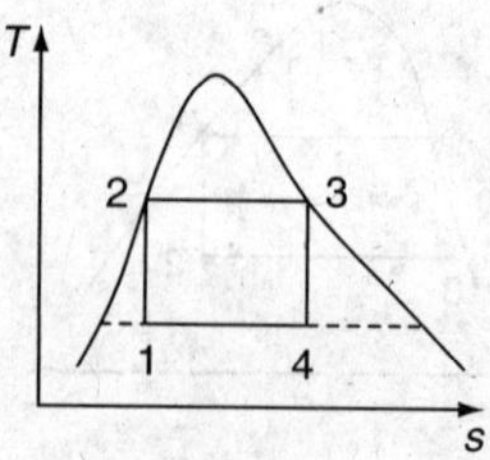

Carnot Cycle

$h_1 = h_5$ + Pump work = 340.6 + 5.09 = 345.69.

Net power output = $(h_3 - h_4)$ – pump work = (2794.2 – 2069.65) – 5.09 = 719.46.

Heat input = $(h_3 - h_1)$ = 2794.2 – 345.69 = 2448.51.

$$\text{Cycle efficiency} = \frac{719.46}{2448.51} = 0.2938 \text{ or } 29.38\%.$$

$$h_3 = 2794.2$$

Since the isentropic expansion 3–4 is the same, we get

$$h_4 = 2069.65$$

$$h_2 = h_f\,(50 \text{ bar}) = 1154.5;\; s_2 = s_f = 2.9206.$$

During isentropic compression 1–2, $s_2 = s_1 = 2.9206$.
We can evaluate the dryness fraction of steam at point 1.

$$s_f + x_1\, s_{fg} = 2.9206.$$

$$x_1 = \frac{2.9206 - 1.0912}{6.5035} = 0.28.$$

$$h_1 = 340.6 + 0.28 \times 2305.4 = 986.112.$$

Net power output = $(h_3 - h_4) - (h_2 - h_1)$ = (2794.2 – 2069.65) – (1154.5 – 986.112)

= 556.162.

Heat supplied = $(h_3 - h_2) = 2794.2 - 1154.5 = 1639.7$

$$\text{Cycle efficiency} = \frac{556.162}{1639.7} = 0.3391 \text{ or } 33.91\%.$$

54. A Rankine cycle works with 150 bar, 500°C inlet and 0.1 bar exit conditions. Determine the cycle efficiency and output per kg/sec of flow. What will be the change in efficiency if the exit pressure is 0.05 bar?

Ans: At 150 bar, 500°C,

$$h_3 = 3308.6;\ s_3 = 6.344$$

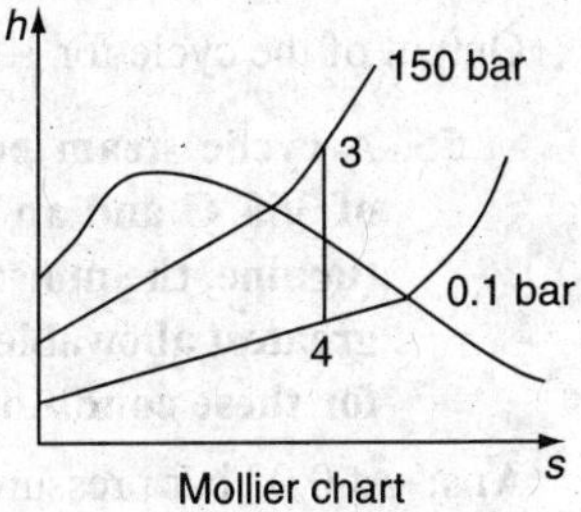

At 0.1 bar pressure,

$$h_f = 191.8;\ h_{fg} = 2392.8;\ h_g = 2584.7$$

$$s_f = 0.649;\ s_{fg} = 7.501;\ s_g = 8.150;\ v_f = 0.00101$$

During isentropic expansion 3 – 4, $s_3 = s_4 = 6.344$.
Since s_4 is less than s_g(8.150), the steam is wet after expansion.

$$x_4 = \frac{6.344 - 0.649}{7.501} = 0.759$$

$$h_4 = h_f + x_4\, h_{fg} = 191.8 + 0.759 \times 2392.8 = 2007.9$$

$$h_5 = h_f = 191.8.$$

Pump work = $(150 - 0.1) \times 100 \times 0.00101 = 15.14$ kJ/kg

$$h_1 = h_5 + \text{pump work} = 191.8 + 15.14 = 206.94.$$

Turbine work = $h_3 - h_4 = 3308.6 - 2007.9 = 1300.7$.
Net output = Turbine work – pump work = 1300.7 – 15.14 = 1285.56.
Heat supplied = $h_3 - h_1 = 3308.6 - 206.94 = 3101.66$.

$$\text{Cycle efficiency} = \frac{\text{Net output}}{\text{Heat supplied}} = \frac{1285.56}{3101.66} = 0.4145 \quad \text{or} \quad 41.45\%.$$

Output of the cycle for 1 kg of flow per sec = 1285.56 kW.
At 0.05 bar pressure,

$$h_f = 137.8;\ h_{fg} = 2423.7;\ s_f = 0.476;\ s_{fg} = 7.919;\ v_f = 0.001005.$$

During isentropic expansion 3 – 4, $s_3 = s_4 = 6.344$.
Since s_4 is less than s_g(0.476 +7.919), the steam is wet after expansion.

$$x_4 = \frac{6.344 - 0.476}{7.919} = 0.741.$$

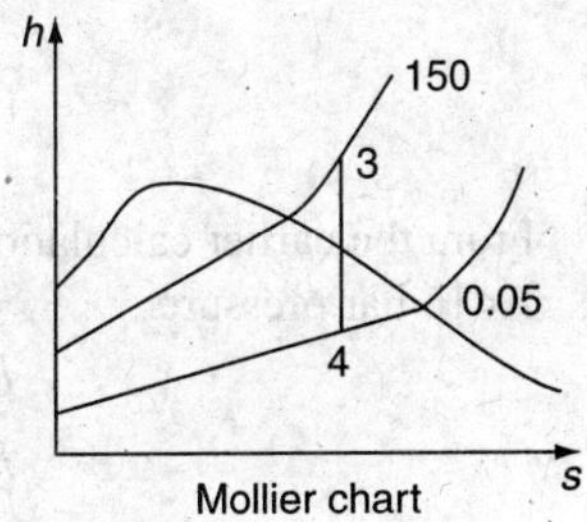

$$h_4 = 137.8 + 0.741 \times 2423.7 = 1933.8.$$

$$h_5 = h_f = 137.8.$$

$$\text{Pump work} = (150 - 0.05) \times 100 \times 0.001005$$
$$= 15.07 \text{ kJ/kg}.$$

$$h_1 = h_5 + \text{pump work} = 137.8 + 15.07 = 152.87.$$

Turbine work = $h_3 - h_4 = 3308.6 - 1933.8$
$= 1374.8.$

Net output = Turbine work – pump work

$= 1374.8 - 15.07 = 1359.73.$

Heat supplied = $h_3 - h_1 = 3308.6 - 152.87 = 3155.73.$

$$\text{Cycle efficiency} = \frac{\text{Net output}}{\text{Heat supplied}} = \frac{1359.73}{3155.73} = 0.4309 \text{ or } 43.09\%.$$

Output of the cycle for 1 kg of flow per sec = 1359.73 kW.

55. A cyclic steam power plant is to be designed for a steam temperature at turbine inlet of 350°C and an exhaust pressure of 8 kPa. After isentropic expansion of steam in the turbine, the moisture content at the turbine exhaust is not to exceed 16%. Determine the greatest allowable steam pressure at the turbine inlet and calculate the Rankine efficiency for these conditions. Estimate also the mean temperature of heat addition.

Ans: At 0.08 bar pressure, $s_f = 0.593$; $s_{fg} = 7.636$

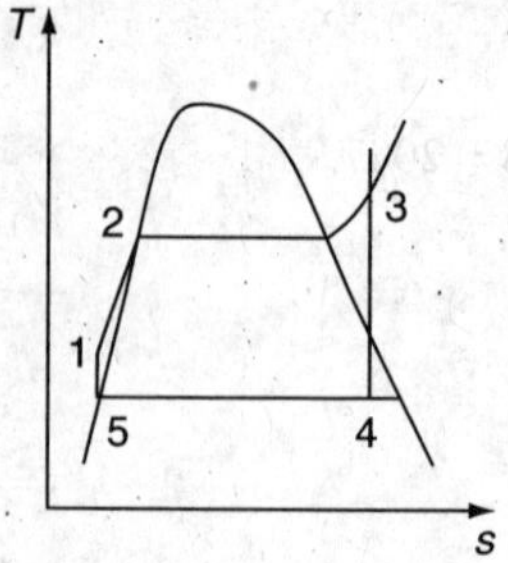

The specific entropy of steam at point 4 (turbine exhaust, 0.84 dry)

$$s_4 = 0.593 + 0.84 \times 7.636 = 7.00724.$$

The specific heat at constant pressure for superheated steam are obtained for the following pressures;

Pressure	t_s	C_p	S_g
15 bar	198.3	2.198	6.4406
16	201.4	2.166	6.4175
17	204.3	2.17	6.3957
18	207.1	2.18	6.3751
19	209.8	2.19	6.3554

From the aforesaid data, we can evaluate the specific entropy of superheated steam at temperature 350°C for three pressures as follows:

$$\text{15 bar: } s = 6.4406 + 2.198 \ln\left(\frac{273+350}{273+198.3}\right) = 7.0539$$

$$\text{16 bar: } s = 6.4175 + 2.166 \ln\left(\frac{273+350}{273+201.4}\right) = 7.0077$$

$$\text{17 bar: } s = 6.3957 + 2.17 \ln\left(\frac{273+350}{273+204.3}\right) = 6.973$$

From the earlier calculations, we can infer that the maximum pressure at turbine inlet is 16 bar. At 16 bar pressure,

$$h_g = 2791.7,\ t_s = 201.4°\text{C}$$

$$h_3 = 2791.7 + 2.166(350 - 201.4) = 3113.56$$

At 0 08 bar pressure,

$$h_f = 173.9;\ h_{fg} = 2403.1;\ v_f = 0.001008$$

$$h_4 = 173.9 + 0.84 \times 2403.1 = 2192.5$$

Pump work = (16 – 0.08) × 100 × 0.001008 = 1.6 kJ/kg

$$h_5 = h_f = 173.9.$$

$$h_1 = h_5 + \text{pump work} = 173.9 + 1.6 = 175.5.$$

Turbine work = $h_3 - h_4 = 3113.56 - 2192.5 = 921.06$.
Heat supplied = $h_3 - h_1 = 3113.56 - 175.5 = 2938.06$.

$$\text{Rankine efficiency} = \frac{W_t - W_p}{h_3 - h_1} = \frac{921.06 - 1.6}{2938.06} = 0.3129 \text{ or } 31.29\%.$$

$$s_3 = 7.007;\ s_5 = s_1 = 0.593.$$

Heat supplied = 2938.06 kJ/kg.
Let T_m be the mean temperature of heat addition.

$$T_m(\Delta s) = \text{heat supplied} = 2938.06$$

$$T_m = \frac{2938.06}{7.007 - 0.593} = 458.06 \text{ K}.$$

Mean temperature of heat addition is 458 K or 185°C.

56. A steam power station uses the following cycle: steam at boiler outlet 150 bar, 500°C, reheat at 4 bar, 500°C; condenser operates at 0.05 bar. Assuming ideal processes, find (i) quality of steam at turbine exhaust, (ii) cycle efficiency and (iii) specific steam consumption.

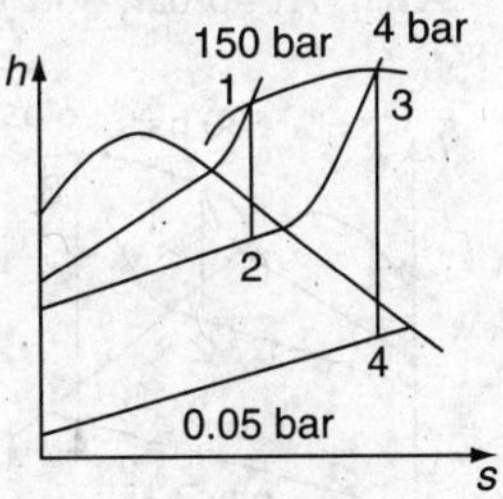

Ans: At 150 bar, 500°C,

$$h_{sup} = 3308.6;\ s_{sup} = 6.344$$

At 4 bar, 500°C,

$$h_{sup} = 3484.9;\ s_{sup} = 8.191$$

At 4 bar pressure,

$$h_f = 604.7;\ h_{fg} = 2133;\ h_g = 2737.7.$$

$$s_f = 1.7764;\ s_{fg} = 5.1179;\ s_g = 6.8943.$$

At 0.05 bar pressure,

$$h_f = 137.8;\ h_{fg}\ 2423.7;\ v_f = 0.001005$$

$$s_f = 0.476;\ s_{fg} = 7.919;\ s_g = 8.395$$

During isentropic expansion 1–2, $s_1 = 6.344 = s_2$

Since s_g (pressure 4 bar) is more than s_2, the condition of steam at point 2 is wet.

$$x_2 = \frac{6.344 - 1.7764}{5.1179} = 0.89;\ h_2 = 604.7 + 0.89 \times 2133 = 2503.$$

During reheating process 2–3 pressure remains the same.

$$h_3 = 3484.9;\ s_3 = 8.191.$$

During isentropic expansion 3 – 4, $s_3 = 8.191 = s_4$
Since s_g(pressure 0.05) is more than s_4, the condition of steam is wet at point 4.

$$x_4 = \frac{8.191 - 0.476}{7.919} = 0.97;\ h_4 = 137.8 + 0.97 \times 2423.7 = 2488.7.$$

Pump work = (150 – 0.05) × 100 × 0.001005 = 15.07 kJ/kg

$$h_5 = 137.8;\ h_6 = h_5 + \text{pump work} = 137.8 + 15.07 = 152.87.$$

Turbine work = $(h_1 - h_2) + (h_3 - h_4)$ = (3308.6 – 2503) + (3484.9 – 2488.7) = 1801.8.
Heat supplied = $(h_1 - h_6) + (h_3 - h_2)$ = (3308.6 – 152.87) + (3484.9 – 2503) = 4137.63.

$$\text{Cycle efficiency} = \frac{(1801.8 - 15.07)}{4137.63} = 0.4318 \text{ or } 43.18\%.$$

$$\text{Specific steam consumption} = \frac{3600}{1801.8 - 15.07} = 2.015 \frac{\text{kg}}{\text{kWhr}}.$$

57. In a thermal plant, steam is supplied at a pressure of 30 bar and temperature of 300°C to the high pressure turbine where it is expanded to 5 bar. The steam is then passed through a reheater at constant pressure to the same temperature. The steam leaves the low pressure turbine at 0.05 bar. Find the efficiency of the cycle with and without reheating.

Ans: At 30 bar, 300°C,

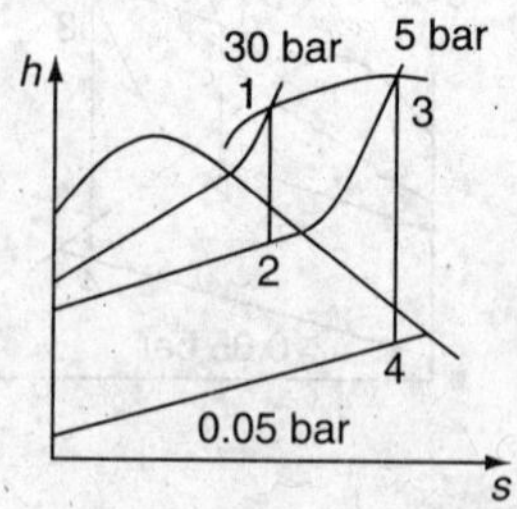

$$h_{sup} = 2993.5;\ s_{sup} = 6.539$$

At 5 bar, 300°C,

$$h_{sup} = 3064.2;\ s_{sup} = 7.460$$

At 5 bar pressure,

$$h_f = 640.1;\ h_{fg} = 2107.4;\ s_f = 1.8604;\ s_{fg} = 4.9588$$

At 0.05 bar pressure,

$$h_f = 137.8;\ h_{fg} = 2423.7;\ v_f = 0.001005$$

$$s_f = 0.476;\ s_{fg} = 7.919;\ s_g = 8.395$$

During the isentropic expansion 1 – 2, $h_1 = 2993.5$ and $s_1 = 6.539 = s_2$
Since s_g(pressure 5 bar) is more than s_2, the condition of steam after expansion is wet.

$$x_2 = \frac{6.539 - 1.8604}{4.9588} = 0.943;\ h_2 = 640.1 + 0.943 \times 2107.4 = 2627.4$$

During reheating process 2 – 3 the pressure remains the same.

$$h_3 = 3064.2;\ s_3 = 7.460$$

During the isentropic expansion 3 – 4, $h_3 = 3064.2$ and $s_3 = 7.460 = s_4$
Since s_g(pressure 0.05 bar) is more than s_4, the condition of steam is wet at point 4.

$$x_4 = \frac{7.460 - 0.476}{7.919} = 0.88;\ h_4 = 137.8 + 0.88 \times 2423.7 = 2270.6$$

Pump work = (30 – 0.05) × 100 × 0.001005 = 3.01 kJ/kg

$$h_5 = 137.8.$$

$$h_6 = h_5 + \text{pump work} = 137.8 + 3.01 = 140.81.$$

(i) *With reheating*

$$\text{Cycle efficiency} = \frac{(h_1 - h_2) + (h_3 - h_4) - \text{pump work}}{(h_1 - h_6) + (h_3 - h_2)}$$

$$= \frac{(2993.5 - 2627.4) + (3064.2 - 2270.6) - 3.01}{(2993.5 - 140.81) + (3064.2 - 2627.4)} = 0.3516 \text{ or } 35.16\%.$$

(ii) *Without reheating*

The steam expands from 30 bar, 300°C to 0.05 bar pressure. During the isentropic expansion 1 – 4, $s_1 = 6.539 = s_4$
Since s_g(pressure 0.05 bar) is more than s_4, the condition of steam is wet.

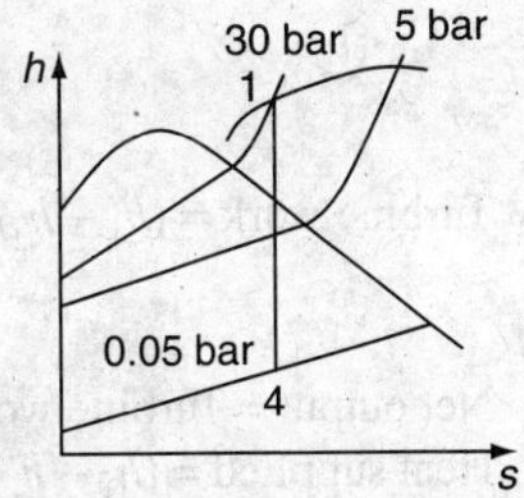

$$x_4 = \frac{6.539 - 0.476}{7.919} = 0.766.$$

$$h_4 = 137.8 + 0.766 \times 2423.7 = 1993.4.$$

$$\text{Cycle efficiency} = \frac{(h_1 - h_4) - \text{pump work}}{(h_1 - h_6)}$$

$$= \frac{(2993.5 - 1993.4) - 3.01}{(2993.5 - 140.81)} = 0.3495.$$

$$= 34.95\%.$$

58. A steam turbine is supplied with steam at a pressure of 1.5 MPa and temperature of 250°C. It is expanded to a pressure of 400 kPa and then reheated at constant pressure to its initial temperature. Further expansion takes place in the turbine. The exhaust pressure is 20 kPa. Calculate the ideal efficiency of the plant and work done.

Ans: At 15 bar, 250°C,

$$h_{sup} = 2923.3;\ s_{sup} = 6.709.$$

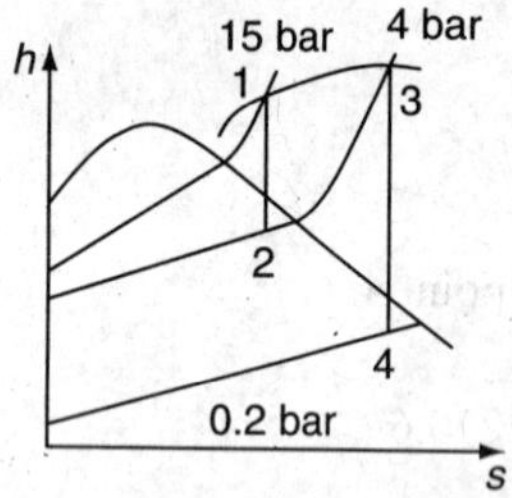

At 4 bar, 250°C,

$$h_{sup} = 2964.2;\ s_{sup} = 7.379.$$

At 4 bar pressure,

$$h_f = 604.7;\ h_{fg} = 2133.0;\ s_f = 1.7764;\ s_{fg} = 5.1179.$$

At 0.2 bar pressure,

$$h_f = 251.5;\ h_{fg} = 2358.4;\ s_f = 0.8321;\ s_{fg} = 7.0773$$

$$v_f = 0.001017.$$

During the isentropic expansion 1 – 2, $h_1 = 2923.3$, and $s_1 = 6.709 = s_2$.

$$x_2 = \frac{6.709 - 1.7764}{5.1179} = 0.96.$$

$$h_2 = 604.7 + 0.96 \times 2133 = 2652.38$$

During reheating operation 2 – 3, the pressure remains the same.

$$h_3 = 2964.2;\ s_3 = 7.379.$$

During the isentropic expansion 3 – 4, $h_3 = 2964.2$ and $s_3 = 7.379 = s_4$

$$x_4 = \frac{7.379 - 0.8321}{7.0773} = 0.925.$$

$$h_4 = 251.5 + 0.925 \times 2358.4 = 2433$$

Pump work = (15 – 0.2) × 100 × 0.001017 = 1.5 kJ/kg.

$$h_5 = 251.5.$$

$$h_6 = 251.5 + \text{pump work} = 251.5 + 1.5 = 253.0.$$

Turbine work = $(h_1 - h_2) + (h_3 - h_4)$ = (2923.3 – 2652.38) + (2964.2 – 2433).

$$= 270.92 + 531.2 = 802.12.$$

Net output = Turbine work – pump work = 802.12 – 1.5 = 800.62.
Heat supplied = $(h_1 - h_6) + (h_3 - h_2)$ = (2923.3 – 253) + (2964.2 – 2652.38).

$$= 2670.3 + 311.82 = 2982.12.$$

Net work done = 800.62 kJ/kg of steam flow.

$$\text{Ideal efficiency} = \frac{800.62}{2982.12} = 0.2685 \text{ or } 26.85\%.$$

59. In a reheat cycle, the inlet steam pressure and temperature are 150 bar and 550°C respectively. The steam is extracted from the turbine when it is just dry and saturated and reheated at constant pressure. It is then expanded to a condenser pressure of 0.1 bar. If the moisture at the condenser inlet is 5%, determine the cycle efficiency.

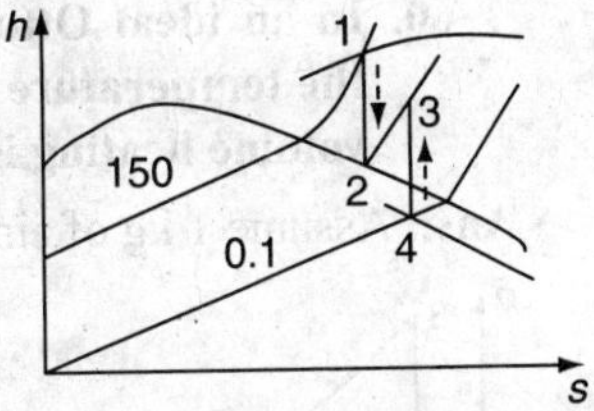

Ans: At 150 bar and 550°C,

$$h_{sup} = 3448.6;\ s_{sup} = 6.520.$$

During iséntropic expansion 1–2, $s_1 = 6.520 = s_2$.

From the steam table, we can find the entropy of dry saturated steam corresponds to a pressure of 12 bar.

At 12 bar pressure,

$$h_g = 2782.7 = h_2.$$

Since the dryness fraction of steam entering the condenser is 0.95 at 0.1 bar pressure, we can find out the entropy of steam at point 4.

At 0.1 bar pressure,

$$h_f = 191.8;\ h_{fg} = 2392.8;\ s_f = 0.649;\ s_{fg} = 7.501;\ v_f = 0.00101\ \text{m}^3/\text{kg}.$$

$$s_4 = 0.649 + 0.95 \times 7.501 = 7.77495.$$

$$h_4 = 191.8 + 0.95 \times 2392.8 = 2464.96.$$

During the isentropic expansion 3–4, $s_3 = s_4$. Since s_4 is known, we can locate the point 3 from entropy calculations as follows:

We can assume $C_p = 2.286$.

At 12 bar pressure,

$$t_s = 188°\text{C};\ s_g = 6.5194.$$

$$s_3 = s_g + C_p \ln\left(\frac{T_3}{T_s}\right);\quad 7.7749 = 6.5194 + 2.286 \ln\left(\frac{T_3}{273+188}\right).$$

Solving the aforesaid equation we get T_3.

$$T_3 = 798.4\ \text{K}.$$

$$h_3 = 2782.7 + 2.286 \times (798.4 - 461) = 3553.9.$$

$$W_{t1} = h_1 - h_2 = 3448.6 - 2782.7 = 665.9.$$

$$W_{t2} = h_3 - h_4 = 3553.9 - 2464.96 = 1088.94.$$

Pump work = $(150 - 0.1) \times 100 \times 0.00101 = 15.1$ kJ/kg.

Net output = $W_{t1} + W_{t2}$ – pump work = $665.9 + 1088.94 - 15.1 = 1739.74$.

$$h_5 = 191.8$$

$$h_6 = h_5 + \text{pump work} = 191.8 + 15.1 = 206.9.$$

Heat supplied = $(h_1 - h_6) + (h_3 - h_2) = (3448.6 - 206.9) + (3553.9 - 2782.7)$

$$= 4012.9.$$

$$\text{Efficiency} = \frac{\text{Net output}}{\text{Heat supplied}} = \frac{1739.74}{4012.9} = 0.4335 \text{ or } 43.35\%.$$

60. In an ideal Otto cycle, at the beginning of compression the pressure is 100 kPa and the temperature is 300 K. The compression ratio is 9.5 and temperature after constant volume heating is 2150 K. Calculate the work output of the cycle.

Ans: Assume 1 kg of air is present in the cycle.

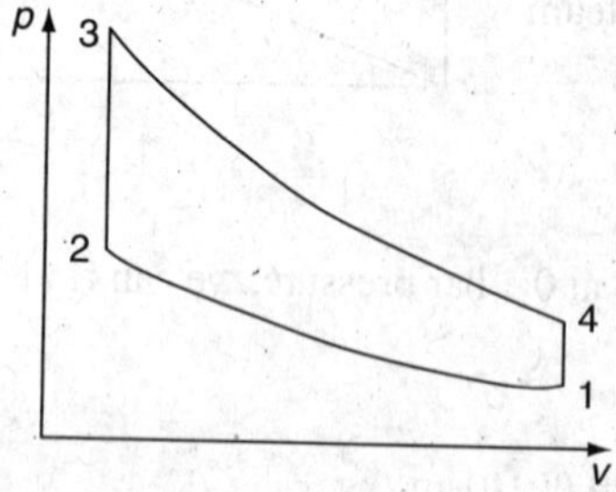

$$T_1 = 300 \text{ K}, p_1 = 1.0 \text{ bar}, T_3 = 2150 \text{ K},$$

Compression ratio, $r_c = 9.5$.

$$\frac{T_2}{T_1} = (r_c)^{\gamma-1}; \quad T_2 = 300(9.5)^{0.4} = 738.3 \text{ K}$$

$$\frac{T_3}{T_4} = \left(\frac{V_4}{V_3}\right)^{\gamma-1}; \quad T_4 = \frac{2150}{(9.5)^{0.4}} = 873.67 \text{ K}.$$

Heat supplied, $Q_s = C_v(T_3 - T_2) = 0.718 \times (2150 - 738.3) = 1013.6$ kJ.

Heat rejected, $Q_r = C_v(T_4 - T_1) = 0.718 \times (873.67 - 300) = 411.895$ kJ.

Work output of the cycle $= Q_s - Q_r = 1013.6 - 411.895 = 601.705$ kN.m.

61. The ratio of the mass of air to fuel supplied to a diesel engine is 50. The oil has a calorific value of 37674 kJ/kg. The temperature of air at the beginning of compression is 20°C and the compression ratio is 14. What is the theoretical efficiency of the engine if it is assumed that the heat is produced and supplied at constant pressure?

Ans: Assume 1 kg of air is present in the cycle.

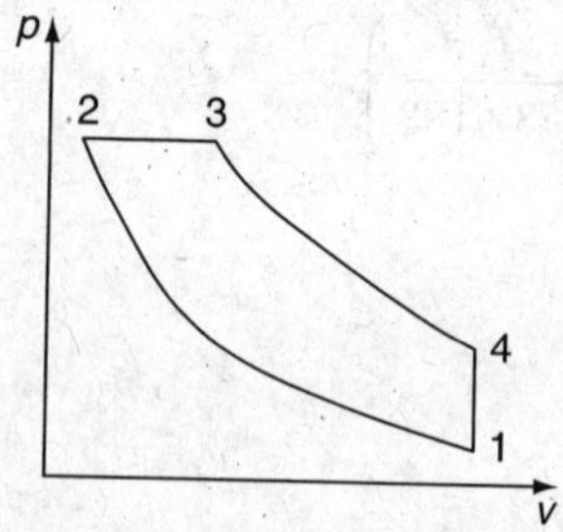

Amount of fuel burnt per cycle = 0.02 kg.

Heat supplied per cycle, $Q_s = 0.02 \times 37674$.

$$= 753.48 \text{ kJ}.$$

$T_1 = 293$ K, Compression ratio, $r_c = 14$

$$\frac{T_2}{T_1} = (r_c)^{\gamma-1}; \quad T_2 = 293 \times (14)^{0.4} = 842 \text{ K}$$

$$Q_s = C_p(T_3 - T_4); \; 753.48 = 1.005 \times (T_3 - 842)$$

Hence, $T_3 = 1591.7$ K.

$$\frac{V_3}{V_2} = \frac{T_3}{T_2} = \frac{1591.7}{842} = 1.89$$

$$\frac{T_4}{T_3} = \left(\frac{V_3}{V_4}\right)^{\gamma-1} = \left(\frac{V_3}{V_2}.\frac{V_2}{V_4}\right)^{\gamma-1} = \left(\frac{V_3}{V_2}\right)^{\gamma-1}\left(\frac{V_2}{V_1}\right)^{\gamma-1} = \frac{(1.89)^{0.4}}{(14)^{0.4}} = 0.\ 44888.$$

$$T_4 = 1591.7 \times 0.44888 = 714.48 \text{ K}.$$

Heat rejected per cycle, $Q_r = C_v(T_4 - T_1) = 0.718 \times (714.48 - 293) = 302.62$ kJ.

$$\text{Thermal efficiency}, \eta = \frac{Q_s - Q_r}{Q_s} = \frac{753.48 - 302.62}{753.48} \times 100 = 59.8\%$$

62. Air enters an ideal diesel cycle at 1 bar and 37°C. The compression ratio is 17. The amount of heat rejected in the cycle is 450 kJ/kg. Find the air standard efficiency and the maximum temperature in the cycle.

Ans:

$$T_1 = 310 \text{ K}, r_c = 17, Q_r = 450 \text{ kJ/kg}$$

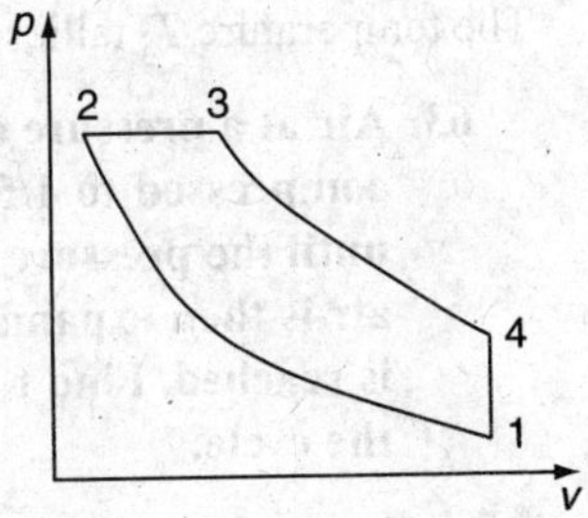

Assume 1 kg of air is present in the cycle.

$$T_2 = T_1 (r_c)^{\gamma-1} = 310(17)^{0.4} = 962.8 \text{ K}$$

$$T_3 = T_2\left(\frac{V_3}{V_2}\right)$$

$$Q_r = m\, C_v (T_4 - T_1)$$

$$450 = 0.718\,(T_4 - 310)$$

Solving the earlier equation, we get

$$T_4 = 936.7 \text{ K}$$

$$\frac{T_3}{T_4} = \left(\frac{V_4}{V_3}\right)^{\gamma-1} = \left(\frac{V_4}{V_2}\right)^{\gamma-1}\left(\frac{V_2}{V_3}\right)^{\gamma-1} = \left(\frac{V_1}{V_2}\right)^{\gamma-1}\left(\frac{T_2}{T_3}\right)^{\gamma-1}$$

$$\frac{T_3}{936.7} = (17)^{0.4}\frac{(962.8)^{0.4}}{T_3^{0.4}}$$

$$T_3^{1.4} = (17)^{0.4} \times (962.8)^{0.4} \times 936.7 = 45414.49.$$

$$T_3 = 2121.1.$$

Maximum temperature in the cycle is 2121.1 K or 1848.1°C.

Heat supplied, $Q_s = C_p (T_3 - T_2) = 1.005 \times (2121.1 - 962.8) = 1164.09$ kJ.

$$\text{Air standard efficiency} = 1 - \frac{Q_r}{Q_s} = 1 - \frac{450}{1164.09} = 0.6134 \text{ or } 61.34\%.$$

As a check, we can calculate the entropy change during the heat addition process 2–3 and during the heat rejection process 4–1 and equate them as as follows:

During the constant pressure process 2–3, $s_3 - s_2 = C_p \ln\left(\frac{T_3}{T_2}\right)$.

During the constant volume process 4–1, $s_1 - s_4 = C_v \ln\left(\frac{T_1}{T_4}\right)$.

For the cyclic process, $\Sigma\, \Delta s = 0 = (s_2 - s_1) + (s_3 - s_2) + (s_4 - s_3) + (s_1 - s_4)$.

We know the entropy change during the processes 1–2 and 3–4 are zeros.

Hence, $C_p \ln\left(\frac{T_3}{T_2}\right) + C_v \ln\left(\frac{T_1}{T_4}\right) = 0$ or $\frac{C_p}{C_v}\ln\left(\frac{T_3}{T_2}\right) = \ln\left(\frac{T_4}{T_1}\right)$.

From the aforesaid relation we get,

$$T_3 = T_2\left(\frac{T_4}{T_1}\right)^{1/\gamma} = 962.8\times\left(\frac{936.7}{310}\right)^{1/1.4} = 2121.1\text{ K.}$$

The temperature T_3 tallies with the value obtained earlier.

63. Air at a pressure of 1.0335 bar and temperature 10°C is drawn into a cylinder, it is then compressed to 1/5 of its previous volume. Heat is then given to it, at constant volume, until the pressure is doubled, and then at constant pressure for 1/8 of piston stroke. The air is then expanded adiabatically to the end of the stroke, i.e. until the original volume is reached. Find the temperature of the air at the end of expansion and the efficiency of the cycle.

Ans:

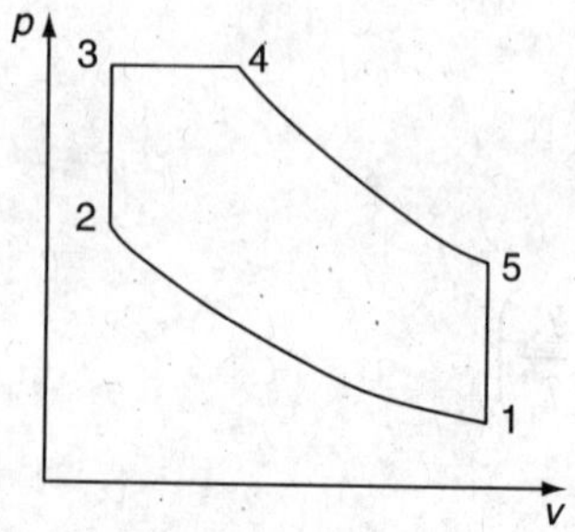

$$T_1 = 283\text{ K};\ p_1 = 1.0335\text{ bar};$$

$$V_2 = \frac{1}{5}V_1;\ p_3 = 2\,p_2;$$

$$(V_4 - V_3) = \tfrac{1}{8}(V_1 - V_2);$$

$$r_c = \frac{V_1}{V_2} = 5;\quad \left(\frac{V_4}{V_3} - 1\right) = \frac{1}{8}\left(\frac{V_1}{V_2} - 1\right)$$

$$\left(\frac{V_4}{V_3} - 1\right) = \frac{1}{8}(5-1);\quad \frac{V_4}{V_3} = 1.5\,.$$

$$p_2 = p_1\,(r_c)^{\gamma} = 1.0335\times(5)^{1.4} = 9.837\text{ bar}$$

$$p_3 = 2\,p_2 = 2\times 9.837 = 19.674\text{ bar}$$

$$T_2 = T_1\,(r_c)^{\gamma-1} = 283\times(5)^{0.4} = 538.73\text{ K}$$

$$T_3 = 2\,T_2 = 2\times 538.73 = 1077.46\text{ K.}$$

$$\frac{T_4}{T_3} = \frac{V_4}{V_3} = 1.5;\quad T_4 = 1.5\times T_3 = 1.5\times 1077.46 = 1616.19\text{ K}$$

$$\frac{T_5}{T_4} = \left(\frac{V_4}{V_5}\right)^{\gamma-1} = \left(\frac{V_4}{V_3}\frac{V_3}{V_5}\right)^{\gamma-1} = \left(1.5\times\frac{V_2}{V_1}\right)^{0.4} = \frac{(1.5)^{0.4}}{(5)^{0.4}} = 0.6178$$

$$T_5 = 1616.19\times 0.6178 = 998.5\text{ K.}$$

The temperature of air at the end of expansion is 998.5 K or 725.5°C.

Heat supplied per cycle, $Q_s = C_v\,(T_3 - T_2) + C_p\,(T_4 - T_3)$

$$= 0.718(1077.46 - 538.73) + 1.005(1616.19 - 1077.46)$$

$$= 928.231\text{ kJ.}$$

Heat rejected per cycle, $Q_r = C_v (T_5 - T_1) = 0.718(998.5 - 283) = 513.729$ kJ

$$\text{Efficiency of the cycle} = \frac{Q_s - Q_r}{Q_s} = \frac{928.231 - 513.729}{928.231} \times 100 = 44.65\%.$$

64. An engine working on the dual combustion cycle has the compression ratio of 10 and the initial pressure and temperature are 1.0 bar and 100°C, respectively. If the maximum pressure is limited to 68.9 bar and the heat supplied per kg of air is 1675 kJ, determine the air standard efficiency.

Ans:

$p_1 = 1.0$ bar; $p_3 = p_4 = 68.9$ bar; $T_1 = 373$ K; $Q = {}_2Q_3 + {}_3Q_4 = 1675$ kJ; $r_c = \frac{V_1}{V_2} = 10.$

$$T_2 = T_1 (r_c)^{\gamma - 1} = 373 \times (10)^{0.4} = 936.9 \text{ K}.$$

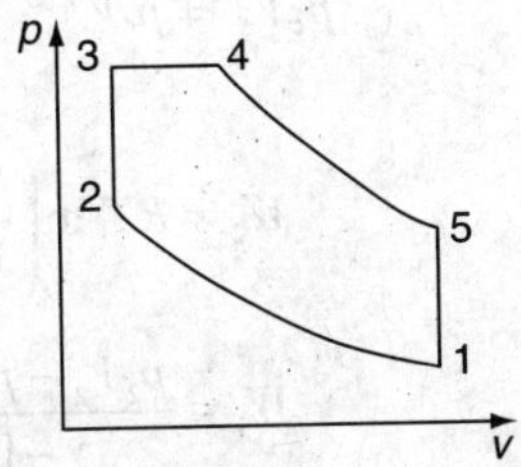

$$p_2 = p_1 (r_c)^{\gamma} = 1 \times (10)^{1.4} = 25.12 \text{ bar}.$$

$$\frac{T_3}{T_2} = \frac{p_3}{p_2} = \frac{68.9}{25.12}; \quad T_3 = T_2 \times 2.7428.$$

$$T_3 = 936.9 \times 2.7428 = 2569.76 \text{ K}.$$

$${}_2Q_3 = 0.718(2569.76 - 936.9) = 1172.4 \text{ kJ}.$$

$${}_3Q_4 = 1675 - 1172.4 = 502.6 \text{ kJ}.$$

$${}_3Q_4 = C_p (T_4 - T_3) = 1.005(T_4 - 2569.76).$$

$T_4 = 3069.85$ K.

$$\frac{V_4}{V_3} = \frac{T_4}{T_3} = \frac{3069.85}{2569.76} = 1.195.$$

$$\frac{T_5}{T_4} = \left(\frac{V_4}{V_5}\right)^{\gamma-1} = \left(\frac{V_4}{V_3}\frac{V_3}{V_5}\right)^{\gamma-1} = \left(1.195 \times \frac{V_2}{V_1}\right)^{0.4} = \frac{(1.195)^{0.4}}{(10)^{0.4}} = 0.4275.$$

$$T_5 = 0.4275 \times T_4 = 0.4275 \times 3069.85 = 1312.36 \text{ K}.$$

Heat rejected per cycle, $Q_r = C_v (T_5 - T_1) = 0.718 \times (1312.36 - 373) = 674.46$ kJ.

$$\text{Air standard efficiency, } \eta = \frac{Q_s - Q_r}{Q_s} = \frac{1675 - 674.46}{1675} \times 100 = 59.73.$$

65. One kilogram of air is first expanded isothermally at a temperature of 250°C from 34.45 bar to 10.335 bar and then adiabatically to 0.689 bar. The air is then cooled at constant pressure and is finally restored to the initial state by adiabatic compression. Calculate the work done by the air per cycle and find the efficiency of the cycle.

Ans:

$p_1 = 34.45$ bar; $T_1 = 523$ K;

$p_2 = 10.335$ bar;

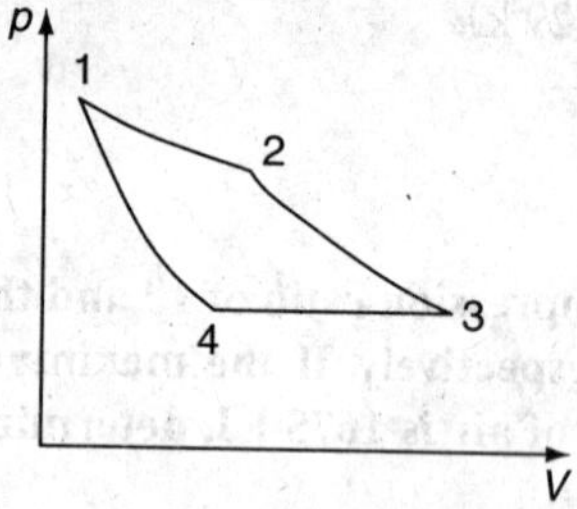

$$p_3 = p_4 = 0.689 \text{ bar}$$

$$V_1 = \frac{287 \times 523}{34.45 \times 10^5} = 0.04357\ m^3$$

$$V_2 = \frac{p_1 V_1}{p_2} = \frac{34.45 \times 0.04357}{10.335} = 0.1452\ m^3$$

$$p_2 V_2^{\gamma} = p_3 V_3^{\gamma};\quad V_3 = V_2 \left(\frac{p_2}{p_3}\right)^{1/\gamma};\ V_3 = 0.1452 \times \left(\frac{10.335}{0.689}\right)^{1/1.4} = 1.004689 \text{ m}^3.$$

$$p_4 V_4^{\gamma} = p_1 V_1^{\gamma};\quad V_4 = V_1 \left(\frac{p_1}{p_4}\right)^{1/\gamma} = 0.04357 \times \left(\frac{34.45}{0.689}\right)^{1/1.4} = 0.71242 \text{ m}^3.$$

$$_1W_2 = RT_1 \ln\left(\frac{V_2}{V_1}\right) = 0.287 \times 523 \times \ln\left(\frac{0.1452}{0.04357}\right) = 180.683 \text{ kN.m.}$$

$$_2W_3 = \frac{p_2 V_2 - p_3 V_3}{\gamma - 1} = \frac{(10.335 \times 0.1452 - 0.689 \times 1.004689)10^5}{(1.4-1)10^3} = 202.103 \text{ kN.m.}$$

$$_3W_4 = p_4 (V_4 - V_3) = 0.689 \times 100 \times (0.71242 - 1.004689) = -20.137 \text{ kN.m.}$$

$$_4W_1 = \frac{p_4 V_4 - p_1 V_1}{\gamma - 1} = \frac{(0.689 \times 0.71242 - 34.45 \times 0.04357)10^5}{(1.4-1)10^3} = -252.532 \text{ kN.m.}$$

Work done per cycle = 180.683 + 202.103 – 20.137 – 252.532 = 110.117 kN.m.
Heat supplied per cycle, $_1Q_2 = {}_1W_2 = 180.683$ kJ.

Efficiency of the cycle, $\eta = \dfrac{110.117}{180.683} \times 100 = 60.94\%.$

66. Find the ideal efficiency of the following air engine cycle: Air is compressed adiabatically from 1.0335 bar and 80°C through a compression ratio of 5; heat is added at constant volume until the pressure is 34.45 bar; the charge then expands adiabatically to 1.0335 bar and finally is brought to its original state by rejecting heat at constant pressure. Compare this efficiency with that of the Otto cycle having the same compression ratio.

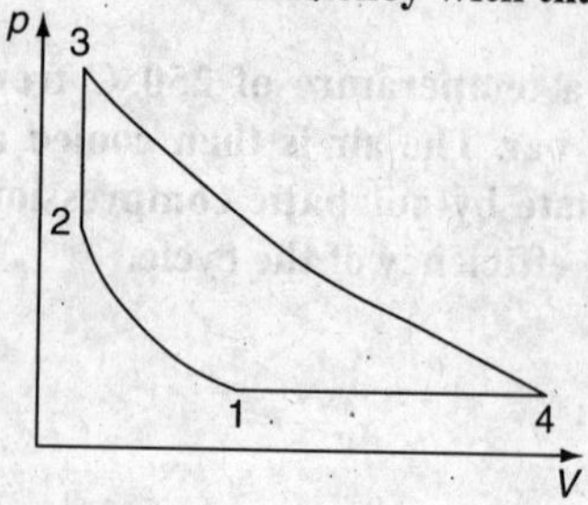

Ans:

$p_1 = 1.0335$ bar, $T_1 = 353$ K;

$p_3 = 34.45$ bar; $p_4 = 1.0335$ bar;

$V_2 = V_3$

$$r_c = \frac{V_1}{V_2} = 5$$

Assume 1 kg of air is present in the cycle.

$$V_1 = \frac{287 \times 353}{1.0335 \times 10^5} = 0.98027 \text{ m}^3.$$

$$T_2 = T_1 \left(\frac{V_1}{V_2}\right)^{\gamma-1} = 353(5)^{0.4} = 671.989 \text{ K}.$$

$$p_2 = p_1 (r_c)^{\gamma} = 1.0335(5)^{1.4} = 9.837 \text{ bar}.$$

$$\frac{T_3}{T_2} = \frac{p_3}{p_2}; \quad T_3 = \frac{34.45}{9.837} \times 671.989 = 2353.36 \text{ K}$$

$$\frac{T_4}{T_3} = \left(\frac{p_4}{p_3}\right)^{\gamma-1/\gamma}; \quad T_4 = 2353.36 \times \left(\frac{1.0335}{34.45}\right)^{0.4/1.4} = 864.132 \text{ K}.$$

Heat supplied, $Q_s = C_v (T_3 - T_2) = 0.718 \times (2353.36 - 671.989) = 1207.224$ kJ.
Heat rejected, $Q_r = C_p (T_4 - T_1) = 1.005 \times (864.132 - 353) = 513.687$ kJ.

$$\text{Ideal efficiency, } \eta = \frac{(1207.224 - 513.687)}{1207.224} \times 100 = 57.448\%$$

$$\eta_{\text{Otto}} = 1 - \frac{1}{(r_c)^{\gamma-1}} = 1 - \frac{1}{(5)^{0.4}} = 0.47469 \text{ or } 47.469\%$$

67. In an air standard Brayton cycle, the air enters the compressor at 1 bar, 15°C. The pressure of air leaving the compressor is 10 bar and the maximum temperature in the cycle is 1100°C. Determine (i) the pressure and temperature at each point in the cycle and (ii) the compressor work, turbine work and cycle efficiency. Assume $C_p = 1.005$ kJ/kg K.

Ans: $p_1 = 1$ **bar;** $p_2 = 10$ **bar;** $p_3 = 10$ **bar;** $p_4 = 1$ **bar;** $T_1 = 288$ **K;** $T_3 = 1373$ **K and** $r_p = 10$.

$$\frac{T_2}{T_1} = \left(\frac{p_2}{p_1}\right)^{\gamma-1/\gamma}; \quad T_2 = T_1 (r_p)^{\gamma-1/\gamma} = 288 \times (10)^{0.4/1.4} = 556 \text{ K}$$

$$\frac{T_3}{T_4} = \left(\frac{p_3}{p_4}\right)^{\gamma-1/\gamma}; \quad T_4 = \frac{T_3}{(r_p)^{\gamma-1/\gamma}} = \frac{1373}{(10)^{0.4/1.4}} = 711 \text{ K}$$

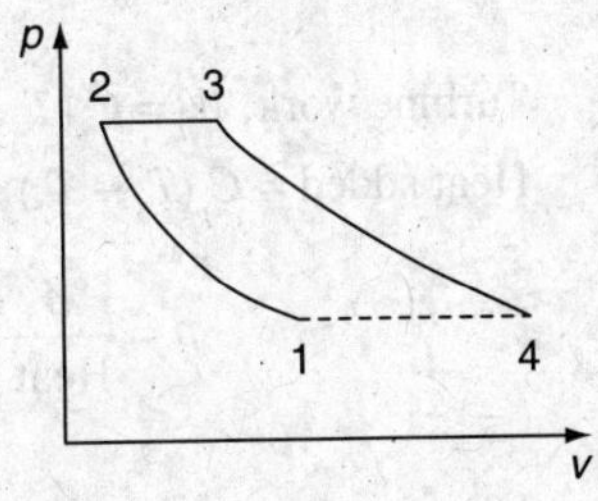

Compressor work, $W_c = C_p (T_2 - T_1)$

$= 1.005 \times (556 - 288) = 269.34$ kJ.

Turbine work, $W_t = C_p (T_3 - T_4)$

$= 1.005 \times (1373 - 711) = 665.31$ kJ.

Heat supplied, $Q_s = C_p\,(T_3 - T_2)$

$$= 1.005 \times (1373 - 556) = 821.085 \text{ kJ.}$$

Net work output $= W_t - W_c = 665.31 - 269.34 = 395.97$ kJ.

$$\text{Cycle efficiency} = \frac{W_t - W_c}{Q_s} = \frac{395.97}{821.085} \times 100 = 48.2\%$$

68. A constant pressure gas turbine works between a lower pressure p_1 and an upper pressure p_2 and there is no exhaust heat regeneration. If the ratio between the maximum and the minimum temperature of the cycle is Z and if the isentropic efficiencies of the turbine and the compressor are η_t and η_c, respectively, show that the cycle efficiency is given by the expression

$$\eta = \left[1 - \frac{1}{x}\right]\frac{(y - x)}{(w - x + 1)}, \text{ in which } w = \eta_c\,(Z - 1);\ x = \left(\frac{p_2}{p_1}\right)^{\gamma-1/\gamma};\ y = \eta_t\,\eta_c\,Z$$

Ans: For compression process 1 – 2,

$p\,V^{\gamma}$ = constant for the process 1 – 2'

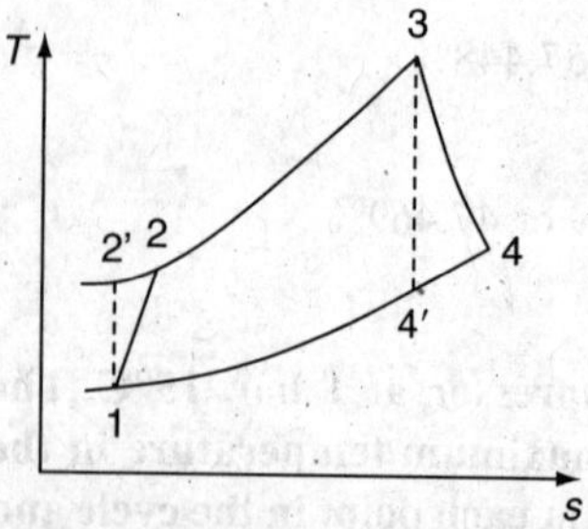

$$\frac{T_{2'}}{T_1} = \left(\frac{p_2}{p_1}\right)^{\gamma-1/\gamma} = x\ ;\quad \eta_c = \frac{T_{2'} - T_1}{T_2 - T_1}$$

$$(T_2 - T_1) = \frac{(T_{2'} - T_1)}{\eta_c} = \frac{T_1}{\eta_c}\left(\frac{T_{2'}}{T_1} - 1\right) = \frac{T_1}{\eta_c}(x - 1)$$

$$T_2 = T_1 + \frac{T_1}{\eta_c}(x - 1) \qquad (1)$$

Compressor work, $W_c = C_p\,(T_2 - T_1)$

For expansion process 3 – 4, $p\,V^{\gamma}$ = constant for the process 3 – 4'

$$\frac{T_3}{T_{4'}} = \left(\frac{p_3}{p_4}\right)^{\gamma-1/\gamma} = \left(\frac{p_2}{p_1}\right)^{\gamma-1/\gamma} = x\,;\eta_t = \frac{T_3 - T_4}{T_3 - T_{4'}};(T_3 - T_4) = \eta_t\,(T_3 - T_{4'}) = \eta_t T_3\left(1 - \frac{1}{x}\right)$$

$$(T_3 - T_4) = \eta_t\,T_3\left(1 - \frac{1}{x}\right) \qquad (2)$$

Turbine work, $W_t = C_p(T_3 - T_4)$

Heat added $= C_p(T_3 - T_2)$

$$\eta = \frac{W_t - W_c}{\text{Heat added}} = \frac{C_p\,(T_3 - T_4) - C_p\,(T_2 - T_1)}{C_p\,(T_3 - T_2)} = \frac{(T_3 - T_4) - (T_2 - T_1)}{(T_3 - T_2)}$$

$$= \frac{T_3\eta_t\left(1-\frac{1}{x}\right)-\frac{T_1}{\eta_c}(x-1)}{T_3-T_1-\frac{T_1}{\eta_c}(x-1)} = \frac{\frac{T_3}{T_1}\eta_t\left(1-\frac{1}{x}\right)-\frac{(x-1)}{\eta_c}}{\frac{T_3}{T_1}-1-\frac{(x-1)}{\eta_c}} = \frac{Z\eta_t\left(1-\frac{1}{x}\right)-\frac{(x-1)}{\eta_c}}{Z-1-\frac{(x-1)}{\eta_c}}$$

$$= \frac{Z\eta_t\eta_c\left(1-\frac{1}{x}\right)-(x-1)}{\eta_c(Z-1)-(x-1)} = \frac{\frac{y}{x}(x-1)-(x-1)}{w-x+1} = \frac{\frac{(x-1)}{x}(y-x)}{w-x+1} = \left(1-\frac{1}{x}\right)\frac{(y-x)}{(w-x+1)}.$$

69. An ideal gas cycle consists of three reversible processes in the following sequence: (i) constant volume pressure rise, (ii) isentropic expansion to *r* times the initial volume and (iii) constant pressure decrease in volume. Sketch the cycle on the *p-v* and *T-s* diagrams. Show that the efficiency of the cycle is

$$\eta_{\text{cycle}} = \frac{r^\gamma - 1 - \gamma(r-1)}{r^\gamma - 1}.$$

Evaluate the cycle efficiency when $\gamma = 1.3333$ and $r = 8$.

Ans: Assume 1 kg of air is present in the cycle.

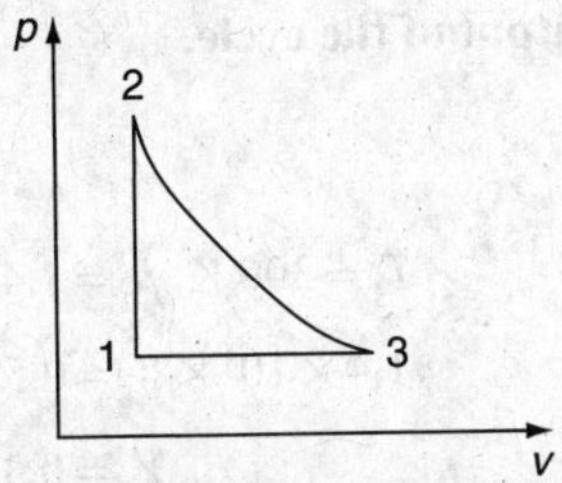

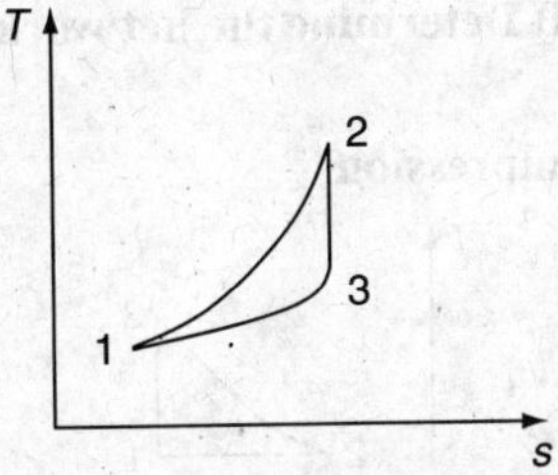

1 – 2 constant volume process (heat addition)

Heat supplied, ${}_1Q_2 = C_v(T_2 - T_1)$

$${}_1W_2 = 0$$

2 – 3 isentropic expansion

$${}_2Q_3 = 0$$

$$p\,v^\gamma = \text{constant}$$

$$\frac{T_3}{T_2} = \left(\frac{V_2}{V_3}\right)^{(\gamma-1)}; \quad T_2 = T_3(r)^{(\gamma-1)} \tag{1}$$

3 – 1 constant pressure cooling (heat rejection)

$$\frac{T_3}{T_1} = \frac{V_3}{V_1} = r; \quad T_3 = T_1 r \tag{2}$$

Heat rejected, ${}_3Q_1 = C_p(T_3 - T_1)$

Substituting for T_3 from equation (2) in equation (1), we get

$$T_2 = T_3\,(r)^{(\gamma-1)} = T_1\,r\,(r)^{(\gamma-1)} = T_1(r)^{\gamma} \quad (3)$$

$$\text{Efficiency, } \eta = \frac{{}_1Q_2 - {}_3Q_1}{{}_1Q_2} = \frac{C_v(T_2 - T_1) - C_p(T_3 - T_1)}{C_v(T_2 - T_1)} = \frac{T_2 - T_1 - \gamma(T_3 - T_1)}{T_2 - T_1}$$

$$= \frac{T_1 r^{\gamma} - T_1 - \gamma(T_1 r - T_1)}{T_1 r^{\gamma} - T_1} = \frac{r^{\gamma} - 1 - \gamma(r-1)}{r^{\gamma} - 1}$$

When $\gamma = 1.3333$ and $r = 8$,

$$\eta = \frac{(8)^{1.3333} - 1 - 1.3333(8-1)}{(8)^{1.3333} - 1} = 0.3777 \text{ or } 37.77\%.$$

70. A cycle using 0.52 kg of air undergoes the following processes: 1 – 2 isothermal compression from 1.4 bar, 27°C to 5.6 bar; 2 – 3 constant pressure heating until the gain in entropy equals the loss in entropy during the isothermal process, 3 – 1 isentropic expansion to the initial state. (i) Sketch the cycle on the *p-v* and *T-s* planes, labelling corresponding points on both diagrams with the same symbol. (ii) Determine the maximum change in entropy in kJ/K. (iii) Determine the net work output of the cycle.

Ans:

1–2 isothermal compression

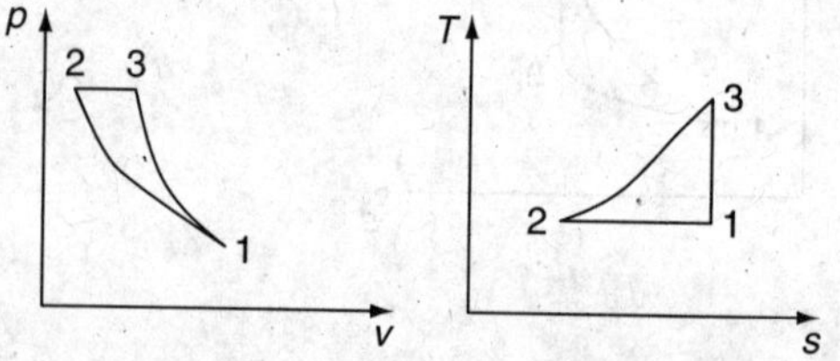

$T_1 = 300$ K; $p_1 = 1.4$ bar; $m = 0.52$ kg.

$$1.4 \times 10^5 \times V_1 = 0.52 \times 287 \times 300$$

$$V_1 = 0.3198 \text{ m}^3.$$

$$p_1 V_1 = p_2 V_2,\ V_2 = \frac{p_1 V_1}{p_2} = \frac{1.4 \times 0.3198}{5.6} = 0.07995 \text{ m}^3.$$

$$S_2 - S_1 = m R \ln\frac{V_2}{V_1} = 0.52 \times 0.287 \times \ln\left(\frac{0.07995}{0.3198}\right) = -0.20689\ \frac{\text{kJ}}{\text{K}}$$

$${}_1W_2 = m R T_1 \ln\left(\frac{V_2}{V_1}\right) = 0.52 \times 0.287 \times 300 \times \ln\left(\frac{0.07995}{0.3198}\right) = -62.067 \text{ kN.m}$$

2–3 constant pressure heating

Given that $S_3 - S_2 = |S_2 - S_1| = 0.20689$

$$S_3 - S_2 = m C_p \ln\left(\frac{V_3}{V_2}\right) = 0.52 \times 1.005 \times \ln\left(\frac{V_3}{0.07995}\right) = 0.20689$$

$$V_3 = 0.11878 \text{ m}^3.$$

$${}_2W_3 = p_3\,(V_3 - V_2) = 5.6 \times 100 \times (0.11878 - 0.07995) = 21.7448 \text{ kN.m.}$$

3–1 isentropic expansion

$$p\,V^{\gamma} = \text{constant}$$

$$_3W_1 = \frac{p_3V_3 - p_1V_1}{\gamma - 1} = \frac{(5.6\times0.11878 - 1.4\times0.3198)10^5}{(1.4-1)10^3} = 54.362 \text{ kN.m.}$$

Maximum entropy change = 0.20689 kJ/K
Net work output of the cycle = –62.067 + 21.7448 + 54.362 = 14.0398 kN.m.

71. The cycle of an engine consists of three stages: an isothermal compression, an increase of pressure at constant volume and an adiabatic expansion. If r is the ratio of expansion and compression, show that the efficiency is

$$1 - \frac{R\ln(r)}{C_v\left(r^{\gamma-1} - 1\right)}$$

In such a cycle air is supplied at 1.0335 bar and 27°C and compressed to 5.1675 bar. Find, per kg of air, the heat supplied and the work done per cycle.

Ans:

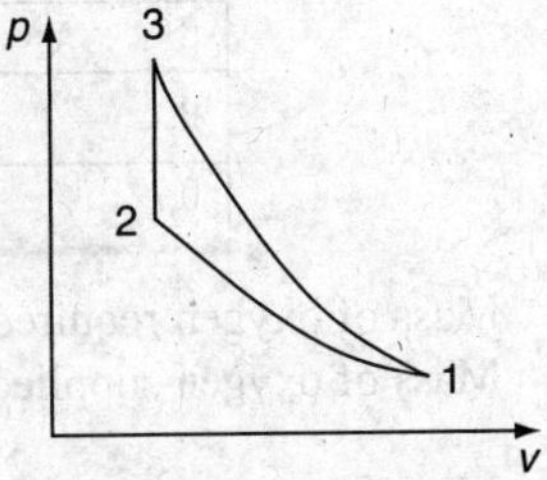

Compression ratio $= \dfrac{V_1}{V_2} = r$

Expansion ratio $= \dfrac{V_1}{V_3} = r$

1–2 isothermal compression

$$_1W_2 = RT_1\ln\left(\frac{V_2}{V_1}\right) = -RT_1\ln\left(\frac{V_1}{V_2}\right) = -RT_1\ln(r) \quad (1)$$

$$_1Q_2 = {}_1W_2 = -RT_1\ln(r) \quad (2)$$

2–3 constant volume heating

$$_2Q_3 = C_v(T_3 - T_2) = C_v(T_3 - T_1) \text{ since } T_2 = T_1$$

3–1 isentropic expansion

$$p\,V^{\gamma} = \text{constant}$$

$$\frac{T_3}{T_1} = \left(\frac{V_1}{V_3}\right)^{\gamma-1} = (r)^{\gamma-1}; \quad T_3 = T_1(r)^{\gamma-1}$$

Heat supplied, $_2Q_3 = C_v(T_3 - T_1)$

Heat rejected, $_1Q_2 = R\,T_1\,\ln(r)$

$$\text{Efficiency}, \eta = \frac{Q_s - Q_r}{Q_s} = 1 - \frac{Q_r}{Q_s} = 1 - \frac{RT_1\ln(r)}{C_v\left(T_1(r)^{\gamma-1} - T_1\right)} = 1 - \frac{R\ln(r)}{C_v\left[(r)^{\gamma-1} - 1\right]}$$

$$p_1V_1 = p_2V_2; \frac{V_1}{V_2} = \frac{p_2}{p_1} = \frac{5.1675}{1.0335} = 5; r = \frac{V_1}{V_2} = 5$$

$$T_1 = 300 \text{ K}; r = 5; p_1 = 1.0335 \text{ bar}; p_2 = 5.1675 \text{ bar}.$$

$$T_3 = T_1(5)^{\gamma-1} = 300 \times (5)^{0.4} = 571.096 \text{ K}.$$

Heat rejected $= R\,T_1 \ln(5) = 0.287 \times 300 \times \ln(5) = 138.57$ kJ.

Heat supplied $= C_v(T_3 - T_2) = C_v(T_3 - T_1) = 0.718 \times (571.096 - 300) = 194.64$ kJ.

Work done per cycle = 194.64 – 138.57 = 56.07 kJ or kN.m.

72. The analysis of the coal used in a boiler is as follows: C – 82%; H_2 – 6%; O_2 – 4%; ash – 6% and moisture – 2%. Determine the theoretical minimum air required for complete combustion of 1 kg of coal.

Ans:

Combustion of 1 kg of Coal

Constituent	Mass (kg)	Chemical Equation	Mass of O_2 Required (kg)
C	0.82	$C + O_2 = CO_2$	2.186
H_2	0.06	$2\,H_2 + O_2 = 2\,H_2O$	0.48
O_2	0.04		

Mass of oxygen required for combustion = 2.186 + 0.48 = 2.666.

Mass of oxygen supplied by air = 2.666 – 0.04 = 2.626.

$$\text{Mass of air required} = \frac{2.626}{0.23} = 11.46 \text{ kg}.$$

73. Producer gas used as fuel has the following volumetric composition: H_2 – 28%; CO – 12%; CH_4 – 2%; CO_2 – 16% and N_2 – 42%. Find the volume of air necessary for complete combustion of one cubic metre of this gas.

Ans:

Combustion of 1 m³ of Producer Gas

Constituent	Volume (m³)	Combustion Equation	O_2 Required (m³)
H_2	0.28	$H_2 + ½\,O_2 = H_2O$	0.14
CO	0.12	$CO + ½\,O_2 = CO_2$	0.06
CH_4	0.02	$CH_4 + 2\,O_2 = CO_2 + 2\,H_2O$	0.04
CO_2	0.16		–
N_2	0.42		

$$\text{Air required to burn 1 m}^3 \text{ of the gas} = \frac{0.14 + 0.06 + 0.04}{0.21} = 1.1428 \text{ m}^3.$$

74. 100 m³ of air per minute at 15°C DBT and 80% RH is heated until its temperature is 22°C. Find (i) the heat added to air per minute, (ii) relative humidity of the heated air and (iii) wet-bulb temperature of the heated air.

Ans:

The following values are obtained from the psychrometric chart

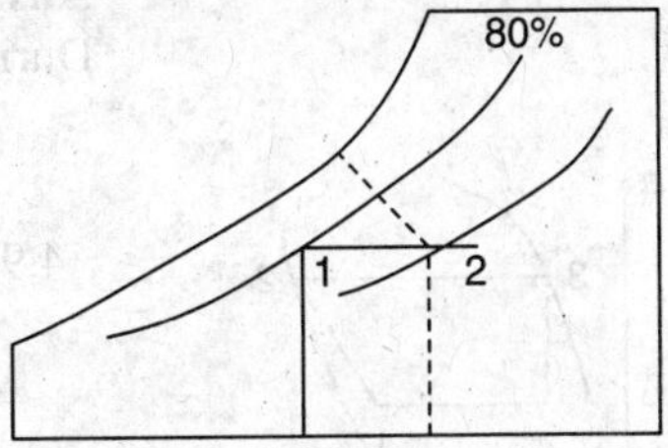

Psychrometric chart

$$h_1 = 38 \text{ kJ/kg of air}$$

$$v_1 = 0.83 \text{ m}^3\text{/ kg}$$

$$h_2 = 45 \text{ kJ/kg of air}$$

$$\varphi_2 = 54\%$$

wet-bulb temperature = 16°C

(i) Heat added per min = $\dfrac{Q}{v_1}(h_2 - h_1) = \dfrac{100}{0.83}(45-38) = 843.37\dfrac{\text{kJ}}{\text{min}}$.

(ii) Relative humidity of the heated air = 54%.

(iii) Wet-bulb temperature = 16°C.

75. A refrigerator works on the Carnot cycle in reverse, between the temperatures at –7°C and 27°C. It makes 500 kg of ice per hour at –5°C from water at 14°C. If the specific heat of ice is 0.5 and latent heat of fusion of ice is 350 kJ/kg, determine (i) the power required to drive the compressor and (ii) the heat rejected by the compressor.

Ans: Heat to be removed from water $= 4.186 \times (14 - 0) + 350 + 0.5 \times [0 - (-5)]$

$$= 411.104 \text{ kJ/kg.}$$

Heat to be removed per hr $= 500 \times 411.104 = 205552$ kJ/hr

$$\text{C.O.P} = \frac{(273-7)}{(273+27)-(273-7)} = \frac{266}{34} = 7.824.$$

Power required to drive the compressor $= \dfrac{205552}{7.824} = 26271.98\dfrac{\text{kJ}}{\text{hr}}$ or

$$= \frac{26271.98}{3600} = 7.297 \text{ kW.}$$

Heat rejected by the compressor $= 205552 + 26271.98 = 231823.98$ kJ/hr.

76. An ammonia refrigerator works between 263 K and 303 K; the vapour being dry at the end of compression which is isentropic. There is no under cooling of liquid ammonia and the liquid is expanded through a throttle valve after leaving the condenser. Calculate the theoretical C.O.P using the properties of ammonia given below:

Temperature K	Specific Enthalpy, kJ/kg			Specific Entropy, kJ/kg K	
	Liquid	Latent Heat	Vapour	Liquid	Vapour
263	135.4	1297.6	1433	0.544	5.477
303	323.1	1145.9	1469	1.204	4.984

Ans:

During isentropic compression 1–2, $s_2 = s_1$.

$$s_2 = 4.984 = s_1$$

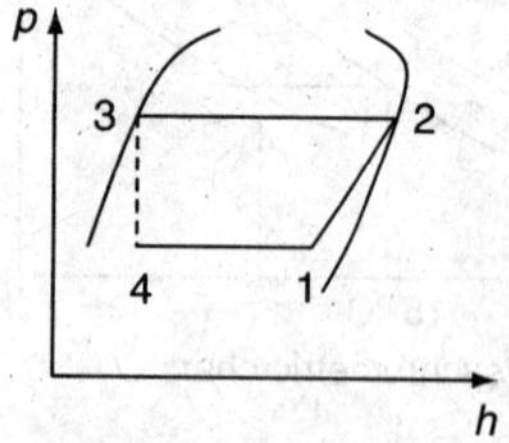

$$4.984 = 0.544 + x_1(5.477 - 0.544)$$

$$x_1 = \frac{4.984 - 0.544}{5.477 - 0.544} = 0.9.$$

$$h_1 = 135.4 + 0.9 \times 1297.6 = 1303.2.$$

$$h_2 = 1469.$$

$$h_3 = 323.1.$$

During throttling process 3 – 4, $h_3 = h_4$.

$$h_4 = 323.1.$$

$$\text{C.O.P} = \frac{(h_1 - h_4)}{(h_2 - h_1)} = \frac{(1303.2 - 323.1)}{(1469 - 1303.2)} = 5.91.$$

12

Additional Practice Problems

1. Air expands at constant pressure from a volume of 0.0849 m^3 at 13.78 bar and 175°C to a volume of 0.2547 m^3. Find the temperature at the end of expansion, the work done during expansion and the change of internal energy.

   (1071°C; 233.98 kN.m; 585.36 kJ)

2. A certain quantity of gas is compressed from an initial state of 0.09 m^3 and 1.0 bar to a final condition of 0.03 m^3 and 4.0 bar. If the specific heat of the gas at constant volume is 0.718 and at constant pressure 1.005, calculate the change of internal energy. Assume the mass of the gas is 0.1 kg.

   (7.49 kJ)

3. The volume, pressure and temperature of 1 kg of air are 0.2676 m^3, 4 bar and 100°C, respectively. Find the temperature and pressure of this air after 240 kJ of heat have been given to it at constant volume. Take the mean specific heat of air at constant volume as 0.718 kJ/kg K.

   (707.3 K; 7.58 bar)

4. One kg of air at a temperature of 10°C is heated to a temperature of 82°C. The specific heat at constant pressure is 1.005 kJ/kg K and that at constant volume is 0.718 kJ/kg K. Find the heat supplied to air if the methods of heating are (i) at constant pressure and (ii) at constant volume.

   (72.36 kJ; 51.6 kJ)

5. Nitrogen at an initial state of 300 K, 150 kPa and 2 m^3 is compressed slowly in an isothermal process to a final pressure of 800 kPa. Determine the work done during the process.

   (–502.2 kJ)

6. A system containing 5 kg of air undergoes an isothermal process in which 200 kJ/kg of air is added to the gas, which is at 200°C. Find the change in enthalpy, work done and change in entropy.

   (0; 1000 kJ; 2.114 kJ/K)

7. Five kg of air expands from a pressure of 13 bar to 1.3 bar. The index of expansion is 1.2. Calculate the change in entropy.

   (1.3757 kJ/K)

8. A system contains 0.15 m^3 of a gas at a pressure of 4 bar and temperature of 150°C. It is expanded adiabatically with an index of 1.4 to a pressure of 1 bar. The gas is then heated at constant pressure till its enthalpy increases by 75 kJ. Taking the specific heat at constant pressure as 1 kJ/kg K, calculate the total work done.

(70.95 kN.m)

9. Two kilograms of an ideal gas are compressed polytropically from a pressure of 1.5 bar and temperature 16°C to 7.5 bar. If the index of compression is 1.3, calculate the work done and heat received or rejected during the process. Assume C_p = 1.005 kJ/kg K and C_v = 0.718 kJ/kg K.

(–248.5 kJ; –61.96 kJ (heat rejection by the gas))

10. A certain mass of gas is compressed in a quasi-static process from 80 kPa, 0.1 m^3 to 0.4 MPa and 0.03 m^3. Assuming that the pressure and volume are related by pV^n = constant, find the work transfer during the process.

(–11.869 kN.m)

11. Air is contained within a piston cylinder device at 1.0335 bar and 15°C. (i) In process A the gas is heated at constant volume until the pressure has doubled. It is then expanded at constant pressure until the volume is three times its initial volume. (ii) In process B the same gas in the same initial state is first expanded at constant pressure until the volume has tripled; then the gas is heated at constant volume until it reaches the same final pressure as in process A. For these two processes determine the net heat effect, the net work effect and the change in internal energy and compare.

(1364.54 kJ; 330.595 kN.m; 1033.949 kJ, 1199.232 kJ; 165.297 kN.m; 1033.949 kJ)

12. Air is compressed adiabatically and reversibly from 1.034 bar, 5°C until the density has tripled in a steady flow process. If the kinetic energy changes are small, calculate the work required in kJ/kg.

(–154.15 kJ/kg)

13. Air at a pressure of 100 kPa and temperature of 30°C is compressed polytropically from a volume of 8 m^3 to 2 m^3 with an index of compression of 1.2. Assume the two specific heats as 1 kJ/kg K and 0.714 kJ/kg K. Calculate the final pressure, final temperature and work transfer.

(5.278 bar; 126.8°C; –1278 kN.m)

14. 0.056 m^3 of carbon dioxide is contained in a cylinder at 310 K and 1.4 bar. During the inward stroke of the piston the gas is compressed to 0.0224 m^3. If the compression process is isothermal, find the change in entropy.

(–0.02317 kJ/K)

15. Steam enters a turbine at a velocity of 40 m/s and specific enthalpy of 2400 kJ/kg and leaves it with a velocity of 90 m/s and specific enthalpy of 1950 kJ/kg. The heat losses from the turbine to the surroundings are 280 kJ/min and the steam consumption is 4500 kg/h. Calculate the power developed by the turbine.

(553.77 kW)

16. The power output of an adiabatic gas turbine is 5 MW and the inlet conditions are p_1 = 2 MPa, T_1 = 1000°C, V_1 = 50 m/sec, Z_1 = 10 m and the exit conditions are p_2 = 115 kPa, T_2 = 354°C, V_2 = 180 m/sec, Z_2 = 6 m. Assume C_p = 1.005 kJ/kg K.
   (i) Compare the magnitudes of Δh, ΔK.E and ΔPE
   (ii) Determine the work done per kg
   (iii) Mass rate of gas flowing through the turbine.

(749.36 kJ/kg; −14.95 kJ/kg; 0.03924 kJ/kg, 734.449 kJ/kg, 6.81 kg/sec)

17. The diffuser of a jet engine draws in air at a pressure of 0.55 bar, a temperature of −7°C and a velocity of 244 m/sec. The air stream leaves the diffuser at a velocity of 122 m/sec. If the flow is adiabatic and quasistatic, what will be the temperature and pressure of the air at the outlet? What is the ratio of outlet area to inlet area?

(15.21°C; 0.7282 bar; 1.636)

18. A gas in steady flow through a frictionless device is compressed from an initial state of 1.034 bar, 0.7798 m^3/kg, to a final pressure of 4.134 bar. During the process the pressure and specific volume are related by $p\,v^{1.3}$ = constant, and the enthalpy of the gas increases by 55.8 kJ/kg. If the changes in kinetic and potential energies are negligible, determine the quantity and direction of heat transfer in kJ/kg.

(−75.68 kJ/kg, from the gas)

19. A steam turbine is reported to deliver 1850 kW if 11000 kg/h of steam passes through the turbine from inlet conditions of 15 bar and 350°C to an exit condition of 1 bar. Are the reported data accurate if the process is adiabatic?

(Inaccurate data)

20. Estimate the amount of heat to be supplied to water initially at 40°C to produce steam at 10 bar and 200°C.

(2660.3 kJ/kg)

21. Steam at a pressure of 1.5 MPa and 300°C expands adiabatically in a turbine to a pressure of 10 kPa. Find the quality of steam leaving the turbine.

(wet and 0.84 dry)

22. Steam at a pressure of 4 bar and dryness fraction 0.75 is heated at constant volume to a pressure of 5 bar. Using steam tables, determine the final condition of steam.

(wet and 0.924 dry)

23. 4 kg of 0.5 dry steam at 6.0 bar pressure is heated so that it becomes (i) 0.95 dry at 6.0 bar pressure, (ii) dry saturated at 6.0 bar pressure and (iii) superheated to 300°C at 6.0 bar pressure. Determine in each case the quantity of heat to be supplied.

(3753 kJ; 4170.4 kJ; 5394.8 kJ)

24. Dry saturated steam at a pressure of 5 bar expands isentropically to a pressure of 0.7 bar. Using steam tables, calculate the change in enthalpy during the process.

(−327.3 kJ)

25. An inventor claims to have a heat engine which is capable of developing 8 kW, while working between the temperatures limits of 130°C and −40°C. It receives 1000 kJ/min of heat. Discuss the possibility of the claim.

(claim incorrect)

26. A heat engine receives reversibly 100 kJ of heat per cycle from a source at 327°C and rejects heat to a sink at 27°C. There are no other heat transfers. For each of the three hypothetical amounts of heat rejected in (i), (ii) and (iii) below, compute the cycle integral $\delta Q/T$. From these results, show which case is reversible, which is irreversible and which is impossible: (i) 50 kJ/cycle, (ii) 75 kJ/cycle and (iii) 25 kJ/cycle.

((i) Reversible; (ii) Irreversible; (iii) Impossible)

27. A Carnot engine develops 2 kW and rejects 3600 kJ/h to a sink at 15°C. Determine the source temperature.

(864 K)

28. A Carnot engine operating between temperature levels of 727°C and 27°C is supplied 500 kJ/cycle. Sixty percent of the work output is used to drive a heat pump, which rejects to the surroundings at 27°C. If the heat pump removes 1050 kJ from the low temperature reservoir per cycle, determine the temperature of that reservoir in °C.

(−23°C)

29. A reversed Carnot engine is used for heating a building. It supplies 3500 kJ/min to the building at 20°C. The outside air is at −5°C. Find the power requirement of the motor.

(4.977 kW)

30. Compare the Carnot cycle efficiency with the Rankine cycle efficiency if the maximum and minimum temperatures are, respectively, 300°C and 72°C, the steam in the case of Rankine cycle being supplied at 20 bar.

(39.79%; 26.62%)

31. A high-pressure boiler delivers steam at 80 bar and 350°C. The steam is expanded in the first section of the turbine to 10 bar and then withdrawn and passed on to the reheater, where it is reheated to the initial temperature. The expansion now takes place in the remaining section of the turbine, down to a condenser pressure of 0.07 bar. Determine (i) efficiency of the ideal cycle for a flow of 1 kg/sec of steam; (ii) the work output and efficiency of the reheat cycle without any pressure drop in the reheater.

(38.23%; 1298.3 kJ, 38.15%)

32. A Rankine cycle works between inlet condition of 120 bar and 500°C and 1 bar of exhaust pressure. If the exhaust pressure is reduced to 0.1 bar, calculate the percentage change in Rankine efficiency.

(33.78%; 40.77%, 20.69%)

33. A steam engine uses steam at 10 bar and 0.9 dry and exhausts at 1.1 bar. Determine the Rankine efficiency. What will be the percentage improvement in efficiency if the initial temperature of steam is 200°C?

(15.85%; 16.07%, 1.388%)

34. The diameter and stroke length of a single cylinder two-stroke engine working on the constant volume cycle are 200 mm and 300 mm, respectively, with clearance volume of 0.0028 m^3. While running at 135 rpm the indicated mean effective pressure is 5.2 bar and the fuel consumption is 8.8 m^3/h. If the calorific value of the fuel is 16350 kJ/m^3, find (i) the air standard efficiency, (ii) the indicated power, (iii) the indicated thermal efficiency and (iv) the relative efficiency.

(44.54%; 11.027 kW; 27.59%; 61.94%)

35. An oil engine with a clearance volume of 20 percent of the stroke volume has an indicated thermal efficiency of 30.7 percent. Find the relative efficiency.

(59.99%)

36. Calculate the percentage increase in efficiency of an engine working on Otto cycle when the compression ratio raises from 5 to 8.

(18.98%)

37. In an engine working on Otto cycle the measured suction temperature was 100°C and the temperature at the end of compression was 300°C. Taking $\gamma = 1.4$ for compression and expansion, find the compression ratio and the ideal efficiency.

(2.925; 34.9%)

38. In an engine working on ideal Otto cycle, the initial pressure and temperature are 1.6 bar and 35°C, respectively. Taking the ratio of specific heat as 1.4 and the thermal efficiency of the engine as 54 percent, calculate the compression ratio and the final condition at the end of compression.

(6.97; 24.25 bar; 669.64 K)

39. Calculate the air standard efficiency of an engine working on the Diesel cycle when the suction pressure is 1 bar and fuel is injected for 5 percent of stroke. The clearance volume is 8 percent of stroke. Assume $\gamma = 1.4$ for air.

(60.73%)

40. In a Diesel cycle having compression ratio of 16, cut off occurs at 6 percent of stroke. Determine the reduction in air standard efficiency if cut off occurs at 7 percent of stroke.

(1.195%)

41. In a Diesel engine cycle, the temperature and pressure at the beginning of compression are 87°C and 100 kPa. If the compression ratio is 14, find the temperature and pressure at the end of compression. If the temperatures at the beginning and end of expansion are 1795°C and 677°C, respectively, determine the efficiency of the cycle.

(59.2%)

42. The compression ratio of an engine is 10 and the pressure and temperature at the beginning of compression are 1.013 bar and 40°C, respectively. The compression and expansion processes are both isentropic and heat is rejected at exhaust at constant volume. The amount of heat added during the cycle is 1500 kJ/kg. Determine the thermal efficiency of the cycle if (i) the maximum pressure is limited to 70 bar and heat is added both at constant volume and constant pressure, and (ii) all heat is added at constant volume.

(59.59%; 60.19%)

43. A reversible heat engine employing an ideal gas with constant specific heat undergoes the following cyclic process: constant volume heating from state 1 to state 2, adiabatic expansion from state 2 to state 3 and then constant pressure compression from state 3 to state 1. Prove that the thermal efficiency of the cycle is

$$\eta = 1 - \gamma \left(\frac{\dfrac{V_3}{V_1} - 1}{\dfrac{p_2}{p_1} - 1} \right)$$

44. A cycle consists of three processes: a constant pressure compression process $a - b$, a constant volume heat addition process $b - c$ and an adiabatic expansion process $c - a$. The internal energies at points a, b and c are, respectively, 80, 40 and 200 kJ. Pressure at a is 1 bar, volume at a is 1 m^3 and volume at b is 0.5 m^3. Calculate the net work done in the cycle.

(70 kN.m)

Bibliography

1. Cengel, Y. A. *Introduction to Thermodynamics and Heat Transfer*, McGraw Hill, New York, 1980.
2. Fast, J. D. *Entropy*, McGraw Hill, New York, 1963.
3. Jones, J. B. and Dugan, R. E. *Engineering Thermodynamics*, Prentice Hall, New Delhi, 1998.
4. Keenan, J. H. *Thermodynamics*, John Wiley, New York, 1941.
5. Lee, J. F. and Sears, F. W. *Thermodynamics*, Addison Wesley, Cambridge, Mass, 1962.
6. Obert, E. F. *Concepts of Thermodynamics*, McGraw Hill, New York, 1960.
7. Reynolds, W. C. and Perkins, H. C. *Engineering Thermodynamics*, McGraw Hill, New York, 1977.
8. Rogers, G. F. C. and Mayhew, Y. R. *Engineering Thermodynamics: Work and Heat Transfer*, Longmans Green, London, 1967.
9. Saad, M. A. *Thermodynamics for Engineers*, Prentice Hall, New Delhi, 1969.
10. Spalding, D. B. and Cole, E. H. *Engineering Thermodynamics*, Edward Arnold, London, 1973.
11. Van Wylen, G. J. and Sonntag, R. E. *Fundamentals of Classical Thermodynamics*, John Wiley, New York, 1976.
12. Wark, K. *Thermodynamics*, McGraw Hill, New York, 1966.
13. Wrangham, D. A. *The Theory and Practice of Heat Engines*, Cambridge University Press, London, 1962.
14. Zemansky, M. W. *Heat and Thermodynamics*, McGraw Hill, New York, 1957.

Index

Locator followed by "*f*" refers to figures in the index.